Analysis of
Bipolar and CMOS
Amplifiers

Analysis of Bipolar and CMOS Amplifiers

Amir M. Sodagar
K. N. Toosi University of Technology
Tehran, Iran
and
University of Michigan
Ann Arbor, MI, U.S.A.

CRC Press
Taylor & Francis Group
Boca Raton London New York

CRC Press is an imprint of the
Taylor & Francis Group, an informa business

CRC Press
Taylor & Francis Group
6000 Broken Sound Parkway NW, Suite 300
Boca Raton, FL 33487-2742

Printed in the United States of America on acid-free paper
10 9 8 7 6 5 4 3 2 1

International Standard Book Number-10: 1-4200-4644-6 (Hardcover)
International Standard Book Number-13: 978-1-4200-4644-1 (Hardcover)

Library of Congress Cataloging-in-Publication Data

Sodagar, Amir M.
Analysis of bipolar and CMOS amplifiers / Amir M. Sodagar.
p. cm.
Includes bibliographical references and index.
ISBN-13: 978-1-4200-4644-1 (alk. paper)
ISBN-10: 1-4200-4644-6 (alk. paper)
1. Transistor amplifiers. 2. Transistor circuits. 3. Metal oxide semiconductors, Complimentary. 4. Bipolar transistors. I. Title.

TK7871.2.S73 2007
621.3815'282--dc22 2007000413

Visit the Taylor & Francis Web site at
http://www.taylorandfrancis.com

and the CRC Press Web site at
http://www.crcpress.com

To the fruits of my life:

Ali and Shokoufeh

Table of Contents

Preface

Several invaluable reference books have been published, which generally introduce the basics of analog circuit analysis and design, and specifically apply these concepts to amplifiers. These books usually describe the classic approach to evaluating small-signal amplifiers, which includes all the contributing factors and precise calculations of the quantities of desired parameters. This technique involves a tedious and sometimes difficult process, however, especially when the number of components increases. Even if a circuit is precisely designed or analyzed using hand analysis, the circuit variables will likely deviate from the expected values in practice. Thus, it is sometimes preferred to roughly determine the quantities of interest with a straightforward method and not bother with spending too much time and effort to precisely analyze the circuit by solving several equations.

After more than 14 years of research and teaching experience in analog and mixed-signal circuits, I decided that aside from the voluminous wide-scope reference books, there should be a book mainly focused on the introduction of a much simpler technique, known as "analysis by inspection," for bipolar and *complementary metal-oxide semiconductor* (CMOS) small-signal amplifiers. This allows a straightforward approach to analyzing a wide variety of small-signal analog circuits quickly, yet within an acceptable accuracy range while still based upon scientific fundamentals. This includes both AC analysis at midband frequencies, as well as analysis at low and high frequencies, for both bipolar and CMOS circuits.

The target audience of this book includes senior undergraduate and first-year graduate students in electrical engineering and related disciplines. It is assumed that the reader has a basic background in the analysis of electric circuits and the theory of signals and systems.

Chapters 4, 5, 7, and 8 discuss the main concepts presented in this book. Chapters 1 through 3 and 6 provide only the minimum basic concepts required to support the flow of the discussion. For example, transistors in Chapter 2 are introduced with their electrical function and not their physical behavior. Moreover, no body or short-channel effects are discussed for MOS transistors, allowing for the simplest possible small-signal model for transistors to be utilized in practice.

I am grateful to the following for their help during the development of this text: Professors A. Ashrafi (University of Alabama), P. Bhatti (Georgia Institute of Technology), M. Ghovanloo (North Carolina State University), R. Lofti (Ferdowsi University of Mashhad), also Dr. Y. Yao (University of Michigan), and University of Michigan graduate students: A. Borna, R.M. Haque, G.E. Perlin, and S. Naraghi. I would also like to express my appreciation to all the indivdiuals who have contributed to the production of this book at Taylor and Francis/CRC Press, especially Nora Konopka, the Editor of Engineering and Environmental Sciences, Marsha Pronin, the Project Coordinator, and Jennifer Smith, the Project Editor.

I also want to take this opportunity to thank Professor G. Roientan Lahiji, who opened my eyes to the exciting world of advanced electronics.

Finally, I thank, from the bottom of my heart, my parents who were my first and best teachers and with my most sincere gratitude I acknowledge my wife's patience, support, and help, without which this work would not have been possible.

Amir M. Sodagar
University of Michigan
Ann Arbor, Michigan

Chapter 1

Basics of Amplifiers

1.1 Introduction

As the name states, an *amplifier* is a circuit that receives a signal at the input and provides a larger or amplified version of that signal at the output. The type of signal and the nature of the amplification are the key factors based on which different types and classes of amplifiers are introduced. This chapter discusses the basic concepts and definitions that will be essential in understanding, analyzing, and designing electronic amplifiers.

1.2 Basic Concepts

First, it is essential to define what a *signal* is and characterize it in such a way that it can facilitate and simplify the analysis methods to be introduced. A signal, according to the *Merriam-Webster* dictionary, is a detectable physical quantity by which messages or information can be transmitted.

Perhaps the simplest and most understandable definition for a signal in general is that:

A "signal" is a quantity that conveys useful information.

For example, the height of the mercury column in a conventional mercury thermometer can be a signal that informs us about the temperature. In electronic circuits and systems, signals are usually voltages or currents. In many applications, signals convey information about real quantities, such as pressure, temperature, light, and voice, which are transduced into a voltage or current by sensors. In some other applications, signals carry virtual information generated or synthesized by manmade systems (e.g., the commands sent from a remote control to a TV or VCR to set the volume or change the channel). Hereafter, the name *signal* refers to a voltage or current of interest that is or has been processed by the circuit under study.

As discussed in section 1.6, the amplifiers studied in this book will be assumed as *linear* circuits. Thus, we need to define when a circuit, or more specifically an amplifier, is linear. A simplified definition for a linear circuit is as follows:

> *Suppose that the responses of a circuit to inputs $x_1(t)$ and $x_2(t)$ are $y_1(t)$ and $y_2(t)$, respectively, then the circuit is said to be a "linear" circuit if its response to $[ax_1(t) + bx_2(t)]$ is $[ay_1(t) + by_2(t)]$.*

Next, we consider the shape of the signals. Throughout the book, it is assumed that all signals are periodic with a fixed frequency. In certain sections, however, the response of the circuits to nonperiodic signals having arbitrary wave shapes will be studied. In addition, as we know from advanced engineering mathematics, every periodic signal can be represented as a linear combination of sine components (Fourier Series Expansion). Thus, the basic waveform will be a sine wave unless otherwise stated. Because the amplifiers studied in this book are assumed to be operating linearly, it will be easy to find the circuit's response to any other periodic waveform by finding and properly superposing the amplifier's responses to the individual sinusoidal frequency components present in the input signal.

1.3 A Signal and Its DC and AC Components

Decomposing a periodic signal into its DC and AC components will allow us to introduce an analysis method that will be significantly simpler than the regular circuit analysis methods studied in basic circuit theory. The *DC component* of a periodic signal is its average or mean value. Mathematically speaking, for the signal x(t) depicted in Figure 1.1, the DC component, X_{DC}, is defined as:

$$X_{DC} = \frac{1}{T}\int_{t_0}^{t_0+T} x(t)dt \tag{1.1}$$

where t_0 is an arbitrary point on the time axis, and T is the period of the signal.

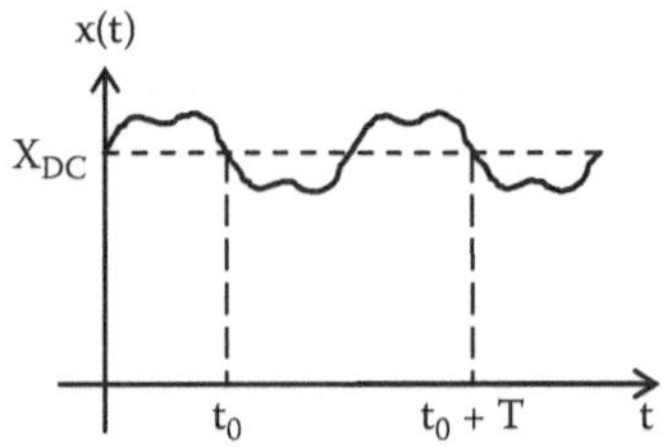

FIGURE 1.1
A periodic signal, x(t), with period T.

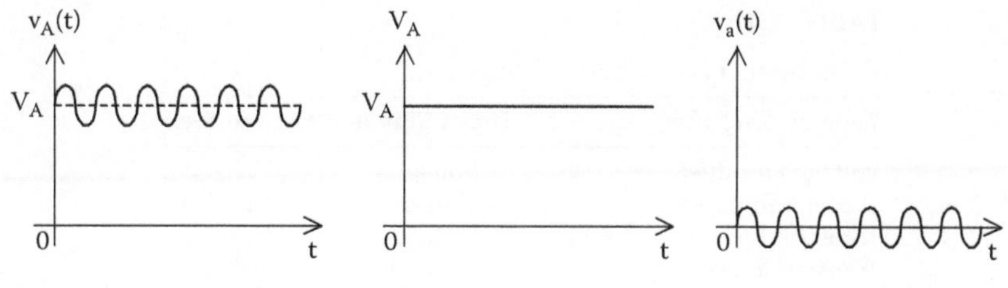

FIGURE 1.2
Convention of naming a signal and its components (a) whole signal, (b) DC component, and (c) AC component.

If the DC component of a signal is extracted out of the signal, the part that remains will be its *AC component*. In other words, for a periodic signal, the DC component represents the mean value, which is constant and has no variations with time, and the AC component will have zero mean value and represents the signal's variations with time.

1.3.1 A Convention to Name a Signal and its Components

To easily refer to a signal and its DC and AC components without any confusion, we will make the following convention to name signals and their components: the entire signal is written with the variable in lower case and capital subscript, the DC component with capital letters and capital subscript, and the AC component with small letter and small subscript. This convention is used in naming the voltage signal depicted in Figure 1.2.

1.4 Basic Types of Amplifiers

Figure 1.3 is the general block representation of an amplifier, with $x_I(t)$ as the input signal, $x_O(t)$ as the output signal, and the amplification factor, hereafter *gain*, of K.

As mentioned before, amplifiers can be classified based on the type of their input and output signals. Depending on whether the input or the output is a voltage or current, four *basic types of amplifiers* would be possible, which are listed in Table 1.1.

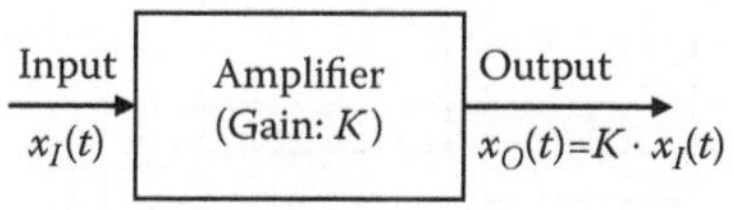

FIGURE 1.3
An amplifier.

TABLE 1.1

Four Basic Types of Amplifiers

Type of Amplifier	Input Signal	Output Signal
Voltage amplifier	Voltage	Voltage
Current amplifier	Current	Current
Transconductance amplifier	Voltage	Current
Transresistance amplifier	Current	Voltage

1.4.1 Voltage Amplifier

A *voltage amplifier* is an amplifier that receives a voltage as the input signal, amplifies it, and provides the amplified signal at the output in voltage form. The two-port model for an ideal voltage amplifier is presented in Figure 1.4, where A_V is the gain of the amplifier, usually referred to as the *voltage gain*.

The input signal to an amplifier is provided by a *signal source*, which can be a laboratory instrument (e.g., a waveform generator), a sensor or transducer converting the quantity of interest into electrical form, or another circuit. Whatever the signal source is, so long as the provided input signal to the amplifier is of voltage type, the input signal source can be modeled by Thevenin's theorem. The Thevenin equivalent circuit consists of a voltage source, v_S, in series with a resistor, R_S. In addition, the amplified signal should be delivered to a device or circuit at the output, called a *load*. Loads can in general be any kind of linear/nonlinear active/passive device or circuit, but the most complex form for a load is usually an impedance. Throughout this book, this impedance is simply a resistive load unless otherwise stated. Termination of an ideal amplifier to a signal source at the input and to a load at the output is illustrated in Figure 1.5.

For the circuit in Figure 1.5, because no current is flowing through R_S, no voltage will drop across it. As a result, the whole v_S is delivered to the amplifier for amplification:

$$v_{IN} = v_S, \tag{1.2}$$

and the amplified voltage, $A_V v_{IN}$, would be completely delivered to the load as:

$$v_{OUT} = A_V v_{IN} = A_V v_S. \tag{1.3}$$

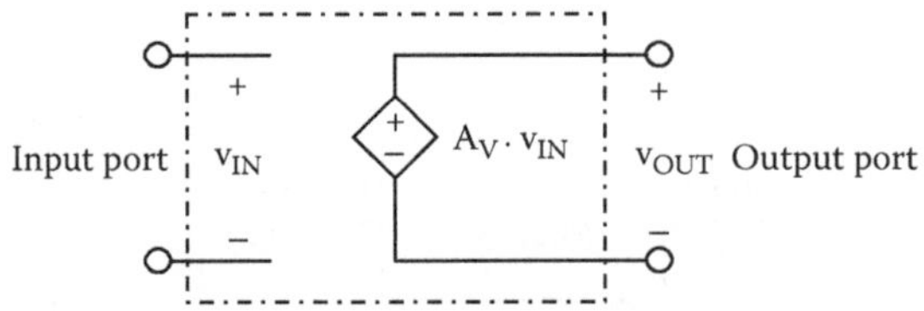

FIGURE 1.4
Two-port model for an ideal voltage amplifier.

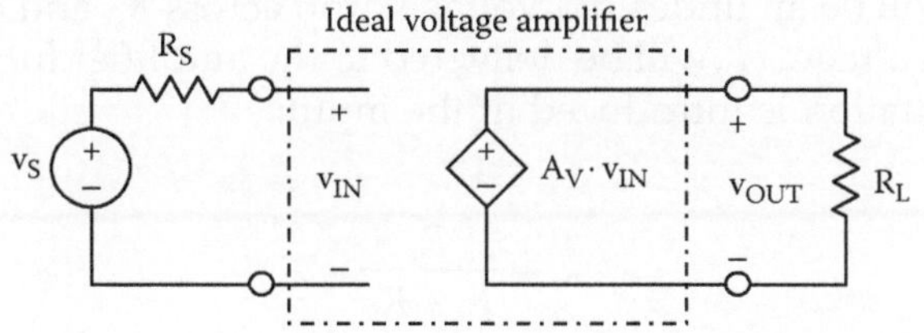

FIGURE 1.5
An ideal voltage amplifier connected to a signal source at the input and a load at the output.

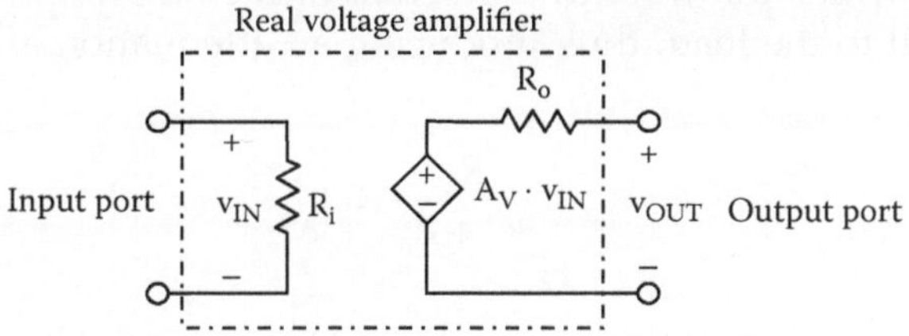

FIGURE 1.6
Two-port model for a real voltage amplifier.

In the real world, no ideal dependent voltage source is available and an amplifier should be implemented using electrical and electronic elements and devices. This will make the amplifier's two-port model slightly different from that of Figure 1.4, although still representing the same basic function. Figure 1.6 is the two-port model for a real amplifier.

Obviously, the two aforementioned models differ in their *input resistance*, R_i, and *output resistance*, R_o. These are simply the Thevenin equivalent resistances seen into the amplifier from the input and output ports, respectively. Connecting a signal source to the input of a real voltage amplifier with finite (not very large) input resistance causes a current to be drawn from the signal source. In other words, R_i and R_s form a voltage divider, as illustrated in Figure 1.7.

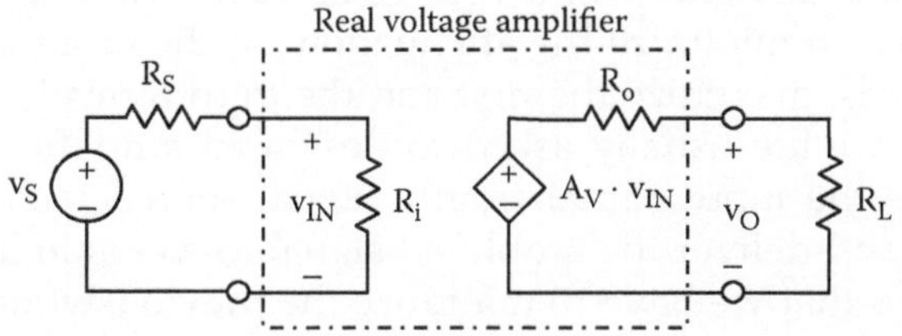

FIGURE 1.7
Connecting a real voltage amplifier to a signal source and a load.

The result will be an undesired voltage drop across R_s, and only a fraction of the source voltage, v_S, will be delivered to the amplifier for amplification. Thus an attenuation is introduced at the input:

$$v_{IN} = \frac{R_i}{R_i + R_S} v_S. \tag{1.4}$$

In addition, as depicted in Figure 1.7, having a nonzero output resistance for a voltage amplifier is also not welcome when it is terminated to a load at the output. This is caused by the voltage division network that is formed by R_o and R_L, which allows only a fraction of the amplified voltage, $A_V v_{IN}$, to be delivered to the load, thus producing an attenuation at the output:

$$v_O = \frac{R_L}{R_L + R_o} A_V v_{IN} \tag{1.5}$$

Combining Equations (1.4) and (1.5) to find the output voltage as a function of the input signal, we have:

$$v_O = \frac{R_L}{R_L + R_o} A_V v_{IN} = \frac{R_L}{R_L + R_o} A_V \frac{R_i}{R_i + R_S} v_S, \tag{1.6}$$

and the *overall voltage gain*, A_{VS}, can be written as:

$$A_{VS} = \frac{v_O}{v_S} = \frac{R_i}{R_i + R_S} A_V \frac{R_L}{R_L + R_o}, \tag{1.7}$$

which consists of the first term representing the attenuation at the input, the second term demonstrating the amplification by the amplifier, and the third term indicating the attenuation at the output. It is obvious that the first and the third terms are smaller than one, but the second term is supposed to be greater than one. Therefore, as long as the overall product is greater than unity, it can be said that the load receives an amplified version of the source signal, v_S. To take advantage of a voltage amplifier as much as possible, it is recommended to minimize the attenuations at the input and the output, or in other words, maximize the first and the third terms in Equation (1.7). As a designer, we are usually asked to design an amplifier with a certain gain to receive the input signal from a signal source with a given input resistance, R_S, and deliver the amplified signal to a certain load. Therefore, the only choices that we have to minimize the previously mentioned attenuations are at the input and the output resistances of our amplifier. This leads to two important recommendations in the design of a voltage amplifier:

make the input resistance much greater than the resistance of the signal source, and the output resistance much smaller than the load resistance:

$$R_i \gg R_S, \tag{1.8}$$

and

$$R_o \ll R_L. \tag{1.9}$$

Meeting these two requirements for a voltage amplifier results in the design of a good voltage amplifier, and the overall voltage gain can be written as:

$$A_{VS} \approx A_V. \tag{1.10}$$

Example 1.1

It is desired to amplify the voltage signal $v_s(t) = 1^{mV} Sin(2\pi 1^{kHz} t)$ provided by a signal source with internal resistance of 1 kΩ, and deliver it to a 1-kΩ resistive load. A voltage amplifier with $R_i = 10$ kΩ, $R_o = 100$ Ω, and $A_V = 100$ is used for this purpose. Figure Ex. 1.1 illustrates the amplifier connected to the signal source at the input, and terminated to the load at the output. Find the overall voltage gain.

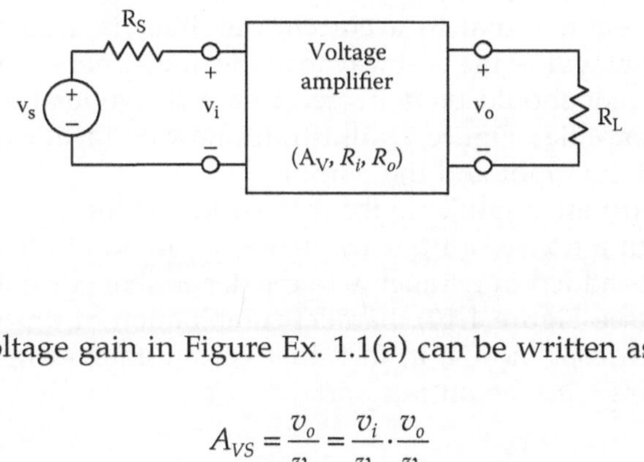

FIGURE EX. 1.1

Solution:

The overall voltage gain in Figure Ex. 1.1(a) can be written as:

$$A_{VS} = \frac{v_o}{v_s} = \frac{v_i}{v_s} \cdot \frac{v_o}{v_i}$$

The first term, v_i/v_s, represents the attenuation at the input. As already illustrated in Figure 1.7, this attenuation is caused by the voltage division between the internal resistance of the signal source and the input resistance of the amplifier:

$$\frac{v_i}{v_s} = \frac{R_i}{R_i + R_S} = \frac{10^{k\Omega}}{10^{k\Omega} + 1^{k\Omega}} \cong 0.91.$$

The second term, v_o/v_i, includes both the pure voltage gain provided by the amplifier and the attenuation at the output. This attenuation is a result of the voltage division between the output resistance of the amplifier and the load resistance:

$$\frac{v_o}{v_i} = A_V \cdot \frac{R_L}{R_L + R_o} = 100 \times \frac{1^{k\Omega}}{1^{k\Omega} + 0.1^{k\Omega}} \cong 91.$$

The overall voltage gain for the amplifier is then achieved:

$$A_{VS} = \frac{v_o}{v_s} = \frac{v_i}{v_s} \cdot \frac{v_o}{v_i} \cong 0.91 \times 91 = 82.81.$$

1.4.2 Current Amplifier

Current amplifiers amplify input current signals and provide the output signal in current form. As illustrated in Figure 1.8, an ideal current amplifier receives a current signal at the input; thus, it is modeled by a short circuit at the input. In addition, as is usual in current-output circuits, to allow the amplified current, $A_I.i_{IN}$, to flow, the output port is shorted. A_I is the current amplifier's gain, which is usually called the *current gain*.

Because the input signal to a current amplifier is a current, a Norton equivalent circuit will be the best way to model the signal source. In addition, the output current should be delivered to a resistive load as was the case for a voltage amplifier. Figure 1.9 illustrates how an ideal current amplifier is connected at the input and the output.

To realize a current amplifier in the real world, the short circuit at the input is replaced with a nonzero input resistance, R_i. In addition, a finite output resistance, R_o, is added in parallel with the dependent current source ($A_I.i_{IN}$) at the output (see Figure 1.10). Parallel combination of this current source and the output resistance is, in fact, the Norton equivalent circuit for the amplifier looking into the output port.

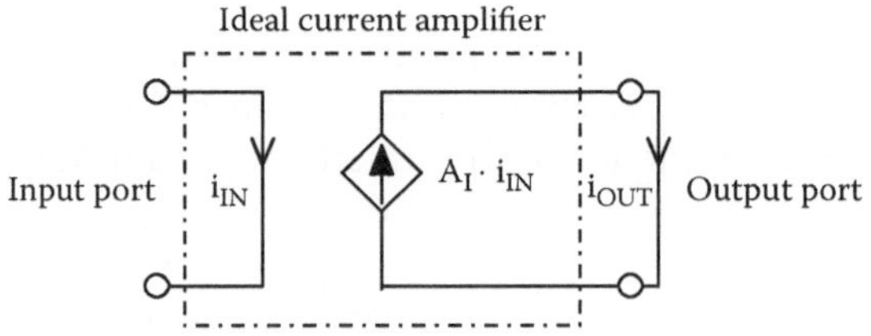

FIGURE 1.8
Two-port model for an ideal current amplifier.

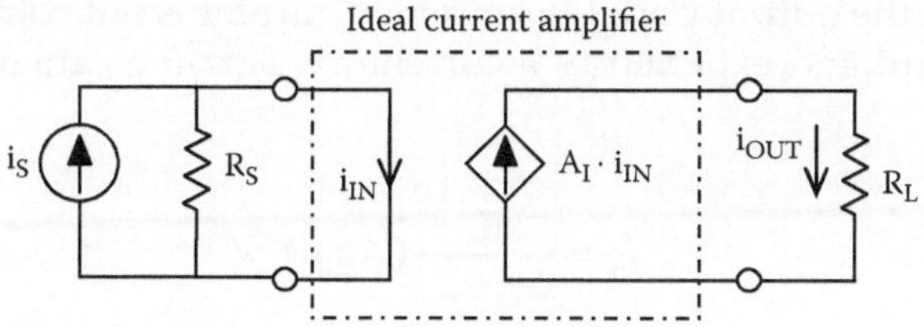

FIGURE 1.9
Termination of an ideal current amplifier to an input signal source and a load.

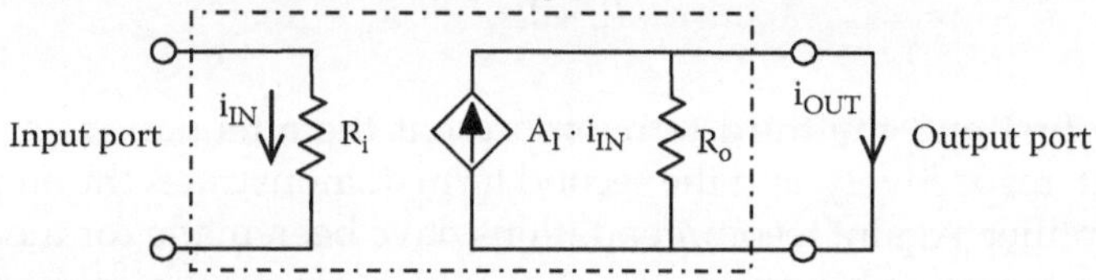

FIGURE 1.10
Two-port model for a real current amplifier.

To analyze the amplification performed by the real current amplifier, a real signal source comprised of i_S, in parallel with its internal resistance, R_S, is connected to the input port, and the output current is delivered to the load, R_L (see Figure 1.11).

Unlike the ideal case where the zero input resistance meant that the amplifier was receiving the entire signal source current ($i_{IN} = i_S$) independently from the internal resistance of the signal source, in the real case the source signal is divided between R_s and R_i, and the input current to the amplifier, i_{IN}, is determined as:

$$i_{IN} = \frac{R_S}{R_i + R_S} i_S \tag{1.11}$$

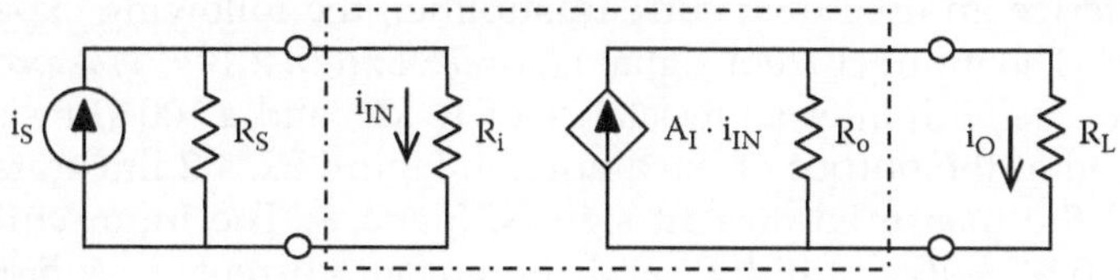

FIGURE 1.11
Termination of a real current amplifier to a real signal source at the input and a load at the output.

Similarly at the output port, the amplified current is not completely delivered to the load and experiences a current division (i.e., attenuation) given by:

$$i_O = \frac{R_o}{R_L + R_o}(A_I i_{IN}) \tag{1.12}$$

As a result, the overall current gain of the amplifier of Figure 1.11 is:

$$A_{IS} = \frac{i_O}{i_S} = \frac{R_S}{R_i + R_S}.A_I.\frac{R_o}{R_L + R_o} \tag{1.13}$$

where the first and the third terms represent the attenuations at the input and output, respectively, and the second term demonstrates the amplification by the amplifier. Again, recommendations have been made for the input and output resistances of a current amplifier, which can minimize the aforementioned attenuations and thus lead to the design of a good current amplifier. When designing a current amplifier, it is recommended to choose the input resistance much smaller than the internal resistance of the signal source, and make the amplifier's output resistance much greater than the load resistance:

$$R_i \ll R_S, \tag{1.14}$$

and

$$R_o \gg R_L. \tag{1.15}$$

Meeting these conditions, the overall gain can be written as:

$$A_{IS} \approx A_I. \tag{1.16}$$

Example 1.2

To characterize an unknown current amplifier, the following experiment is performed. The input current signal $i_s(t) = 1^{\mu A} Sin(2\pi \times 10^{kHz} t)$ is provided by a signal source with internal resistance of 10 kΩ, and a 100-Ω resistive load is connected at the output of the amplifier. Figure Ex. 1.2 illustrates the test setup and the measured current signals, i_i and i_o. The input current, i_i, is measured $0.8^{\mu A} Sin(2\pi \times 10^{kHz} t)$, and the output current, i_o, is $56^{\mu A} Sin(2\pi \times 10^{kHz} t)$. The load is then replaced with a 1-kΩ resistor, which reduces the output current to $40^{\mu A} Sin(2\pi \times 10^{kHz} t)$. Find the pure current gain, input, and output resistances for the amplifier.

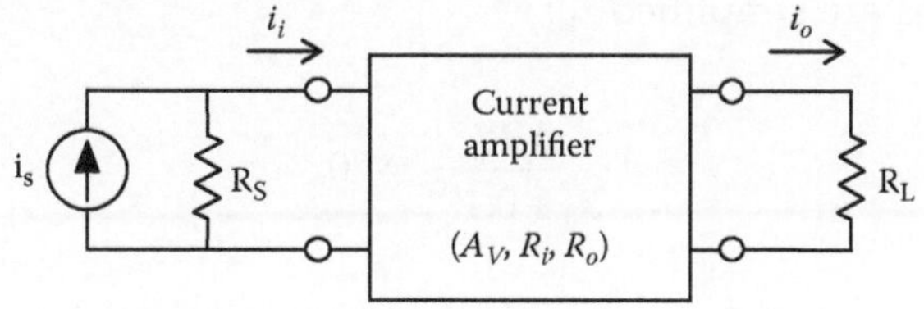

FIGURE EX. 1.2

Solution:

The input resistance can be found by writing the current division relationship at the input:

$$i_i = \frac{R_S}{R_i + R_S}.i_s$$

which if solved for R_i gives:

$$R_i = R_S\left(\frac{i_s}{i_i} - 1\right)$$

$$= 10^{k\Omega}\left(\frac{1^{\mu A}}{0.8^{\mu A}} - 1\right)$$

$$= 2.5^{k\Omega}.$$

The output current is also determined by current division between the output resistance of the amplifier and the load resistance:

$$i_o = \frac{R_o}{R_o + R_L} A_I i_i.$$

The problem is that the above equation includes two unknowns: A_I and R_o. The two values for the output current with different load resistances help us to find the output resistance and the pure gain simultaneously:

$$R_L = 100^{\Omega}: \quad i_o = 56^{\mu A} Sin(2\pi \times 10^{kHz} t) = \frac{R_o}{R_o + 0.1^{k\Omega}}.A_I 0.8^{\mu A} Sin(2\pi \times 10^{kHz} t)$$

and

$$R_L = 1^{k\Omega}: \quad i_o = 40^{\mu A} Sin(2\pi \times 10^{kHz} t) = \frac{R_o}{R_o + 1^{k\Omega}}.A_I 0.8^{\mu A} Sin(2\pi \times 10^{kHz} t).$$

These equations are simplified to:

$$\frac{A_I R_o}{R_o + 0.1^{k\Omega}} = 70$$

and

$$\frac{A_I R_o}{R_o + 1^{k\Omega}} = 50,$$

and give:

$$R_o = 2.15^{k\Omega}$$

and

$$A_I = 73.26 .$$

1.4.3 Transconductance Amplifier

A *transconductance amplifier* is an amplifier that amplifies the voltage applied to its input port and provides a current at its output port, which is proportional to the input voltage. Figure 1.12 is the two-port model for a transconductance amplifier.

Unlike the gain of voltage and current amplifiers, which was dimensionless, the gain of a transconductance amplifier is of conductance nature and will be in units of A/V (Siemens or Mho). This is because such an amplifier converts a voltage to a current, and based on the Ohm's law, it functions as a kind of conductance. For a regular conductance, however, the application of a voltage causes a proportional current to flow across the same terminal. Here, the voltage is applied between two nodes and the proportional current is taken from two different nodes — thus the prefix "trans." The gain of a transconductance amplifier is referred to as the amplifier's *transconductance*.

Figure 1.13 and Figure 1.14 depict a real transconductance amplifier's two-port model and its termination to a real signal source at the input and to a nonzero load at the output. It is clear that nothing is new about this amplifier;

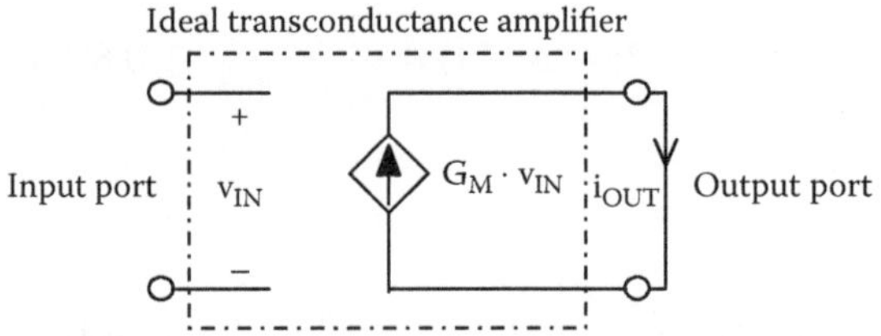

FIGURE 1.12
Two-port model for an ideal transconductance amplifier.

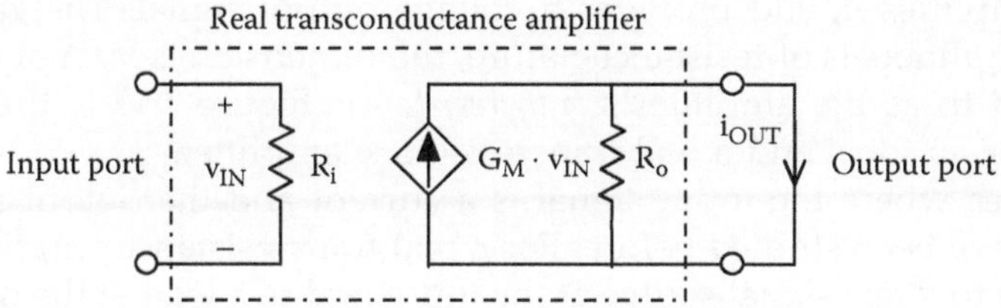

FIGURE 1.13
Two-port model for a real transconductance amplifier.

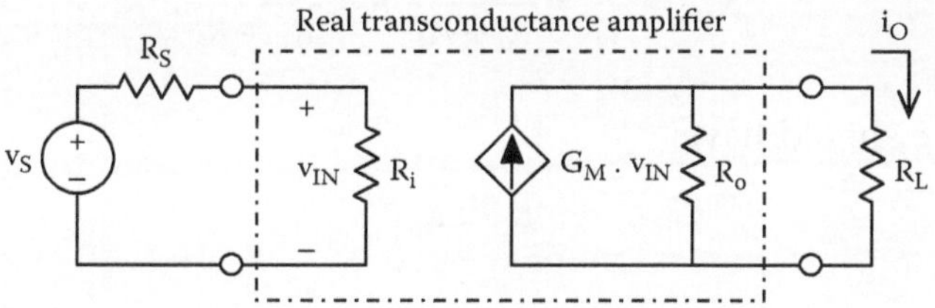

FIGURE 1.14
Termination of a real transconductance amplifier to a real signal source and a load.

we have already studied the cases where the input signal is a voltage and the output signal is a current (in voltage amplifiers and current amplifiers, respectively). So, for the sake of simplicity, without going into the details, the overall transconductance of the circuit in Figure 1.14 is presented as:

$$G_{MS} = \frac{i_O}{v_S} = \frac{R_i}{R_i + R_S} . G_M \cdot \frac{R_o}{R_L + R_o} \tag{1.17}$$

It can be concluded easily that to design a good transconductance amplifier, one should have:

$$R_i \gg R_S, \tag{1.18}$$

and

$$R_o \gg R_L. \tag{1.19}$$

Meeting these conditions, it can be written:

$$G_{MS} \approx G_M. \tag{1.20}$$

1.4.4 Transresistance Amplifier

Based on what we learned about the transconductance amplifier, it can be guessed that a *transresistance amplifier* is a circuit that receives a current input signal, amplifies it, and provides a voltage output signal. The gain of this type of amplifiers is of resistance nature, the dimension is V/A or Ohm, and is referred to as the amplifier's *transresistance*. Figure 1.15 is the two-port models for an ideal and a real transresistance amplifier.

The cases where the input signal is a current and the output signal is a voltage have been studied before. For a real transresistance amplifier that is connected to a real signal source at the input and to a load at the output (see Figure 1.16), we know how the overall transresistance becomes:

$$R_{MS} = \frac{v_O}{i_S} = \frac{R_S}{R_i + R_S}.R_M.\frac{R_L}{R_L + R_o}, \tag{1.21}$$

and why we should have:

$$R_i \ll R_S, \tag{1.22}$$

and

$$R_o \ll R_L. \tag{1.23}$$

to achieve an overall transresistance almost equal to the amplifier's transresistance. Then it can be said:

$$R_{MS} \approx R_M. \tag{1.24}$$

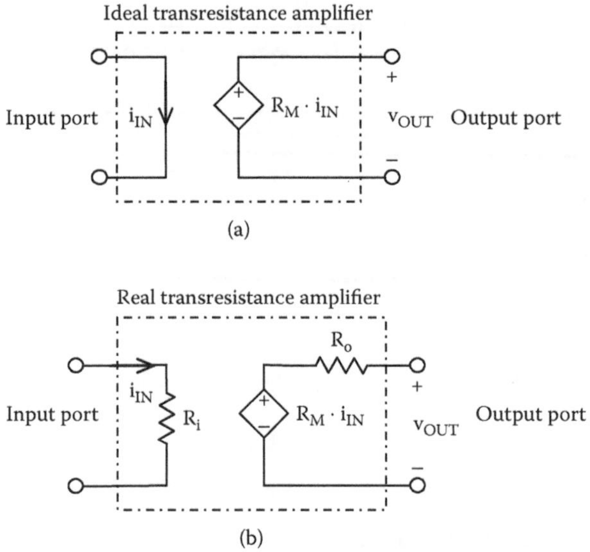

FIGURE 1.15
Two-port models for a transresistance amplifier (a) ideal (b) real.

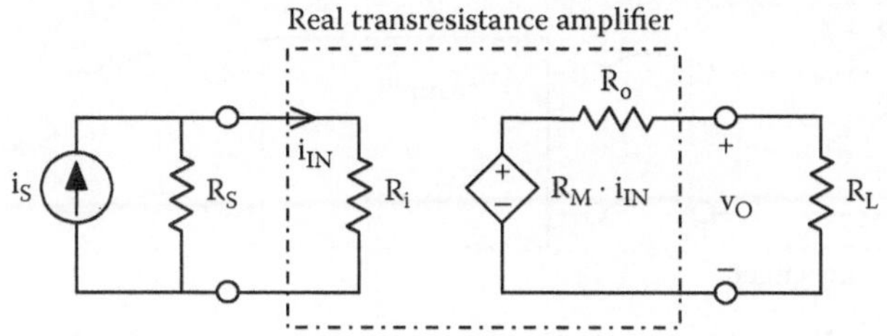

FIGURE 1.16
A real transresistance amplifier connected to a real input signal source and a load.

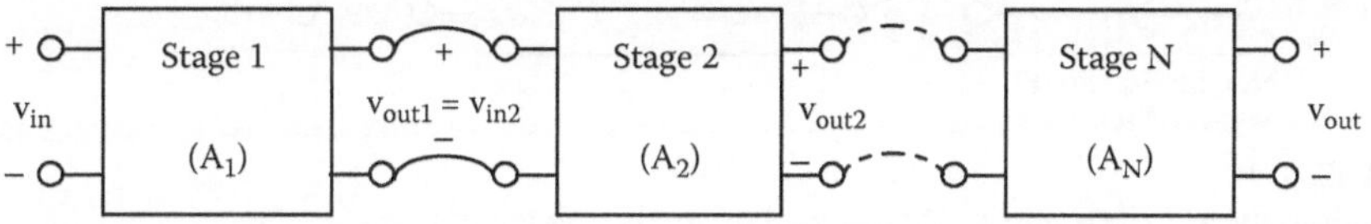

FIGURE 1.17
A cascade of N amplifier stages.

1.5 Cascaded Amplifiers

Different implementations for amplifiers will be studied in Chapter 4. It will be seen that the amount of the gain that each amplifier is capable of providing is limited. In addition, none of the amplifier configurations can exhibit a good gain and excellent input and output resistances at the same time. As will be extensively studied in Chapter 5, in most applications, it is preferred to cascade two or more amplifiers to obtain the desired performance. Figure 1.17 illustrates a cascade of N amplifier stages. In multistage amplifiers, each stage amplifies the signal and delivers it to the next stage. When studying the amplification by the *i-th* stage, the $(i-1)$-th stage performs as the signal source, and the $(i+1)$-th stage is considered as the load. The easiest way to analyze each stage is to model the previous stage by its Thevenin or Norton equivalent circuit, depending on whether the signal is a current or a voltage, and represent the next stage by only a resistance as the load. This is illustrated in Figure 1.18.

1.6 Small-Signal and Large-Signal Amplifiers

As will be discussed in Chapters 4 and 5, amplifiers are comprised of *transistors*, which in general are not linear devices. They have different modes

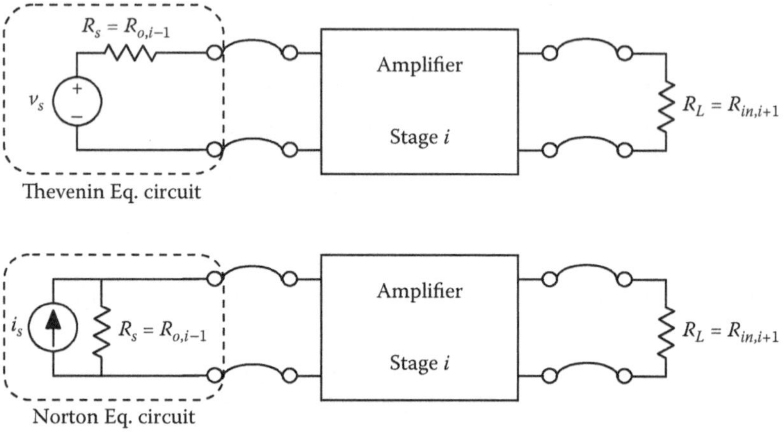

FIGURE 1.18
Modeling the previous and next stages when analyzing stage i.

of operation, which enable them to be used in both linear applications such as amplification and nonlinear applications such as digital circuits and systems. When a transistor is intended to be used as a linear amplifier, it is usually biased in the proper operation mode, and the input signal should be either directly or indirectly applied to one or more terminals of the transistor to be amplified. For the transistor characteristics to be linearized with negligible error, the input signal should be kept very small. A linear model is valid with a small-signal input because the transistor stays in the desired operating mode and the signal-dependent variations in the transistor's currents and voltages are small compared with the established DC values. Under this condition, the transistor is assumed *semilinear,* and the amplifier can be simply analyzed using the rules and methods that are known for linear circuits and systems, of course with minor modifications and considering some assumptions. Such an operation is called *small-signal operation,* and the amplifier will be referred to as a *small-signal amplifier.* If the signal is not small enough to be considered a "small signal" based on the preceding definition, then the circuit's behavior will be called *large-signal operation.*

1.7 A Fundamental Question

As stated, to function as an amplifier, a transistor should be properly biased using a well-designed bias network usually consisting of one or more DC voltage sources and some other elements as required. Therefore, a real amplifier circuit can be better described by Figure 1.19, assuming that it is biased using two DC power supplies, V_{DD} and $-V_{SS}$.

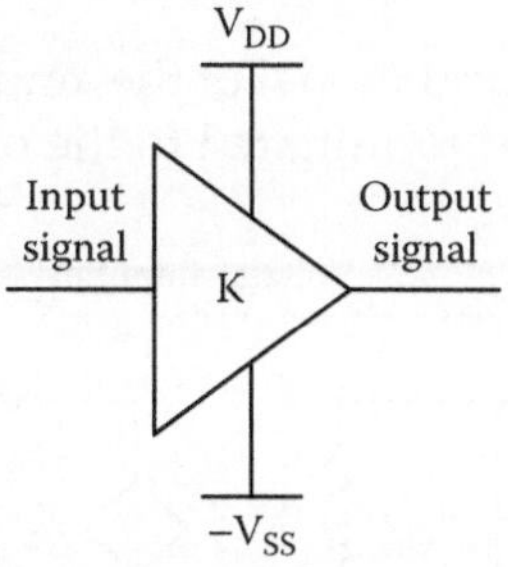

FIGURE 1.19
An amplifier symbol including the DC power sources.

Now, the fundamental and critically important question that so far has probably been forming in your mind can be answered using the illustration of an amplifier in Figure 1.19. The question is:

> Isn't the basic concept of amplification in conflict with the physical principle of "energy conservation"?

In simpler words,

> When amplifying a weak signal and providing a much larger version of it at the output, where does the added energy come from?

The answer is that an amplifier's function is not really taking the same input signal through the amplification path and delivering it at the output. The art of an amplifier is, in fact, sensing the input signal, getting some energy from the DC power supplies, and delivering it to the output in voltage or current (or sometimes electric power) form proportional to the input signal with a proportionality factor usually greater than unity. Therefore,

> An amplifier's function is only getting the electrical energy from the DC power supplies and delivering it at the output as dictated by the input signal.

1.8 Simulation Examples

Example 1.3

A voltage amplifier is used to amplify the voltage signal $v_s(t) = 1^{\mathrm{mV}} Sin(2\pi 1^{\mathrm{kHz}} t)$, provided by a signal source with an internal resistance of 1 kΩ, and deliver the output signal to a 1-kΩ resistive load. The amplifier has a gain of 100 and output resistance of 100 Ω. Simulate the amplifier with the input resistance of 1 kΩ, 10 kΩ, and 100 kΩ, and find the overall gain.

Solution:
Figure Ex. 1.3(a) is the electrical model of the amplifier connected to the input signal source at one end and terminated to the output load at the other end.

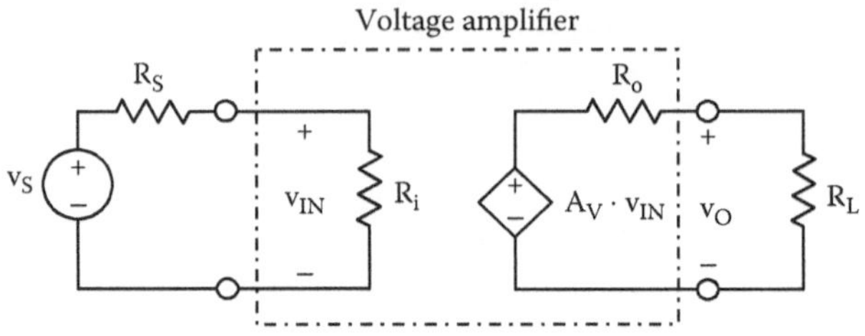

FIGURE EX. 1.3(a)

```
VS S 0 sin(0 1mV 1k)
RS S in 1k
XAMP in 0 out 0 VAMP PARAMS: Ri={Ri} Ro={Ro}
RL out 0 1k
.PARAM Ri=1k Ro=0.1k
.STEP PARAM Ri LIST 100k 10k 1k
.TRAN 1u 5m
.PROBE
.SUBCKT VAMP inp inn outp outn PARAMS: Ri=10k Ro=0.1k
Ri inp inn {Ri}
E int outn inp inn 100
Ro outp int {Ro}
.ENDS
.END
```

The resulting input and output waveforms are presented in Figure Ex. 1.3(b). The upper part illustrates the input voltage and the lower traces are the output voltage. The input voltage results from a voltage division between the internal resistance of the input signal source, R_S, and the input resistance of the amplifier, R_i. For the case where $R_i = R_S = 1$ kΩ, only half of the 1-mV input source signal is delivered to the amplifier, but as the input resistance of the amplifier increases to higher values, a larger portion of the signal is given to the amplifier. The attenuation at the input, v_{in}/v_s, which is 0.5 for $R_i = 1$ kΩ, increases to 0.9 and 0.99 for $R_i = 10$ kΩ and 100 kΩ, respectively.

The input voltage, v_{in}, is then amplified by a gain of 90. This includes the pure gain of the amplifier (100), and the attenuation at the output (0.9). The latter is the result of a voltage division between the 100-Ω output resistance and the 1-kΩ load.

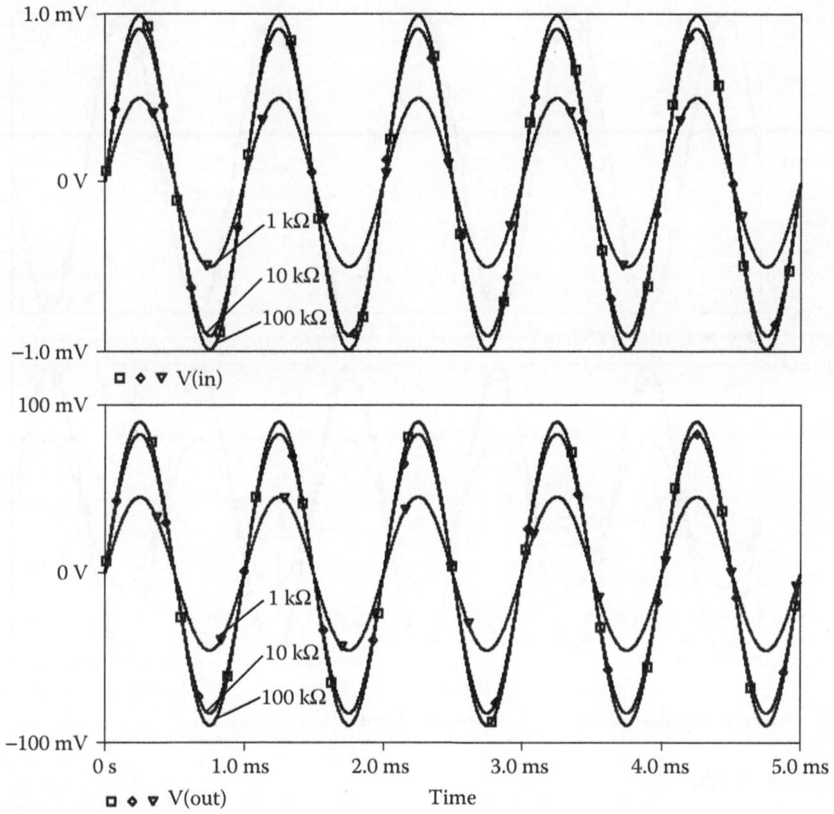

FIGURE EX. 1.3(b)

Example 1.4

A current amplifier with $R_i = 1\text{k}\Omega$, $R_o = 10\ \text{k}\Omega$, and $A_I = 100$ is used to amplify the input signal $i_s(t) = 1^{\text{mA}}Sin(2\pi 1^{\text{kHz}}t)$, provided by a signal source with the internal resistance of 100 kΩ. Simulate the circuit, given in Figure Ex. 1.4(a), and find the overall current gain for the load resistances 0.1 kΩ, 1 kΩ, and 10 kΩ.

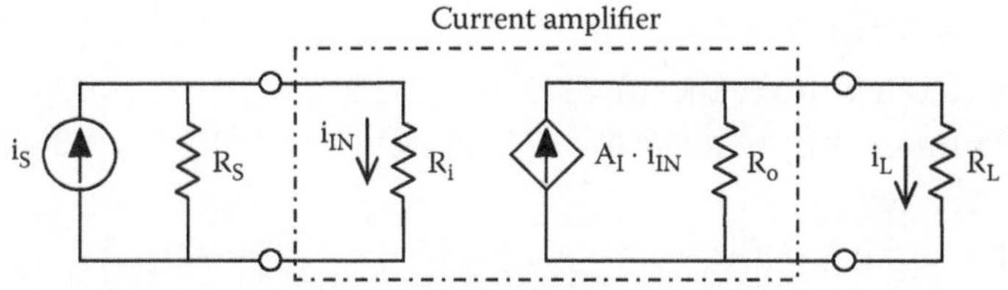

FIGURE EX. 1.4(a)

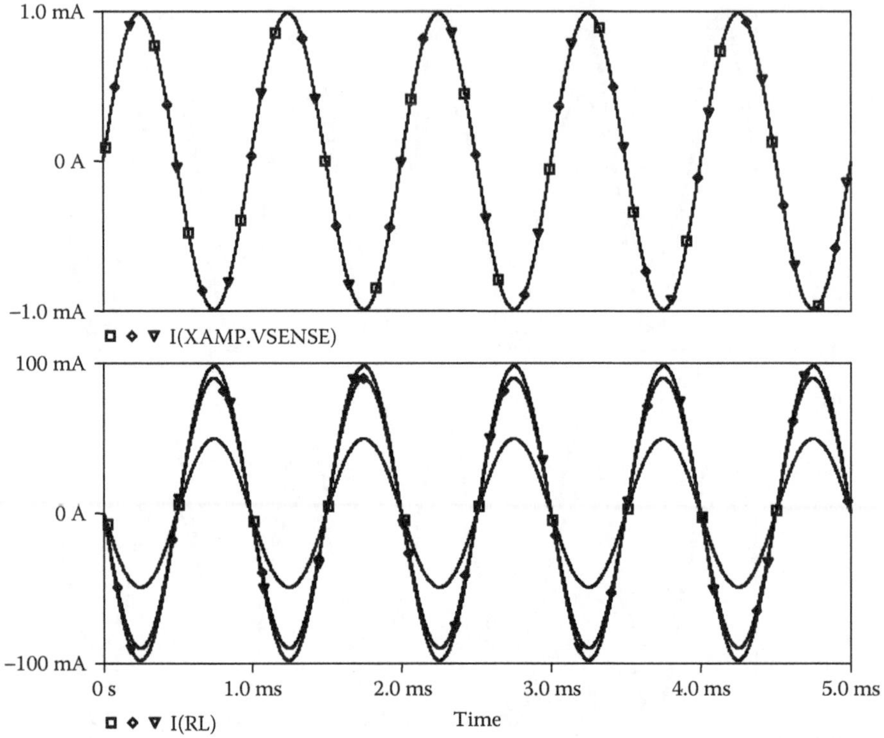

FIGURE EX. 1.4(b)

Solution:

The output current of a current-controlled current source in SPICE is defined to be a function of the current passing through a voltage source somewhere in the circuit. Thus, to implement the dependent current source, a voltage source is necessary at the input. The trick is to consider a 0-V voltage source (V_{SENSE}, equivalent to a short circuit) in series with the input resistance to sense the current that enters the amplifier.

```
IS 0 in sin(0 1m 1k)
RS in 0 100k
XAMP in 0 out 0 IAMP PARAMS: Ri={Ri} Ro={Ro}
RL out 0 {RL}
.PARAM Ri=1k Ro=10k RL=1k
.STEP PARAM RL LIST 10k 1k 0.1k
.TRAN 1u 5m
.PROBE
.SUBCKT IAMP inp inn outp outn PARAMS: Ri=1k Ro=10k
Ri inp ins {Ri}
VSENSE ins inn 0
```

```
F outp outn VSENSE 100
Ro outp outn {Ro}
.ENDS
.END
```

Figure Ex. 1.4(b) on the previous page shows the signals resulting from simulation. The input current is $i_{in} = 0.99^{mV}Sin(2\pi 1^{kHz}t)$. This means that nearly all the signal is delivered to the amplifier, due to its much smaller input resistance compared with the internal resistance of the signal source. This signal is then amplified with a gain of 100.

Different values for the load resistance causes different amplitudes for the output current, i_L, resulting in different values for the overall gain. The overall gain measured 98, 90, and 49.5 for R_L = 100 Ω, 1 kΩ, and 10 kΩ, respectively.

Example 1.5

Two identical voltage amplifiers with R_i = 10 kΩ, R_o = 100 Ω, and A_V = 100 are cascaded to realize a voltage amplifier with a gain of 10^4. This large voltage gain is needed to amplify a very weak neural signal modeled in Figure Ex. 1.5(a). The input signal $v_s(t)$ is provided by a signal source with an internal resistance of 1 kΩ, and the output signal is delivered to a load resistance of 100 Ω. The circuit is illustrated in Figure Ex. 1.5(b). Simulate the circuit, determine how much the signal is amplified by each stage, and find the overall voltage gain.

Solution:

```
VS S 0 PWL REPEAT FOR 5 (0 0V 0.1m 0V 0.15m -10uV 0.2m 0V
+0.5m 400uV 0.9m -100uV 1.2m -150uV 1.5m -140u 1.9m -10uV
+2.1m 0 20m 0) ENDREPEAT
RS S in 1k
XAMP1 in 0 out1 0 VAMP PARAMS: Ri={Ri} Ro={Ro}
XAMP2 out1 0 out 0 VAMP PARAMS: Ri={Ri} Ro={Ro}
RL out 0 0.1k
.PARAM Ri=10k Ro=0.1k
.TRAN 1u 100m
.PROBE
.SUBCKT VAMP inp inn outp outn PARAMS: Ri=10k Ro=0.1k
Ri inp inn {Ri}
E int outn inp inn 100
Ro outp int {Ro}
.ENDS
.END
```

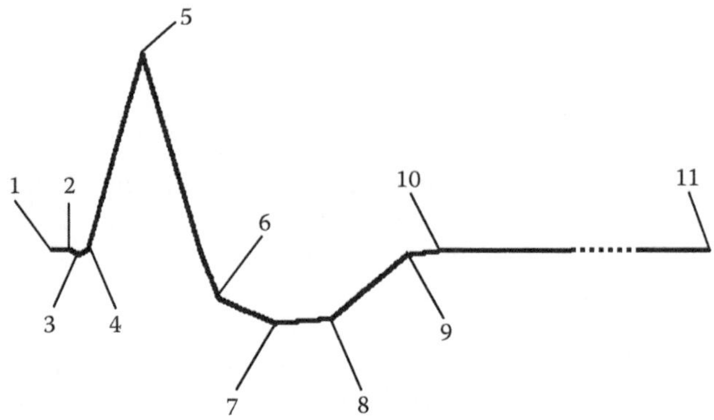

Corner point	Time (ms)	Voltage (μV)
1	0	0
2	0.1	0
3	0.15	−10
4	0.2	0
5	0.5	400
6	0.9	−100
7	1.2	−150
8	1.5	−140
9	1.9	−10
10	2.1	0
11	20	0

FIGURE EX. 1.5(a)

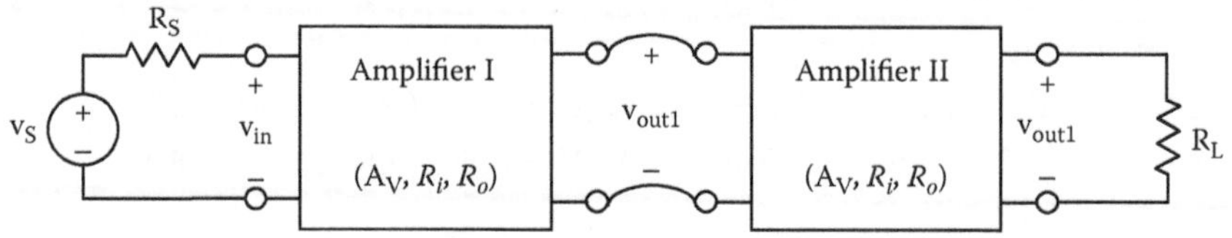

FIGURE EX. 1.5(b)

Figure Ex. 1.5(c) depicts the signals at the input and output of both amplifier stages. Peak-to-peak amplitude of the input source signal (v_s) is 550 μV. The voltage at the input of the first stage (v_{in}) has a peak-to-peak amplitude of 497 μV, which is 90% of the input source signal, indicating the attenuation at the input. This signal is then amplified and appears at the output of the first stage as (v_{out1}) with peak-to-peak amplitude of 49.17 mV. This indicates that the gain of the first stage is nearly 98.9. This includes the amplification with a gain of 100 and a slight attenuation at the output of the first stage

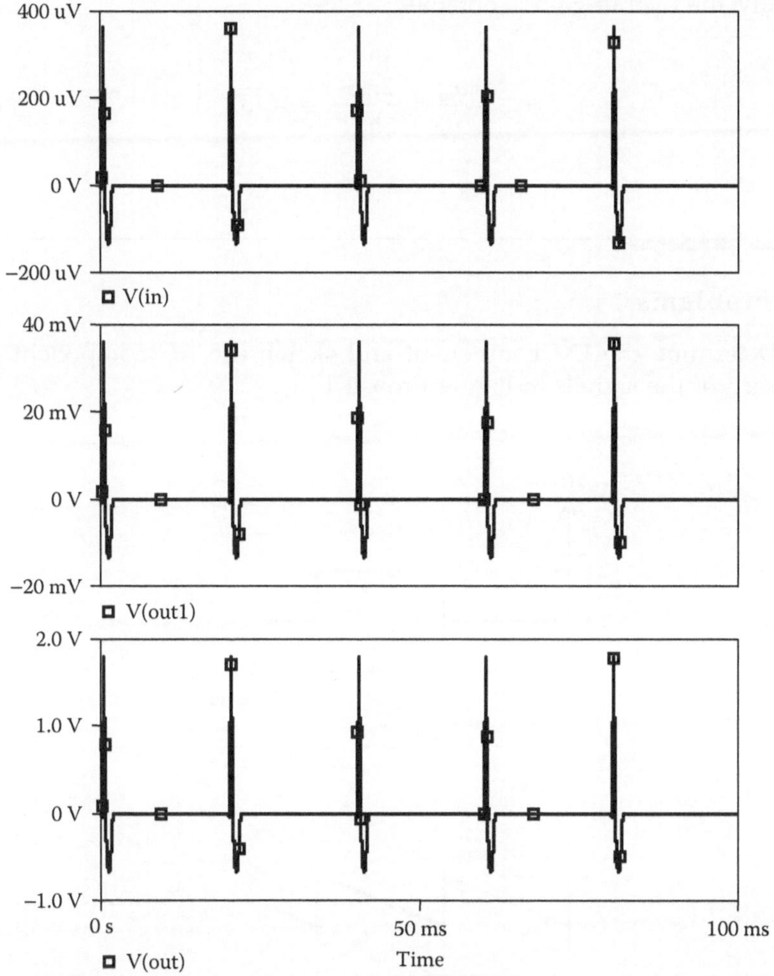

FIGURE EX. 1.5(c)

(0.989). This attenuation demonstrates the loading effect of the second stage on the first stage. Because the input resistance of the second stage is 100 times larger than the output resistance of the first stage, this loading is so small, thus negligible. The voltage v_{out1} is then amplified by the second stage with a total gain of 50, resulting in peak-to-peak amplitude of 2.46 V for the output voltage v_{out}. The gain of the second stage is half of its pure gain ($A_V = 100$) because it is loaded by a load resistance, which is equal to the output resistance:

$$A_{V2} = A_V \cdot \frac{R_L}{R_L + R_o} = 100 \times \frac{100^{\Omega}}{100^{\Omega} + 100^{\Omega}} = 50$$

Finally, the overall gain is obtained:

$$A_{VS} = \frac{v_{out}}{v_s} = \frac{2.46^V}{550^{\mu V}} \cong 4473 .$$

1.9 Problems

1.1 Determine the DC component and sketch the AC component for each of the signals in Figure Prob. 1.1.

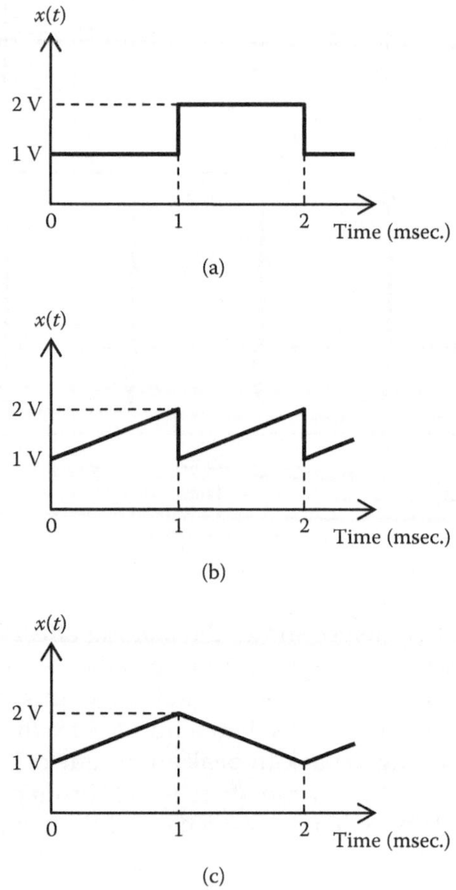

FIGURE PROB. 1.1

1.2 A voltage amplifier is designed to reject the DC component of the input signal, and amplify its AC component by a gain of 100. What is the output of this amplifier when the voltage $2.5^{V} + 1^{mV}\text{Sin}(120\pi t)$ is applied at the input?

1.3 To obtain a voltage gain of 2×10^4, two voltage amplifiers are cascaded as illustrated in Figure Prob. 1.3. Specifications of the amplifiers are given in the following table:

	A_V	R_i	R_o
Amplifier I	100	100 kΩ	1 kΩ
Amplifier II	200	5 kΩ	100 Ω

(a) Determine the overall gain.

(b) Explain why the expected overall gain is not achieved using thses amplifiers.

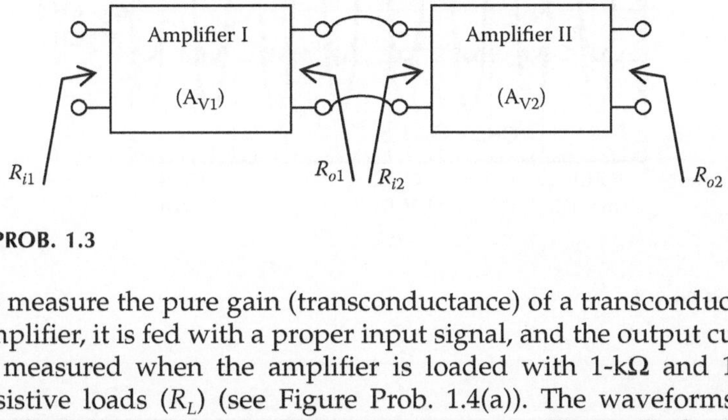

FIGURE PROB. 1.3

1.4 To measure the pure gain (transconductance) of a transconductance amplifier, it is fed with a proper input signal, and the output current is measured when the amplifier is loaded with 1-kΩ and 10-kΩ resistive loads (R_L) (see Figure Prob. 1.4(a)). The waveforms that resulted from this experiment are given in Figure Prob. 1.4(b). Assuming that the trace on CH.1 depicts the input voltage, and CH.2 and CH.3 depict the output voltage for $R_L = 10$ kΩ and 1 kΩ, respectively, determine:

(a) The overall transconductance of the circuit, $G_{MS} = i_o/v_i$, for each value of R_L

(b) The pure transconductance of the amplifier, G_M

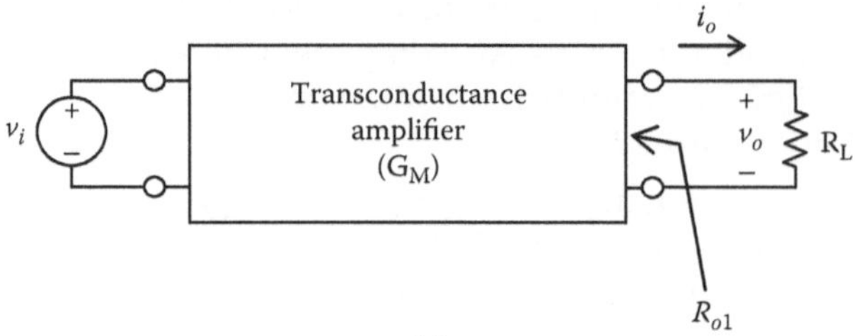

(a)

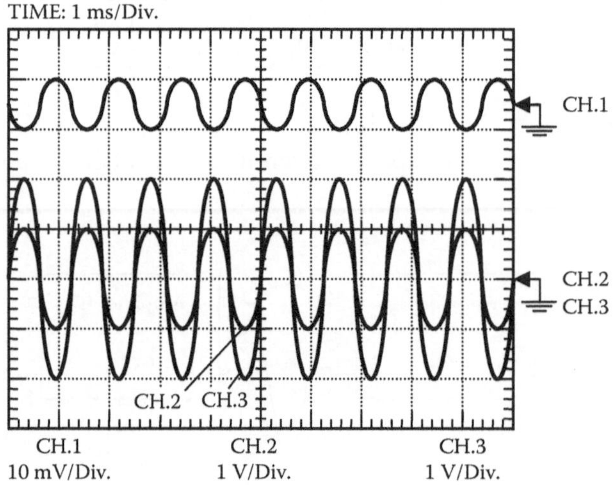

(b)

FIGURE PROB. 1.4

Chapter 2

Transistors

2.1 Introduction

After less than 60 years since the invention of the first transistor, it is horrible to even imagine what will happen if all the transistors in the world suddenly refuse to work. Computers and computerized systems, medical equipment, transportation, navigation systems, communication systems, security systems, and so many other aspects of daily activities are affected by solid-state electronic circuits that involve devices called transistors.

Common types of transistors are introduced in this chapter, and their basic function and modes of operation are studied.

2.2 Basic Concepts

A transistor has several good and precise descriptions, but a transistor can also be simply considered as a device that provides a controllable path for electric current to flow. Unlike regular switches, relays, and vacuum tubes that sometimes demonstrate a similar functionality, a transistor is a semiconductor device that can be fabricated in a very small size. This unique feature of semiconductor technology allows for the integration of millions of transistors on a fairly small area.

As illustrated in Figure 2.1, the current, i, flowing from A to B is controlled by the third terminal of the transistor, C.

A transistor is usually employed in one of the following two major applications:

- In *switching applications*, the channel *AB* is either shorted or opened by the control terminal, C. In this case, the transistor performs as a switch. Although the switch realized by a transistor is not ideal, in many applications it is good enough to be approximated as an ideal

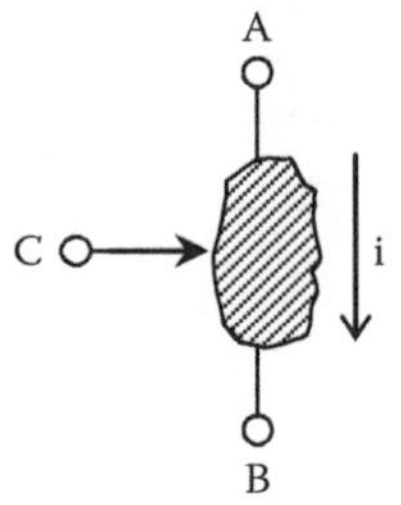

FIGURE 2.1
General illustration of a transistor.

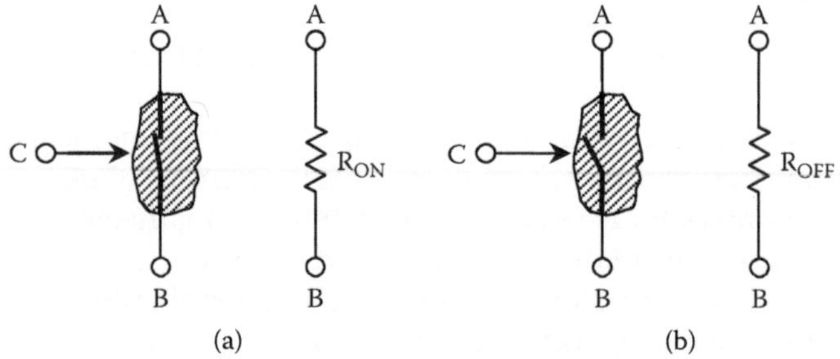

FIGURE 2.2
Transistor as a switch (a) closed (b) open.

switch. Simplifying its role, a transistor in the closed position is equivalent to a resistor with a relatively small resistance, R_{ON}. In addition, when it is open, it can be simply assumed as a resistor with a very high resistance, R_{OFF} (see Figure 2.2).

- In *linear applications*, the transistor always allows the current to flow from *A* to *B*, but the channel *AB* will be under the influence of the control terminal, *C*. In other words, in these applications the current flowing from *A* to *B*, *i*, is a function of the *control variable*, which is the control terminal's voltage or current. This is illustrated in Figure 2.3.

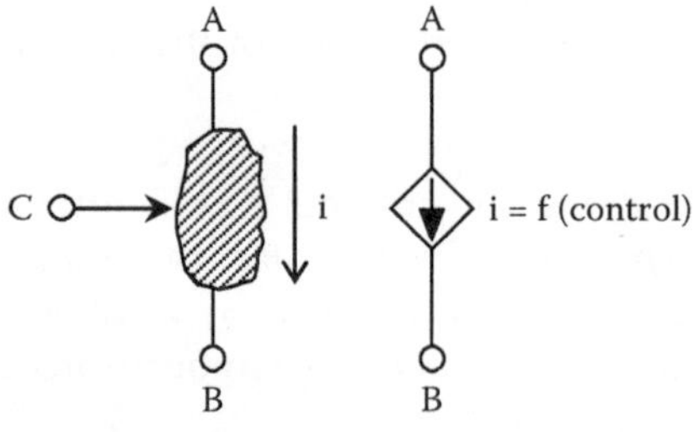

FIGURE 2.3
Transistor in linear applications.

In general, the behavior of a transistor is nonlinear, but, as will be explained in Chapter 4, certain conditions are usually applied to its terminals under which the current i will be a linear function of the control variable.

In semiconductor technology, transistors can be implemented in different ways and still have the same functionality described previously. Bipolar junction transistors (BJTs or simply bipolar transistors) and metal-oxide semiconductor field-effect transistors (MOSFETs or simply MOS transistors) are the two most popular types of transistors in academia and industry, and are discussed throughout this book.

Because the focus of this book is mainly on the introduction of amplifier analysis methods, transistors are studied here only in terms of their simplified functionality and the relationship between the associated currents and voltages and not their physical structure and behavior. The interested reader is referred to the textbooks that discuss these aspects of transistors in more detail [3]–[5].

2.3 Metal-Oxide Semiconductor Field-Effect Transistor

MOSFETs are known as the dominant type of transistor in the industry that are currently used to make electronic integrated circuits (ICs). The structure of this class of transistors is based on three layers: *metal*, *oxide*, and *semiconductor*. As mentioned earlier, a transistor is a device that provides a controllable path for electric current to flow. In MOS transistors, the conductive path through which the current flows is called the *channel*. In this family of transistors, an electric field induced by the control signal affects the electric characteristics of the channel, thus the term: *field-effect*.

Two major classes of MOS transistors, with different types of current-conducting channels, can be described. *Depletion* MOS transistors are made to conduct current without applying any effort to establish the channel, as it is already available after the transistor is fabricated. Depletion transistors were widely used in early days of MOS technology. For a long time now, the dominant class of MOS transistors is the *enhancement* MOS transistor, in which the channel is formed by applying certain voltages to the proper terminals of the transistor. Throughout this book, the enhancement MOS transistor is the default type of MOS transistor.

Two types of MOS transistors are available, which are identical in general physical structure, basics of operation, and functionality, and different in some of the electrical aspects. Figure 2.4 presents the circuit symbols for these two types of MOS transistors, known as *negative-channel metal-oxide semiconductor* (NMOS) and *positive-channel metal-oxide semiconductor* (PMOS) transistors, with the three terminals called the drain (D), the gate (G), and the source (S). The channel connects the drain and the source, and is controlled by the voltage applied to the gate. MOS transistors are known as

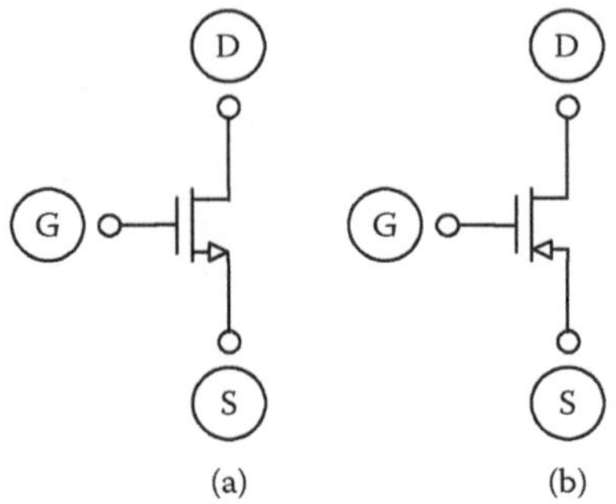

FIGURE 2.4
Circuit symbols for (a) NMOS and (b) PMOS transistor.

voltage-controlled devices because it is the voltage applied to the gate that controls the characteristics of the channel and, consequently, the operation of the transistor.

MOS transistors are primarily described for the NMOS case in this section, and then the differences between NMOS and PMOS transistors will be studied in the next section.

2.3.1 NMOS Transistor

The key variables describing an NMOS transistor are depicted in Figure 2.5. The gate current for an MOS transistor is almost always zero*:

$$i_G = 0, \tag{2.1}$$

and that the source and the drain currents are equal:

$$i_D = i_S. \tag{2.2}$$

Three basic modes** of operation are defined for an NMOS transistor: *cutoff*, *triode*, and *saturation* modes. In *cutoff mode*, the transistor performs as an open switch, which does not allow any current to flow. In this mode, all the currents will be zero:

$$i_D = i_S = i_G = 0. \tag{2.3}$$

An NMOS transistor will be in cutoff mode when the gate-source voltage, v_{GS}, is less than a specific voltage level called the *threshold voltage*, V_{TH}.

* This is true for both DC conditions and AC mode, provided that the signal frequency is not so high that the internal parasitic capacitances of the transistor are not negligible anymore. The latter will not happen unless the transistor is operated in the "high-frequency" band, which will be studied in Chapter 8.

** These are just the *basic* modes of operation, which are simplified enough for a basic course in MOS circuit analysis.

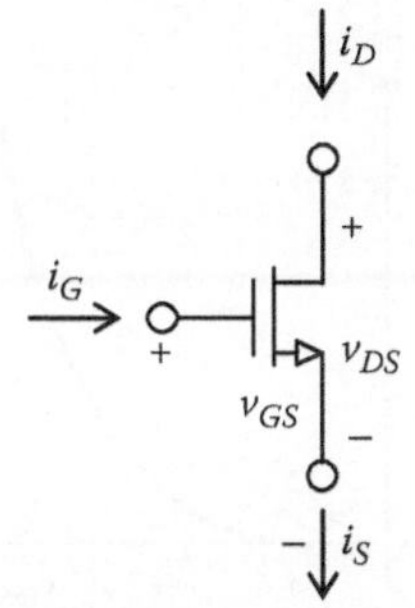

FIGURE 2.5
Key voltages and currents for an NMOS transistor.

Threshold voltage is one of the specifications of the transistor, with a value that is determined by a variety of contributing factors, from physical parameters, which remain almost constant after it is fabricated, to the factors that depend on how the transistor is electrically connected and conditioned. To simplify the discussions and analyses, it is assumed in this book that the threshold voltage has a constant value independent of electrical conditions of the transistor.*

If the gate-source voltage is equal to or greater than V_{TH}, the transistor can be in either *saturation mode* or *triode mode* depending on the amount of drain-source voltage, v_{DS}. If v_{DS} is equal to or greater than a certain voltage level, called "the minimum drain-source voltage," $V_{DS}(Sat.)$, the transistor will be in *saturation mode*.

For a transistor in saturation mode, drain current is a quadratic function of gate-source voltage:

$$i_D = \frac{\beta}{2}(v_{GS} - V_{TH})^2, \tag{2.4}$$

in which β is the transistor's voltage-to-current conversion factor,** with the dimension of A/V^2. The parameter β itself, is a function of some physical and technology-related parameters and the size of the transistor's channel:

$$\beta = k'\frac{W}{L}, \tag{2.5}$$

where k' represents the physical and technology-related parameters, and W and L are the width and length of the transistor's channel. The parameter k' has the same dimension as β does as the W/L ratio is dimensionless. The

* Interested readers are encouraged to study "body effect," "body bias," or "back bias" concepts for the contribution of electrical issues to the threshold voltage of an MOS transistor [6]–[7].

** It is also called the *transconductance factor*, which should not be mistaken for the *transconductance* of the transistor to be introduced in Chapter 4.

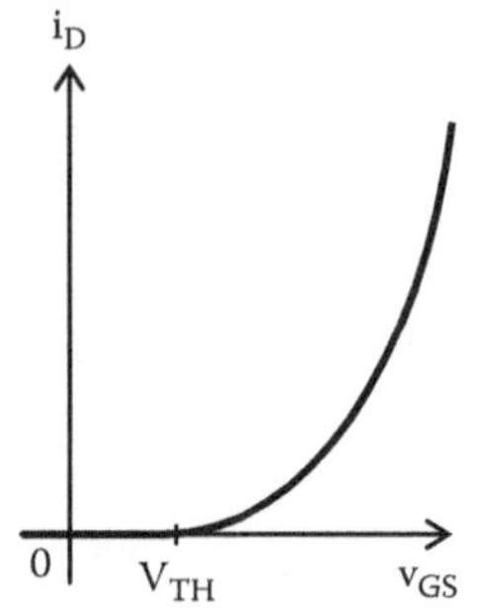

FIGURE 2.6
$i_D - v_{GS}$ characteristic curve for an NMOS transistor.

square-law $i_D - v_{GS}$ characteristic curve for an NMOS transistor expressed by Equation (2.4) is plotted in Figure 2.6.

The voltage $V_{DS}(Sat.)$ is a function of both the transistor's transconductance factor (β) and the DC component of its drain current:

$$V_{DS}(Sat.) = \sqrt{\frac{2I_D}{\beta}} \tag{2.6}$$

Now, one can ask how a transistor with such a nonlinear behavior can play an important role in linear applications. In fact, as will be explained in Chapter 4, in such cases the transistor will be operated only on a very small portion of the quadratic characteristic curve. This way, the curve can be linearized (i.e., approximated to a line segment) at and around a certain point. As will be explained later, the transistor can be operated only on a very small portion of the characteristic curve that can be approximated to a line segment, where the transistor will be able to demonstrate linear operation.

The gate-source voltage of an MOS transistor is comprised of two parts: one is the transistor threshold and the other is a term that should be determined according to the current flowing through the transistor. If Equation (2.4) is rewritten for v_{GS} we will have:

$$v_{GS} = V_{TH} + \sqrt{\frac{2i_D}{\beta}}. \tag{2.7}$$

The above terms can be described as follows: for an NMOS transistor to have a drain current of i_D, the gate-source voltage should be high enough so that the channel can form (above V_{TH}), and then increased further to make the intended current flow through the transistor. The latter term is called the *overdrive voltage*, V_{OD}, which gives the drain current on the semiparabolic part of the device's $i_D - v_{GS}$ characteristic curve. If a transistor is to have a drain current of I_D, its overdrive voltage is determined as:

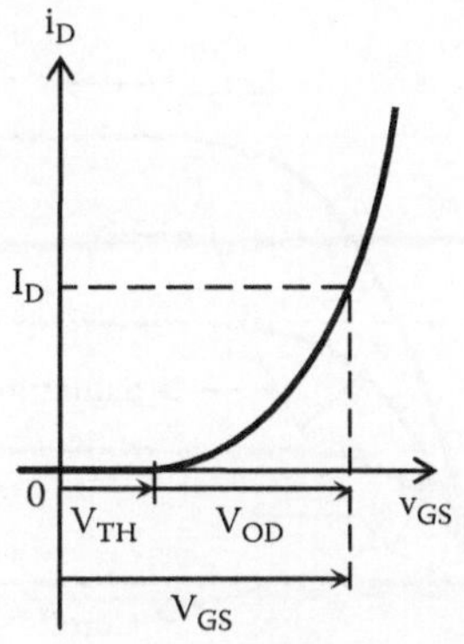

FIGURE 2.7
Illustration of the two terms in the gate-source voltage of an NMOS transistor.

$$V_{OD} = \sqrt{\frac{2I_D}{\beta}}. \tag{2.8}$$

This concept is illustrated in Figure 2.7. It is interesting that the overdrive voltage, V_{OD}, has the same relationship that the minimum drain-source voltage, $V_{DS}(Sat.)$, does.

Example 2.1

An NMOS transistor with W = 10 μm, L = 1 μm, V_{TH} = 0.8 V, and k' = 120 $\mu A/V^2$ is intended to operate in saturation. Find the required gate-source voltage for a drain current of 1 mA.

Solution:

The parameter β is calculated as:

$$\beta = k'\frac{W}{L} = 120^{\mu A/V^2} \times \frac{10^{\mu m}}{1^{\mu m}} = 1.2^{mA/V^2}$$

The required gate-source voltage is determined using Equation (2.7):

$$V_{GS} = V_{TH} + \sqrt{\frac{2I_D}{\beta}}$$

$$= 0.8^V + \sqrt{\frac{2\times 1^{mA}}{1.2^{mA/V^2}}}$$

$$= 2.09^V.$$

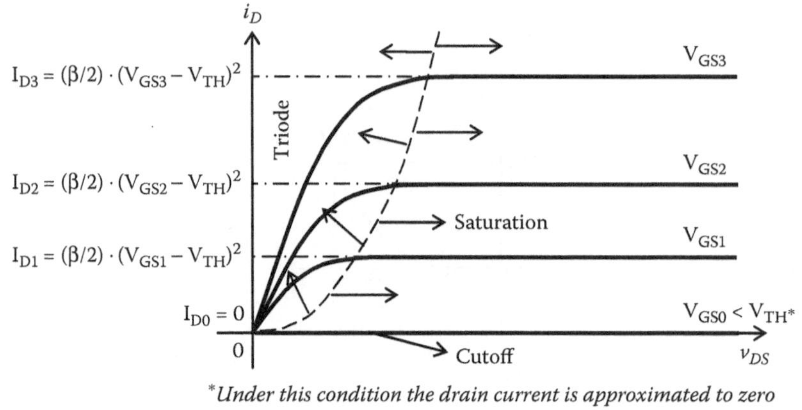

FIGURE 2.8
$i_D - v_{DS}$ characteristic curves for an NMOS transistor.

An MOS transistor also has a second characteristic curve that expresses the relationship between the drain current and the drain-source voltage, called $i_D - v_{DS}$ characteristics, which is illustrated in Figure 2.8. The $i_D - v_{DS}$ characteristic curve is plotted based on the assumption that in the saturation mode the drain current is not a function of v_{DS}, as stated by Equation (2.4). Instead, i_D is considered as a function of v_{GS}, which is treated as a parameter (not the main variable) when plotting this curve.

In practice, electrical behavior of the channel is affected by the voltage across it, which is v_{DS}. The physical phenomenon that causes this effect is known as *channel-length modulation*. Because of this phenomenon, the drain current in saturation mode proportionally follows the variations in the drain-source voltage. To include the channel-length modulation effect, the characteristic equation of an MOS transistor becomes:

$$i_D = \frac{\beta}{2}(v_{GS} - V_{TH})^2(1 + \lambda v_{DS}), \tag{2.9}$$

where λ is the channel-length modulation parameter. This parameter is one of the specifications of an MOS transistor. It is a small number typically in the range of 0.01 V^{-1} to 0.02 V^{-1}. The $i_D - v_{DS}$ characteristics for a real transistor, which includes the channel-length modulation effect, is illustrated in Figure 2.9.

For an NMOS transistor to be in *triode mode*, the gate-source voltage should be greater than the threshold voltage ($v_{GS} > V_{TH}$), and the drain-source voltage should be smaller than the minimum drain-source voltage ($v_{DS} < V_{DS}(Sat.)$). For an NMOS transistor in triode mode, the drain-source can accept any voltage level between zero and $V_{DS}(Sat.)$. When in triode mode, the transistor is still capable of conducting current, but with a different physical mechanism and of course a different relationship for the drain current:

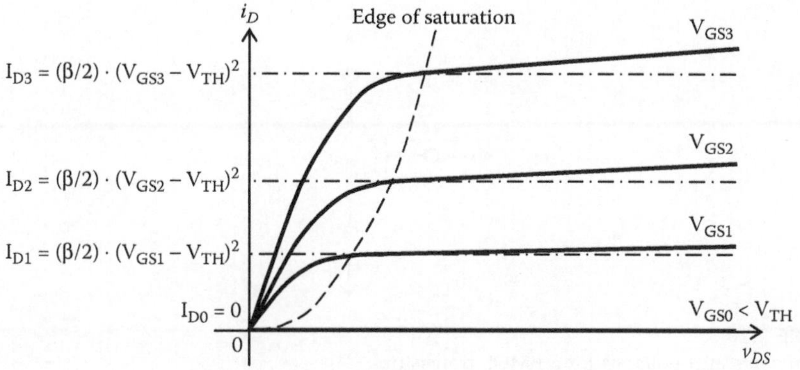

FIGURE 2.9
Adding the channel-length modulation effect to the $i_D - v_{DS}$ characteristic curves for an NMOS transistor.

$$i_D = \beta.\left((v_{GS} - V_{TH})v_{DS} - \frac{v_{DS}^{\ 2}}{2} \right), \tag{2.10}$$

which expresses the characteristic curvatures in triode mode depicted in Figure 2.8. This relationship holds so long as the drain-source voltage, v_{DS} is smaller than $V_{DS}(Sat.)$, and turns out to become Equation (2.4) at the onset of saturation (i.e., when v_{DS} reaches $V_{DS}(Sat.)$). This can be simply shown by replacing the drain-source voltage in Equation (2.10) by $V_{DS}(Sat.)$ defined in Equation (2.6), and derive the drain current in saturation mode presented in Equation (2.4).

Electrical conditions for an NMOS transistor to operate in cutoff, triode, and saturation modes are summarized as follows:

$$\begin{cases} v_{GS} < V_{TH} \Rightarrow \text{Cutoff Mode} \\ v_{GS} \geq V_{TH} \begin{cases} v_{DS} > V_{DS}(Sat.) \Rightarrow \text{Saturation Mode} \\ v_{DS} \leq V_{DS}(Sat.) \Rightarrow \text{Triode Mode} \end{cases} \end{cases} \tag{2.11}$$

2.3.2 PMOS Transistor

As mentioned earlier, the PMOS transistor is considered as the complementary counterpart of the NMOS transistor. In other words, because of the semiconductor materials used in its physical structure, current flow through a transistor is dominantly determined by negative electric charges in NMOS transistors and by positive charges in PMOS transistors. As a result, and remembering the conventions for the direction of currents in electric circuits,

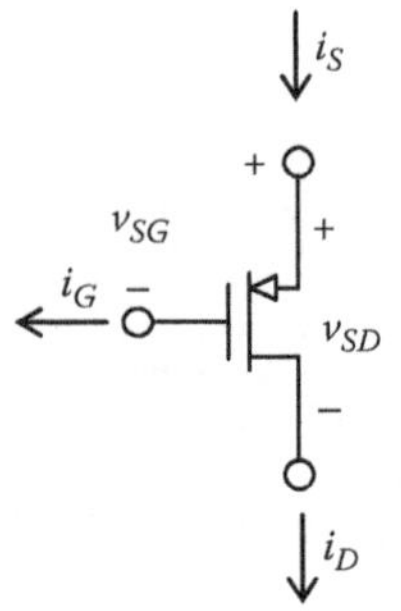

FIGURE 2.10
Key currents and voltages for a PMOS transistor.

for a PMOS transistor all the currents flow in opposite directions compared to those of an NMOS transistor (i.e., from source to drain). Another consequence of the difference between the materials used in their structure is the opposite polarity of the voltages applied between their terminals. Key voltages and currents describing a PMOS transistor's functionality are depicted in Figure 2.10. To use the same descriptive sentences that were used for NMOS devices and avoid negating the voltages, it is preferred to embed the negative sign in how the signal is named, thus using v_{SG} and v_{SD} as opposed to $-v_{GS}$ and $-v_{DS}$, respectively. It should also be noted that the threshold voltage for a PMOS transistor is always negative.* To avoid confusion, the NMOS transistor's formulas will also be used for the PMOS transistor in which the threshold voltage is replaced by the absolute value of the PMOS transistor's threshold voltage.

With the terminology in Figure 2.10 for a PMOS transistor, we have the $i_D - v_{SG}$ characteristic equation in saturation mode as:

$$i_D = \frac{\beta}{2}\left(v_{SG} - |V_{TH}|\right)^2, \tag{2.12}$$

and in triode mode as:

$$i_D = \beta\left(\left(v_{SG} - |V_{TH}|\right).v_{SD} - \frac{v_{SD}{}^2}{2}\right). \tag{2.13}$$

Electrical conditions for a PMOS transistor to operate in cutoff, saturation, and triode modes are also summarized as follows:

* This is for enhancement PMOS transistors, as explained earlier.

$$\begin{cases} v_{SG} < |V_{TH}| \Rightarrow \text{Cutoff Mode} \\ v_{SG} \geq |V_{TH}| \begin{cases} v_{SD} > V_{SD}(Sat.) \Rightarrow \text{Saturation Mode} \\ v_{SD} \leq V_{SD}(Sat.) \Rightarrow \text{Triode Mode} \end{cases} \end{cases} \tag{2.14}$$

where

$$V_{SD}(Sat.) = V_{OD} = \sqrt{\frac{2I_D}{\beta}}. \tag{2.15}$$

2.4 Bipolar Junction Transistor

Bipolar junction transistors are known as the first successful generation of transistors, having demonstrated superb quality in performance over a couple of decades. Similar to what was described for MOS transistors, two types of bipolar transistors are available, which are identical in general physical structure, basics of operation, and functionality, yet different in some of the electrical aspects, (e.g., the direction of current flow and the polarity of voltages). Figure 2.11 presents the circuit symbols for these two versions of bipolar transistors, known as *NPN* and *PNP* transistors. The names NPN and PNP come from the physical structure of the transistors not studied in this book.

A bipolar transistor has three terminals, called *emitter (E)*, *base (B)*, and *collector (C)*. Referring to the basic illustration of Figure 2.1, the current that passes through the transistor flows from collector to emitter for an NPN transistor and from emitter to collector for a PNP transistor. This current is controlled by the control signal applied to the base terminal. In bipolar

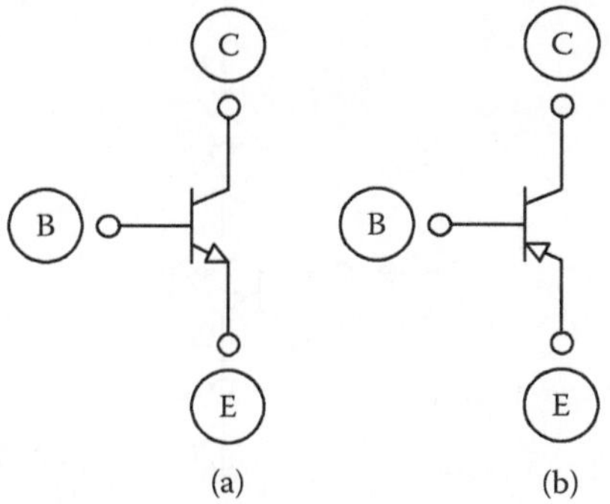

FIGURE 2.11
Circuit symbols for (a) NPN and (b) PNP bipolar transistor.

transistors, it is indeed the base current that controls the current flowing through the transistor. Later in this section, it will be seen that the collector and emitter currents are both functions of the base current. This is why bipolar transistors are sometimes known as *current-controlled* devices. It should be noted, however, that based on the relationship between the base-emitter voltage and the collector current, one can also simply control the behavior of the transistor by the base-emitter voltage. As will be discussed later, this approach is even more widely used in the majority of circuits. It can be shown that these two concepts are interchangeable, and thus there will be no intention to emphasize one and avoid the other.

Because the two types of bipolar transistors, NPN and PNP, are similar in a majority of aspects, the NPN transistor is hereafter taken as the default type of transistor. Similar to MOS transistors, everything is first explained for the NPN transistor in detail, and then the differences between an NPN transistor and its *complementary* counterpart, the PNP transistor, will be learned afterward.

2.4.1 NPN Transistor

The key voltages and currents usually used to express the transistor's functionality are presented in Figure 2.12. Analyzing a bipolar transistor is in general similar to its MOS counterpart, except that the base current is not zero. The current that passes through the transistor is the collector current, i_C, and is controlled by the base current, i_B.

Based on the *cut-set rule* in the analysis of electric circuits, the summation of the currents that arrive at the transistor equals the summation of the currents that leave the transistor:

$$i_C + i_B = i_E \tag{2.16}$$

A bipolar transistor has three useful modes of operation: *cutoff*, *saturation*, and *active* modes, which conceptually correspond to the cutoff, triode, and saturation modes of an MOS transistor, respectively.

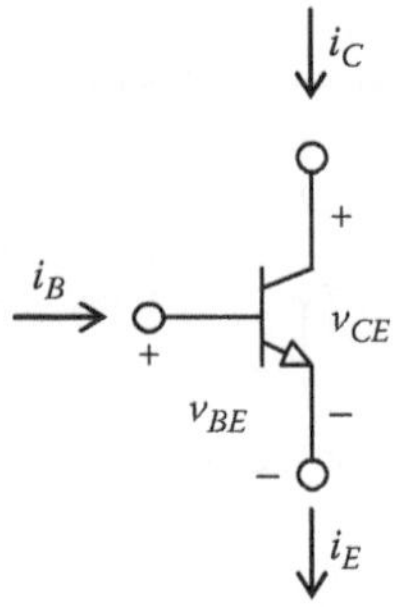

FIGURE 2.12
Key currents and voltages for an NPN transistor.

In *cutoff mode*, the transistor performs as an open switch, which does not allow the collector current to flow. In this mode, base and emitter currents will also be zero:

$$i_C = i_E = i_B = 0. \tag{2.17}$$

An NPN transistor will be in cutoff mode when the base-emitter voltage, v_{BE}, is less than a specific voltage level, the "base-emitter voltage at the onset of current flow," $V_{BE}(ON)$.* Similar to an MOS transistor, $V_{BE}(ON)$ plays the same role for a bipolar transistor as the threshold voltage, V_{TH}, does for an MOS transistor.** If the base-emitter voltage is equal to or greater than $V_{BE}(ON)$, the transistor can be in either *active mode* or *saturation mode*, depending on the amount of collector-emitter voltage, v_{CE}. If v_{CE} is equal to or tends to be less than a certain voltage level called "collector-emitter voltage at saturation," $V_{CE}(Sat.)$, the transistor will be in *saturation mode*. Unlike $V_{DS}(Sat.)$ for an MOS transistor, which was a function of the bias current, $V_{CE}(Sat.)$ for a bipolar transistor is one of the transistor parameters and fixed in value. In this mode, unlike the MOS transistor in the triode mode, the collector-emitter voltage stays at $V_{CE}(Sat.)$ and it can be said that the transistor provides a good path for the current to flow from the collector terminal to the emitter terminal. For an ideal transistor with $V_{CE}(Sat.) = 0$, the collector terminal is shorted to the emitter terminal as if an ideal switch between the collector and the emitter is closed. For a real bipolar transistor,*** typical ranges for $V_{BE}(ON)$ and $V_{CE}(Sat.)$ are 0.6^V–0.7^V and 0.1^V–0.2^V, respectively. *In saturation mode, like a short circuit or a voltage source, the amount of the current passing through the transistor is determined by the external circuitry and not the transistor itself.* In switching applications, the transistor switches between cutoff and saturation modes, whereas the proper mode of operation for a transistor in linear applications is the active mode.

It should be noted that the words "saturation" in the operation modes of bipolar and MOS transistors are chosen to refer to different concepts and have not been intended to point to a specific common property or behavior. The saturation mode for an MOS transistor, indeed, corresponds to the active mode for a bipolar transistor, in which the transistor is capable of linear operation.

* To be conceptually correct, it should be mentioned that the "base-emitter voltage at the onset of current flow," called $V_{\gamma,BE}$, is indeed different from and slightly smaller than the "base-emitter voltage when the transistor is turned on in active mode," $V_{BE}(ON)$, to be dealt with in the next few pages. In this book, for the sake of simplifying the analysis, it is assumed that $V_{BE}(ON)$ is approximately equal to $V_{\gamma,BE}$, which can be mistaken for it.

** It should be emphasized again that in this chapter, transistors are known by the electrical relationships between their currents and voltages, and not the actual physical mechanisms that occur in the transistors. Thus, this analogy is valid only when relating transistors' voltages with their operating modes.

*** Assuming that the transistor is fabricated in silicon technology for small-signal applications.

A transistor will be in the *active mode* if the base-emitter voltage is equal to or greater than $V_{BE}(ON)$ and the collector-emitter voltage is greater than $V_{CE}(Sat.)$. For a transistor in active mode, collector current is an exponential function of the base-emitter voltage:

$$i_C = I_S \exp\left(\frac{v_{BE}}{V_T}\right) \tag{2.18}$$

in which I_S is the transistor's reverse saturation current and V_T is the thermal voltage. I_S is one of the transistor's specifications and is typically in the range of 10^{-15}A to 10^{-14}A. V_T is a physical parameter and a function of temperature:

$$V_T = \frac{kT}{q} \tag{2.19}$$

where k is the Boltzman constant, T is temperature in Kelvin, and q is the electron charge. In room temperature, V_T is approximately 25 mV. The exponential $i_C - v_{BE}$ characteristic curve for an NPN transistor is plotted in Figure 2.13. This is sometimes referred to as the transistor's input characteristic curve as it expresses the relationship between the signal applied to the input terminal (v_{BE}) and the collector current.

One of the major differences between a bipolar and an MOS transistor is the rate at which the transistor current (i_C or i_D) grows when the associated control voltage, (v_{BE} or v_{GS}) increases. It is obvious from the characteristic Equation (2.5) and Equation (2.18), as well as from the plots given in Figure 2.6 and Figure 2.13, that the collector current of a bipolar transistor generally grows with v_{BE} much more rapidly than the drain current of an MOS transistor does with v_{GS}. In other words, based on Equation (2.18), the base-emitter voltage in the active mode can be written as a logarithmic function of the collector current, but because of the very sharp slope of the $i_C - v_{BE}$

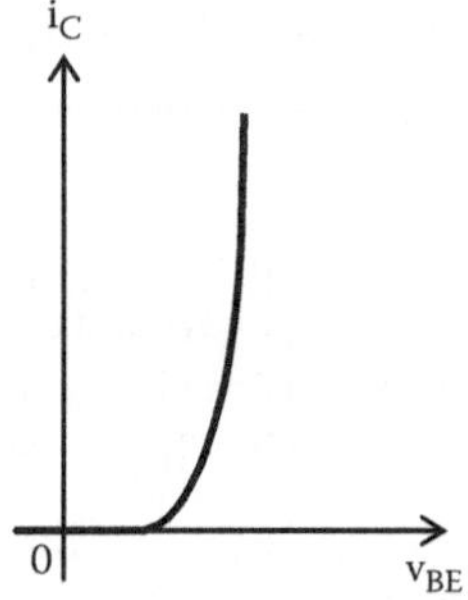

FIGURE 2.13
$i_C - v_{BE}$ characteristic curve for an NPN transistor.

characteristic curve, base-emitter voltage is usually approximated to a constant value, introduced earlier as $V_{BE}(ON)$, typically around 0.7V. This is because with the aforementioned logarithmic function between i_C and v_{BE}, a relatively wide range of i_C corresponds to a so narrow range of v_{BE} that can be approximated to a fixed value. As will be seen later, this simplifies the analysis of a bipolar transistor in DC mode. It is worth noting, however, that one is allowed to do this approximation only when the base-emitter voltage is to be compared to, added to, or subtracted from much larger voltages. In situations where the precise value of base-emitter voltage is used, or small signal-dependent variations of this voltage is of interest, this approximation is not valid.

Example 2.2

Simulate the $i_C - v_{BE}$ characteristic curve for the NPN transistor 2N3904 using SPICE program.

Solution:

The main point of this example is to observe that for a wide range of collector current, the base-emitter voltage can be approximated to 0.7 V. As seen in Figure Ex. 2.2, with a precision of ±10%, it can be said that v_{BE} is almost 0.7 V when the collector current varies by more than two orders of magnitude. The range considered for the sweeping voltage, v_{BE}, in this simulation is restricted to the values that return reasonable values for the collector current (i.e., $0 < v_{BE} < 0.8$ V).

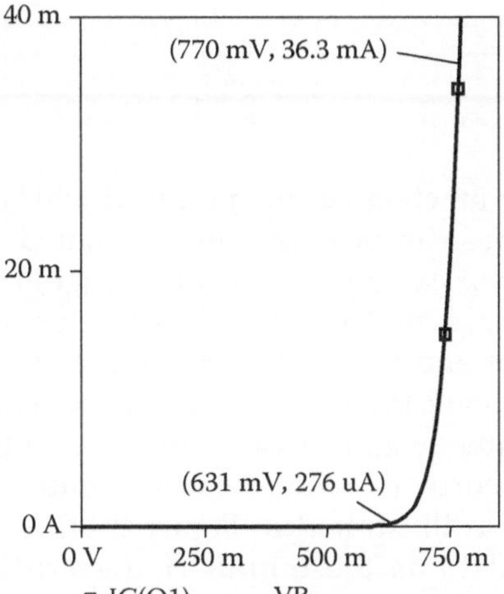

FIGURE EX. 2.2

```
VC C 0 5V
VB B 0 0.5V
Q1 C B 0 Q2N3904
.DC lin VB 0 0.8 0.01
.LIB nom.lib
.PROBE
.END
```

Like the drain current of an MOS transistor, it is obvious that the collector current of a bipolar transistor is also a nonlinear function of the base-emitter voltage. To linearize the function of a bipolar transistor, it will be operated only on a very small portion of the exponential characteristic curve.

In the active mode, collector current is related to the base current as:

$$i_C = \beta .\, i_B, \tag{2.20}$$

in which β is a constant, and one of the transistor's specifications that is determined by its physical structure.* This parameter has a typical range from 70 up to 200 for an NPN transistor, and 50 to 100 for a PNP transistor. Replacing Equation (2.20) in Equation (2.16), one can write:

$$i_E = (\beta + 1)\, i_B, \tag{2.21}$$

and

$$i_C = \alpha\, i_E \tag{2.22}$$

where

$$\alpha = \frac{\beta}{\beta + 1}. \tag{2.23}$$

Because β is usually much larger than unity, α will be somewhere between 0.98 and 0.999 and thus can be easily approximated to 1. Therefore, it can be said that the collector current will be almost equal to the emitter current. It should be emphasized that this is just a matter of approximation, and assuming the collector and the emitter currents to be equal does not mean that, based on Equation (2.16), the base current will be exactly zero.

A transistor has another characteristic curve that indicates the relationship between the collector current and the collector-emitter voltage, called $i_C - v_{CE}$ characteristics, which is illustrated in Figure 2.14. This characteristic curve is sometimes referred to as the output characteristic as it expresses the

* The parameter β is sometimes referred to as the transistor's "common-emitter current gain."

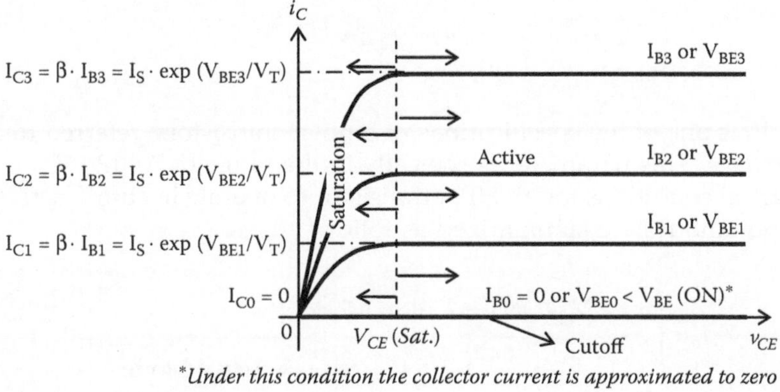

FIGURE 2.14
$i_C - v_{CE}$ characteristic for an NPN transistor.

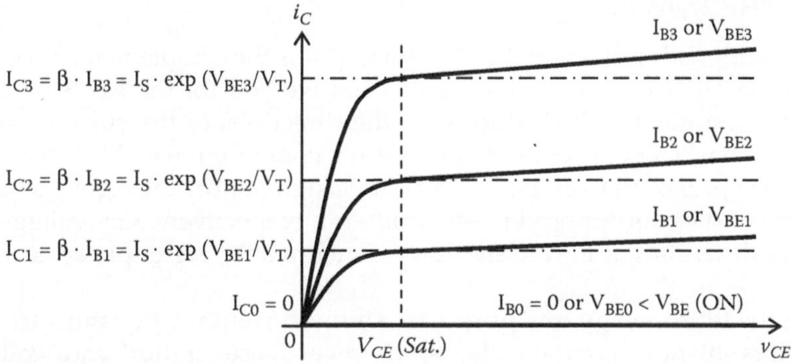

FIGURE 2.15
Adding the base-width modulation effect to $i_C - v_{CE}$ characteristic for an NPN transistor.

behavior of the collector current as a function of the voltage at the output side (i.e., collector-emitter). It will make a clearer sense when, in Chaper 4, it is seen that the output of an amplifier is usually taken from either the collector or the emitter of the amplifying transistor.

The $i_C - v_{CE}$ characteristic curve is plotted based on the assumption that in the active mode the collector current is not a function of v_{CE}, as stated by Equation (2.18). Instead, i_C is considered as a function of either i_B (Equation (2.20)) or v_{BE} (Equation (2.18)), which are treated as parameters (not the main variables) when plotting this curve.

Similar to what was seen for an MOS transistor, in practice, the collector current of a bipolar transistor is also a function of the collector-emitter voltage. This phenomenon is known as *base-width modulation*. As illustrated in Figure 2.15, this makes the collector current proportionally follow the variations in the collector-emitter voltage:

$$i_C = I_S e^{v_{BE}/V_T}\left(1+\frac{v_{CE}}{V_A}\right), \tag{2.24}$$

where V_A is one of the specifications of bipolar transistors, referred to as the *Early voltage*. This parameter is typically in the range of 70–120 V.

Electrical conditions for an NPN transistor to operate in cutoff, active, and saturation modes are summarized as follows:

$$\begin{cases} v_{BE} < V_{BE}(ON) \Rightarrow \text{Cutoff Mode} \\ v_{BE} \geq V_{BE}(ON) \begin{cases} v_{CE} > V_{CE}(Sat.) \Rightarrow \text{Active Mode} \\ v_{CE} \leq V_{CE}(Sat.) \Rightarrow \text{Saturation} \end{cases} \end{cases} \tag{2.25}$$

2.4.2 PNP Transistor

It was mentioned earlier that the PNP transistor is the complementary counterpart of the NPN transistor. Similar to what we had for the MOS transistor, when talking about a PNP transistor, the directions of the currents should be reversed and all the voltages are negated, compared to an NPN transistor. Again, it is preferred to embed the negative sign in how the signal is named, using v_{EB} and v_{EC} as opposed to $-v_{BE}$ and $-v_{CE}$, respectively. Key voltages and currents describing a PNP transistor's functionality are depicted in Figure 2.16.

With the terminology in Figure 2.16, all the current relationships for NPN transistors still hold. For the voltages, wherever "base-emitter" and "collector-emitter" voltages are used, they should be simply replaced with "emitter-base" and "emitter-collector" voltages, that is:

$$v_{BE} \to v_{EB},$$

$$V_{BE}(ON) \to V_{EB}(ON),$$

$$v_{CE} \to v_{EC},$$

and

$$V_{CE}(Sat.) \to V_{EC}(Sat.).$$

Let us summarize the electrical conditions for a PNP transistor to operate in cutoff, active, and saturation modes as follows:

$$\begin{cases} v_{EB} < V_{EB}(ON) \Rightarrow \text{Cutoff Mode} \\ v_{EB} \geq V_{EB}(ON) \begin{cases} v_{EC} > V_{EC}(Sat.) \Rightarrow \text{Active Mode} \\ v_{EC} \leq V_{EC}(Sat.) \Rightarrow \text{Saturation} \end{cases} \end{cases} \tag{2.26}$$

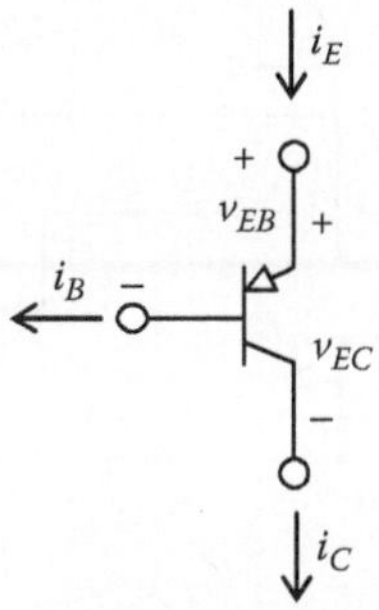

FIGURE 2.16
Key currents and voltages for a PNP transistor.

2.5 Simulation Examples

Example 2.3

In this example, input characteristic curves of an NPN bipolar transistor and an NMOS transistor are plotted. Simulate the $i_C - v_{BE}$ characteristic curve for an NPN transistor with I_S = 1.5 *E*-15A, and β = 120, and the $i_D - v_{GS}$ characteristic curve for an NMOS transistor with V_{TH} = 0.7 *V* and k' = 100 $\mu A/V^2$.

Solution:

```
VDC CC 0 5V
VSWEEP SWP 0 0.5V
Q CC SWP 0 NPNmodel
M CC SWP 0 0 NMOSmodel W=20u L=0.5u
.MODEL NPNmodel NPN (IS=1.5E-15 BF=120)
.MODEL NMOSmodel NMOS (VTO=0.7 kp=100u)
.DC lin VSWEEP 0 1 0.01
.PROBE
.END
```

To simulate the input characteristic curves of the transistors studied in this example, the emitter and the source are grounded, the collector and the drain are connected to a constant voltage of 5 V, and the base and the gate are supplied with a variable voltage sweeping from 0–1 V (see Figure Ex. 2.3(a)).

The resulting characteristics from the simulation are depicted in Figure Ex. 2.3(b). As expected, the exponential curve for $I_C(Q)$ rises more sharply than $I_D(M)$.

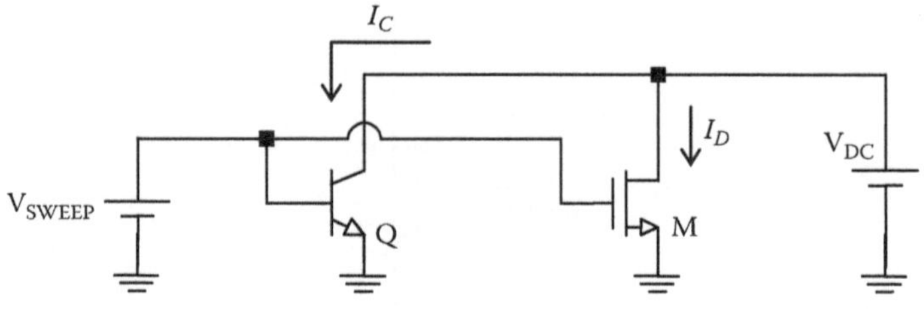

FIGURE EX. 2.3(a)

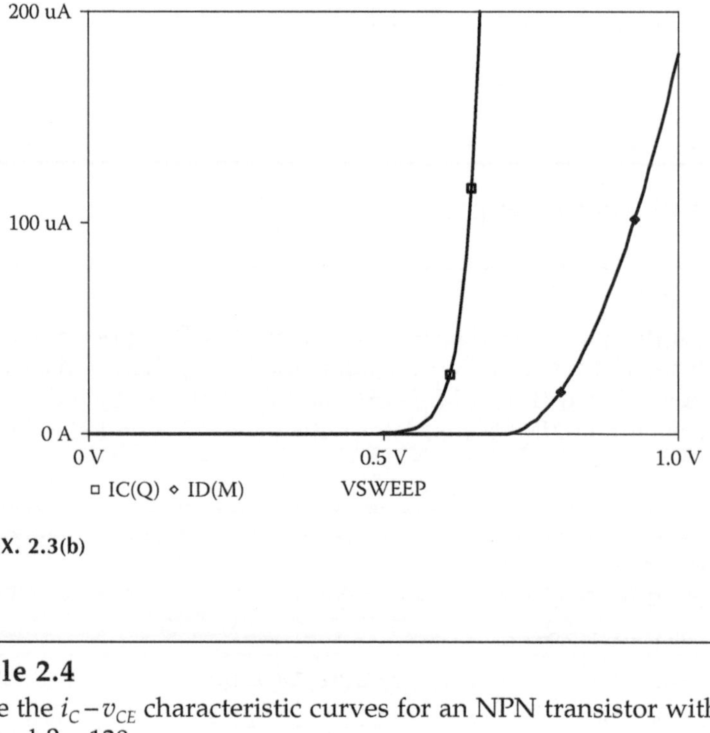

FIGURE EX. 2.3(b)

Example 2.4

Simulate the $i_C - v_{CE}$ characteristic curves for an NPN transistor with $I_S = 1.5$ E-15A, and $\beta = 120$.

Solution:

```
VCE C 0 5V
IB 0 B {IBB}
Q C B 0 NPNmodel
.MODEL NPNmodel NPN (IS=1.5E-15 BF=120)
.PARAM IBB 10u
.DC lin VCE 0 6 0.01
.STEP PARAM IBB 10u 50u 10u
```

```
.PROBE
.END
```

To simulate the output characteristic curves of a transistor, the emitter is grounded, the collector is supplied with a variable voltage sweeping from 0–5 V, and a current source is connected from the ground to the base to feed a certain amount of base current (see Figure Ex. 2.4(a)).

The simulation starts with $I_B = 10\ \mu A$ and an $i_C - v_{CE}$ curve is plotted. Then, using a .STEP command, I_B is changed to 20 μA and another $i_C - v_{CE}$ curve is plotted. This is repeated until I_B reaches $50\ \mu A$. The traces drawn in Figure Ex. 2.4(b) indicate that the collector current is exactly β times the base current, and is absolutely independent of the collector-emitter voltage. To observe the base-width modulation effect, the interested reader can add the Early voltage, VAF = 100V, to the transistor model and repeat the simulation.

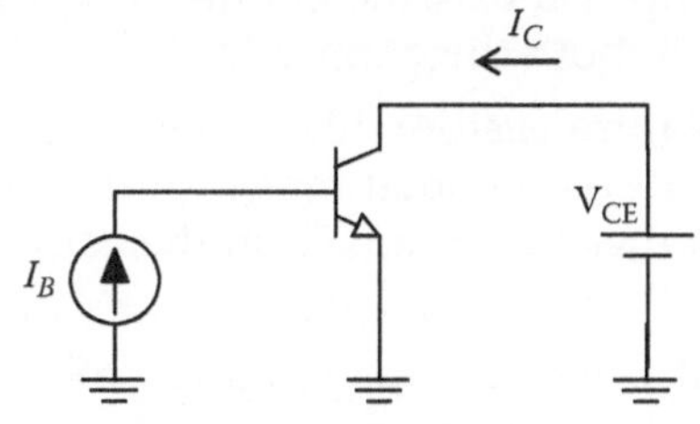

FIGURE EX. 2.4(a)

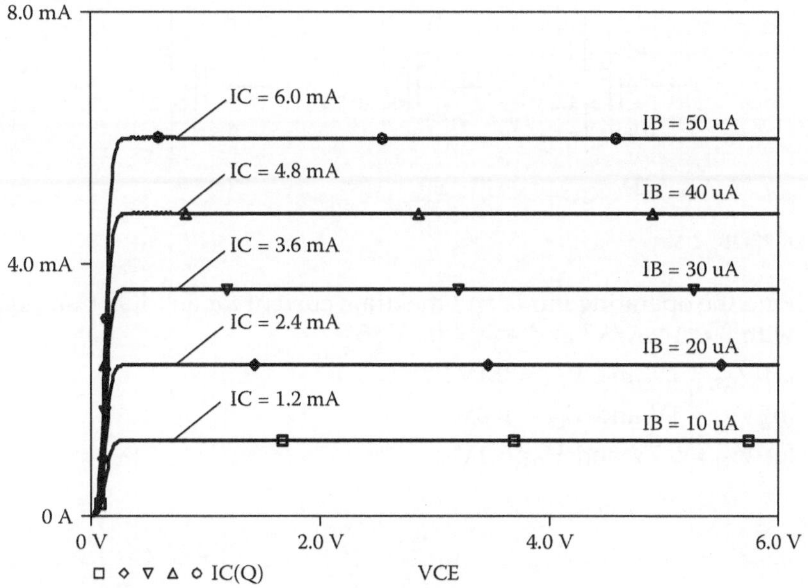

FIGURE EX. 2.4(b)

2.6 Problems

2.1 An NPN bipolar transistor has $\beta = 80$, and its base current is given $100\ \mu A$. What are the collector and emitter currents?

2.2 For a PNP bipolar transistor, find the emitter-base voltages corresponding to 1 mA and 10 mA. Let $I_S = 1.0E - 15A$ and $V_T = 25$ mV.

2.3 Suppose that the emitter-base voltage of the transistor of Problem 2.2 is set to 0.7 V.

(a) Calculate the transistor's collector current.

(b) How much the collector current will change if the emitter-base voltage is changed by ±2%?

(c) Why, in part (b), the collector current decrement is not as high as its increment, whereas the emitter-base variations are of the same amount in both directions?

2.4 For an NPN bipolar transistor with $V_{BE}(ON) = 0.65$ V and $V_{CE}(Sat.) = 0.1$ V, what is the recommended range for its collector-base voltage in order for the transistor to remain in the active mode?

2.5 Assuming $k' = 120\ \mu A/V^2$, W = 50 μm, L = 1 μm, and $|V_{TH}| = 0.8\ V$ for each of the MOS transistors in Figure Prob. 2.5,

(a) Determine the transistor's mode of operation.

(b) Calculate the drain current.

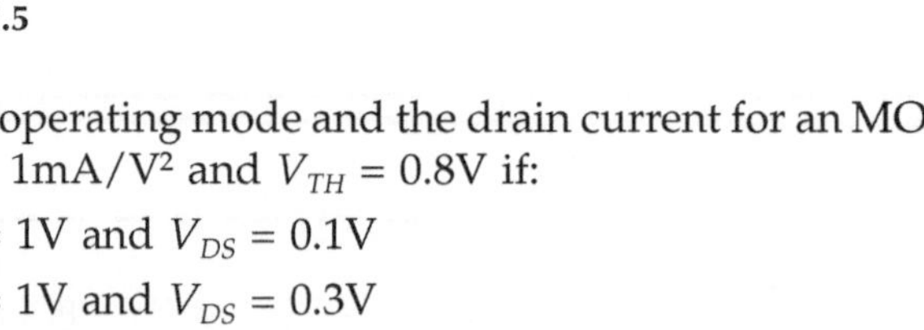

FIGURE PROB. 2.5

2.6 Find the operating mode and the drain current for an MOS transistor with $\beta = 1 mA/V^2$ and $V_{TH} = 0.8V$ if:

(a) $V_{GS} = 1V$ and $V_{DS} = 0.1V$

(b) $V_{GS} = 1V$ and $V_{DS} = 0.3V$

(c) $V_{GS} = 0.7V$ and $V_{DS} = 1V$

2.7 For the PMOS transistor in Figure Prob. 2.7 let $k' = 250\mu A/V^2$, W = 10μm, L = 10μm, and $V_{TH} = -0.6V$. Neglecting the channel-length modulation effect, find V_G.

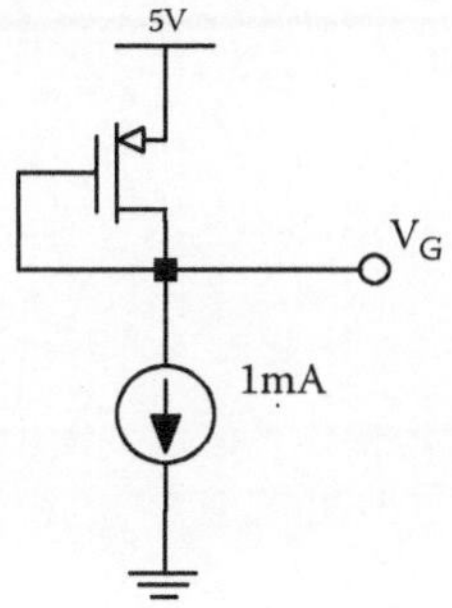

FIGURE PROB. 2.7

2.8 For an NMOS transistor let $k' = 100\mu A/V^2$, $V_{TH} = 0.7V$, and $\lambda = 0.01V^{-1}$. If $V_{GS} = 0.9V$ and $V_{DS} = 1V$, find the proper W/L ratio required for a drain current of 0.1mA.

Chapter 3

Biasing

3.1 Introduction

As explained in the previous chapter, a transistor has different operating modes, which are each suitable for specific applications. To be used as an amplifier or for any other linear application, a transistor needs to be *biased* in a certain mode: bipolar transistors in the active mode and metal-oxide semiconductor (MOS) transistors usually in the saturation mode. Biasing a transistor is usually a preliminary step to make the transistor ready for its main function. To do this, DC voltage and current sources are used to establish proper voltages and currents for the transistor to stay at certain points on its input and output characteristic curves, called *operating points*.

This chapter introduces various approaches and circuit configurations for biasing a transistor in the proper mode for linear applications.

3.2 Biasing Bipolar Transistors

It was mentioned in the previous chapter that for an NPN bipolar transistor to be biased in the active mode, the following voltage conditions should be satisfied:

$$V_{BE} > V_{BE}(ON) \tag{3.1(a)}$$

and

$$V_{CE} > V_{CE}(Sat.), \tag{3.1(b)}$$

which implies applying proper voltages to the three terminals of the transistor. Under these conditions, the emitter, collector, and base currents with the directions depicted in Figure 2.12 all are expected to be positive.

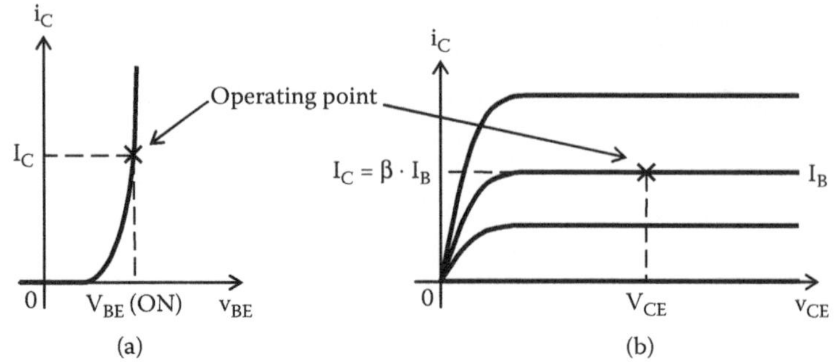

FIGURE 3.1
Operating points on (a) $i_C - v_{BE}$ characteristics (b) $i_C - v_{CE}$ characteristics.

3.2.1 Operating Point

To describe electrical conditions of a transistor, it is not necessary to name all three of the voltages and currents associated with the three terminals of the transistor. For the currents, this is simply because the emitter, collector, and base currents are related to each other by equations (2.16) through (2.18), and thus, knowing one of them will give us the other two. It is the collector current, I_C, that usually represents the transistor currents because it is one of the key variables in both $i_C - v_{BE}$ and $i_C - v_{CE}$ characteristics. As for the voltages, it is the relative potentials of the three terminals that are important in biasing a transistor. Therefore, one of the terminals is taken as the reference, and the voltages of the other two terminals are relatively expressed. For an NPN bipolar transistor, the emitter is taken as the reference terminal and the base and collector relative voltages, V_{BE} and V_{CE}, will be the voltages that express the transistor's voltage conditions. As explained earlier, for an NPN transistor biased in the active mode with a reasonable current, the base-emitter voltage is taken approximately constant, $V_{BE}(ON)$. Thus, the collector-emitter voltage, V_{CE}, will be the voltage that expresses the transistor's voltage conditions. To conclude, electrical conditions of an NPN transistor can be determined by its collector current and collector-emitter voltage, (I_C, V_{CE}), which are capable of representing the operating points of the transistor on both $i_C - v_{BE}$ and $i_C - v_{CE}$ characteristic curves. This is depicted in Figure 3.1.

3.2.2 Biasing a Transistor

To bias an NPN transistor in active mode, even if the ground potential is used to set the voltage for one of the leads, such as the emitter, two other voltage sources are still needed to bias the transistor (see Figure 3.2).

Knowing that the collector and base terminals draw current and the emitter terminal sources current, one can use resistors to convert these currents to the required voltage levels to properly bias the transistor. This way, one

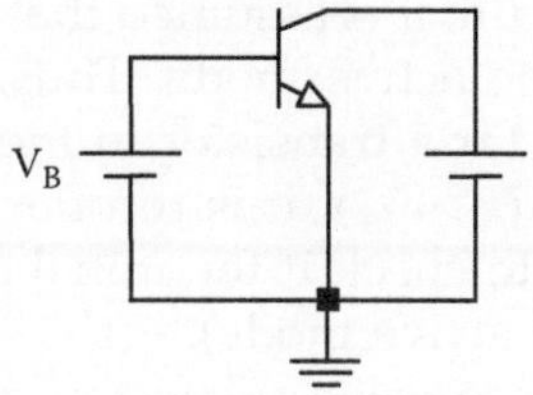

FIGURE 3.2
Illustration of biasing a bipolar transistor using two voltage sources.

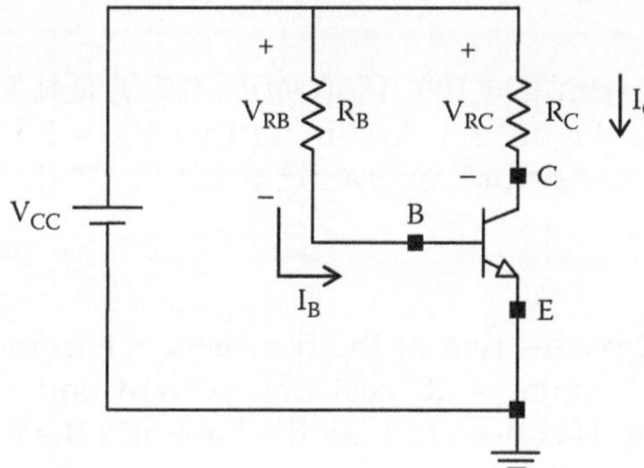

FIGURE 3.3
Biasing by one voltage source.

voltage source (and the required resistors) will be enough for biasing. Figure 3.3 is an example of biasing an NPN transistor using one voltage source and two resistors.

The voltage drops across the resistors are

$$V_{R_B} = R_B I_B \tag{3.2(a)}$$

and

$$V_{R_C} = R_C I_C, \tag{3.2(b)}$$

and the voltages of the three terminals of the transistor in reference to ground become:

$$V_E = 0, \tag{3.3(a)}$$

$$V_B = V_{CC} - V_{RB}, \tag{3.3(b)}$$

and

$$V_C = V_{CC} - V_{RC}. \tag{3.3(c)}$$

The proper choice of the voltage source and the resistances according to the transistor's specifications will result in biasing the transistor in the active mode.

When analyzing a bias circuit, it is assumed that the whole idea of biasing is to put the transistor in the active mode. Thus, the voltage and current relationships and equations for a transistor in the active mode are used to achieve the operating point (I_C, V_{CE}). It is recommended, at the end of calculations, to use the results to check if the initial assumption is true (i.e., if the transistor is really in the active mode).

To better understand how to solve transistor circuits for their operating points, let us continue with the following examples.

Example 3.1

An NPN transistor with $\beta = 100$, $V_{BE}(ON) = 0.7$ V, and $V_{CE}(Sat.) = 0.2$ V is used in the circuit of Figure 3.3. Assuming that $V_{CC} = 5$ V, $R_B = 430$ $k\Omega$, and $R_C = 3$ $k\Omega$, find the transistor's operating point (I_C, V_{CE}).

Solution:

Finding a loop that passes two of the transistor's terminals will result in a KVL* equation that contains at least one current and one voltage of the transistor. If the loop is chosen to pass the base and the emitter, the voltage variable will be v_{BE}, which can be approximated to $V_{BE}(ON)$ with a given constant value. Therefore, the only unknowns of the equation will be one or two of the transistor's currents, which can be converted to each other. In this example, writing KVL equation for the loop in Figure Ex. 3.1(a) gives:

$$\text{KVL 1: } V_{CC} = V_{RB} + V_{BE}(ON)$$

or

$$V_{CC} = R_B I_B + V_{BE}(ON).$$

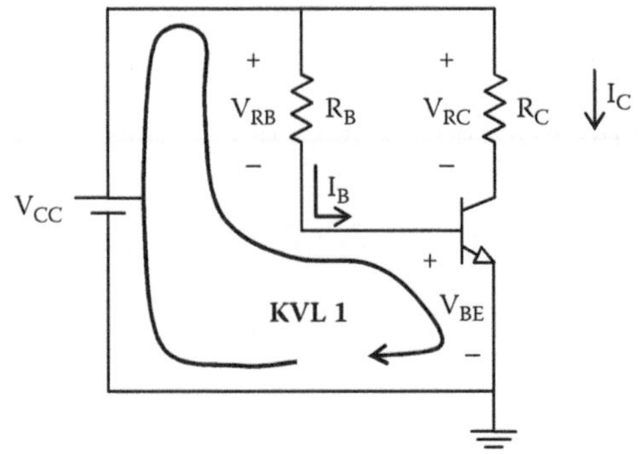

FIGURE EX. 3.1(a)

* Kirchhoff's voltage law.

Replacing the known parameters with their numeric values we have:

$$5\ V = (430^{k\Omega} I_B) + 0.7\ V,$$

which, if solved for I_B, gives:

$$I_B = \frac{5^V - 0.7^V}{430^{k\Omega}} = \frac{4.3^V}{430^{k\Omega}} = 10\ \mu A.$$

Finally, the collector current is found as

$$I_C = \beta I_B = 100 \times 10\ \mu A = 1\ mA.$$

Now, to find the collector-emitter voltage, a second loop will be needed that passes both the collector and the emitter nodes. This loop is depicted in Figure Ex. 3.1(b), which gives:

$$\text{KVL 2: } V_{CC} = V_{RC} + V_{CE}$$

or

$$V_{CC} = R_C I_C + V_{CE}.$$

Replacing the known parameters with their numeric values we have:

$$5\ V = (3^{k\Omega} \times 1^{mA}) + V_{CE},$$

which, if solved for V_{CE}, gives:

$$V_{CE} = 5\ V - (3^{k\Omega} \times 1^{mA}) = 2V.$$

Because the collector-emitter voltage is well above the transistor's $V_{CE}(Sat.)$, it can be said that the initial assumption of the transistor being in the active mode is true.

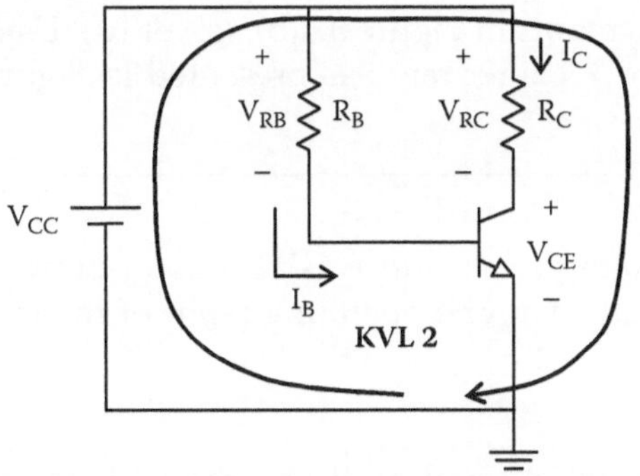

FIGURE EX. 3.1(b)

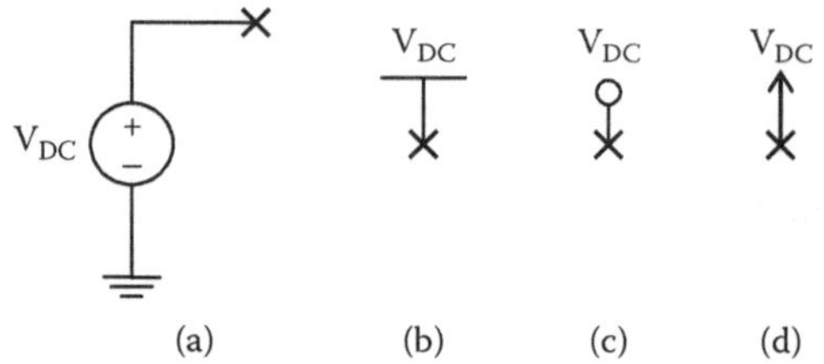

FIGURE 3.4
(a) A DC voltage source; (b), (c), and (d) equivalent symbols.

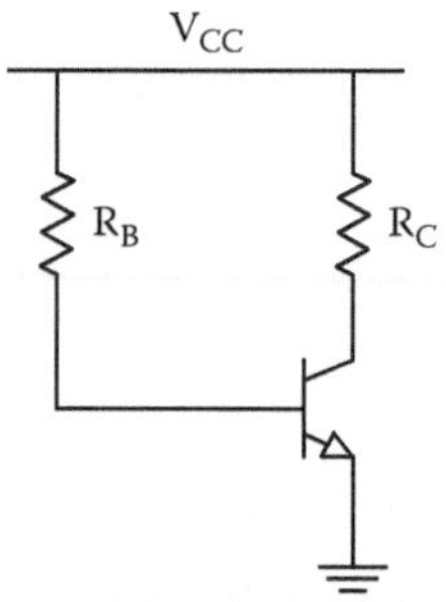

FIGURE 3.5
Simplified schematic of the circuit of Figure 3.3.

3.2.3 A Circuit Drawing Convention

To keep the circuit drawings simple, especially when the number of transistors increases, the circuit-drawing convention depicted in Figure 3.4 has become a default in drawing electronic circuit schematics. As seen, the voltage source and the ground symbol connected to it (Figure 3.4(a)) are not drawn anymore, although still present in the circuit, and are replaced with one of the simple symbols in Figure 3.4(b), (c), or (d). Using this convention, the circuit of Figure 3.3 is redrawn as presented in Figure 3.5.

Example 3.2

In the circuit of Example 3.1, add a 1-$k\Omega$ resistor between the emitter and the ground and recalculate the operating point of the transistor.

Solution:

By adding the resistor, R_E, and replacing the voltage source, V_{CC}, with its simplified symbol, the circuit of Figure Ex. 3.2(a) is resulted.

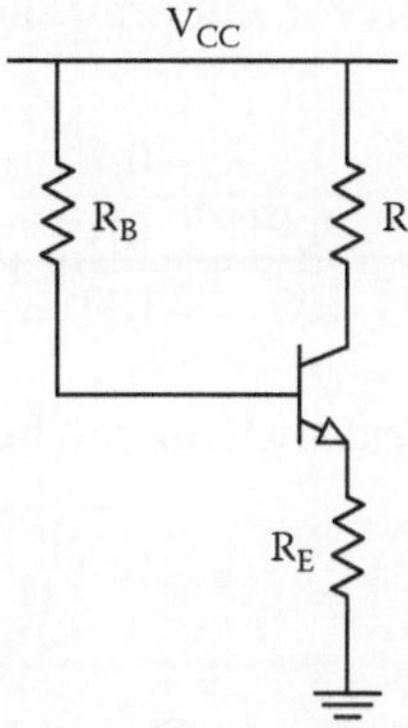

FIGURE EX. 3.2(a)

The two paths for which the required KVL equations should be written are also depicted on the circuit in Figure Ex. 3.2(b).
Writing the first KVL equation helps find the collector current:

$$\text{KVL 1: } V_{CC} = V_{RB} + V_{BE}(ON) + V_{RE}$$

or

$$V_{CC} = R_B I_B + V_{BE}(ON) + R_E I_E.$$

In this equation, I_B and I_E are unknown variables, but both can be rewritten in terms of the collector current, I_C:

$$V_{CC} = \left(R_B \frac{I_C}{\beta}\right) + V_{BE}(ON) + R_E \frac{\beta+1}{\beta} I_C.$$

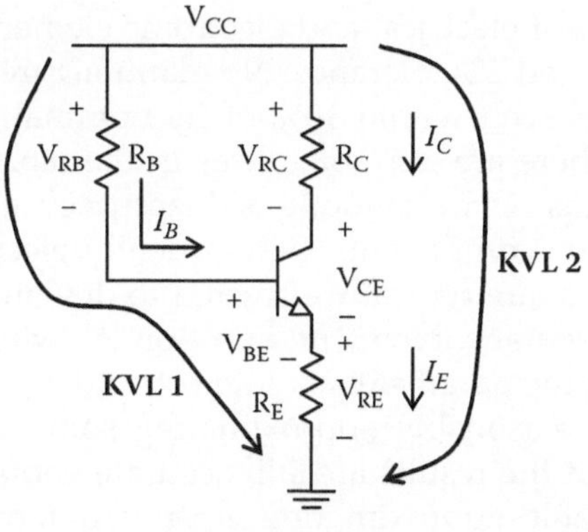

FIGURE EX. 3.2(b)

Solving the preceding equation for the only unknown variable, I_C, gives:

$$I_C = \frac{V_{CC} - V_{BE}(ON)}{\frac{R_B}{\beta} + \left(\frac{\beta+1}{\beta}.R_E\right)} = \frac{5^V - 0.7^V}{\frac{430^{k\Omega}}{100} + \left(\frac{101}{100} \times 1^{k\Omega}\right)} = \frac{4.3^V}{5.31^{k\Omega}} = 0.81^{mA}$$

Collector-emitter voltage of the transistor can be found by solving the second KVL equation:

$$\text{KVL 2: } V_{CC} = V_{RC} + V_{CE} + V_{RE}$$

or

$$V_{CC} = R_C I_C + V_{CE} + R_E I_E.$$

Solving for V_{CE} and replacing the known parameters with their numeric values we have:

$$V_{CE} = V_{CC} - \left(R_C + \frac{\beta+1}{\beta} R_E\right) I_C$$

and

$$V_{CE} = 5^V - \left(3^{k\Omega} + \frac{101}{100} \times 1^{k\Omega}\right) \times 0.81^{mA} = 5^V - 3.25^V = 1.75^V.$$

The collector-emitter voltage is well above the transistor's $V_{CE}(Sat.)$, and thus, the initial assumption of the transistor being in the active mode is true.

3.2.4 Approximation in the Analysis of Electronic Circuits

When analyzing or designing a circuit on paper, we usually use a scientific calculator to precisely find the target values, but in practice, a tolerance exists for the specifications of electrical and electronic elements. Resistors usually come with 10%, 5%, and 2% tolerance. No manufacturer guarantees β value of a transistor to be even within 10% of its nominal value. In integrated circuit technology, there are also tolerances in the fabrication of active and passive devices. Thus, some amount of inaccuracy always occurs in the specifications of circuit components. Consequently, electrical quantities (e.g., currents and voltages) are typically expected to deviate from the calculated values by some percentage. Now, the question is "why do we bother with precisely calculating the parameters of interest, and not simplify the analysis and calculations by reasonably approximating some values, within a safe range, and claim that the results are still accurate enough for practical purposes?" The acceptable approximation error is determined by where the circuit is employed. Throughout this book and in many applications, ±10% approximation error is chosen to be acceptable. Therefore,

- Each quantity can be approximated within ±10% of its precise value.
- When adding or subtracting two numbers, the smaller one can be neglected if it is at least 10 times smaller.

As a result, if two resistors are in series, the smaller one can be neglected if it is at least 10 times smaller than the other, and similarly, for two resistors in parallel, the bigger one can be neglected if it is at least 10 times larger than the smaller one. It should be emphasized, however, that this approximation rule applies only on the cases where a quantity has a negligible contribution compared with some other quantities.

An example of a case where the approximation rule is often applied, but *not always*, is: For a transistor with a large enough β (i.e., typically larger than 50), when determining the emitter current the base current can be ignored and the emitter current can be approximated to the collector current:

$$\beta >> 1 \Rightarrow I_E \cong I_C.$$

It should be emphasized, however, that this does not mean that the base current, I_B, equals zero. For instance, in Example 3.2 where $\beta = 100$, it was seen that the emitter current was $1.01I_C$, which could be easily approximated to I_C. If the approximation rule had been applied, a collector current of 0.811^{mA} would have been calculated. In this case, the base current cannot be approximated to zero as the result of approximating I_E to I_C, because a zero base current would result in zero collector and emitter currents. This is true for both Example 3.1 and Example 3.2 where the base current, however small it is, is the quantity that sets the collector and emitter currents.

Example 3.3

In the bias circuit in Figure Ex. 3.3(a) assume: $\beta = 80$, $V_{BE}(ON) = 0.7$ V, $V_{CE}(Sat.) = 0.1$ V, $V_{CC} = 5$ V, $R_B = 100$ $k\Omega$, and $R_C = 1$ $k\Omega$. Find the transistor's operating point (I_C, V_{CE}).

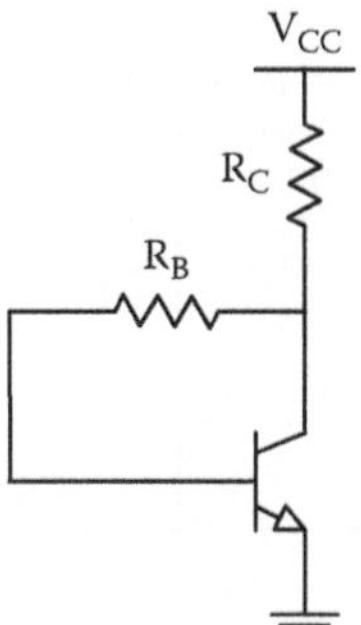

FIGURE EX. 3.3(a)

Solution:

Figure Ex. 3.3(b) illustrates the circuit along with the two paths for which the required KVL equations are written. Similar to the previous examples, writing the first KVL equation for the path that passes through the base and the emitter nodes helps find the collector current:

$$\text{KVL 1: } V_{CC} = V_{RC} + V_{RB} + V_{BE}(ON)$$

or

$$V_{CC} = R_C I_X + R_B I_B + V_{BE}(ON),$$

where $I_X = I_C + I_B$. Solving this equation for the collector current, I_C, we have:

$$I_C = \frac{V_{CC} - V_{BE}(ON)}{\frac{R_B}{\beta} + \left(1 + \frac{1}{\beta}\right).R_C}$$

Because $\beta >> 1$, the term $1/\beta$ in the denominator can be neglected when added to 1:

$$I_C = \frac{V_{CC} - V_{BE}(ON)}{\frac{R_B}{\beta} + R_C} = \frac{5^V - 0.7^V}{\frac{100^{k\Omega}}{80} + 1^{k\Omega}} = \frac{4.3^V}{2.25^{k\Omega}} = 1.91^{mA}$$

The collector-emitter voltage of the transistor can be found by solving the second KVL equation:

$$\text{KVL 2: } V_{CC} = V_{RC} + V_{CE}$$

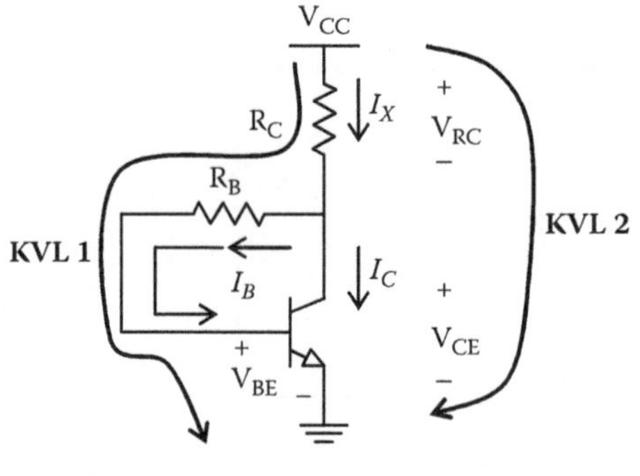

FIGURE EX. 3.3(b)

or

$$V_{CC} = R_C (I_C + I_B) + V_{CE}.$$

Solving for V_{CE} and replacing the known parameters with their numeric values we have:

$$V_{CE} = V_{CC} - R_C \left(1 + \frac{1}{\beta}\right) I_C$$

and

$$V_{CE} = 5^V - 1^{k\Omega} \left(1 + \frac{1}{80}\right) \times 1.91^{mA} = 5^V - 1.93^V = 3.07^V.$$

Because the collector-emitter voltage is still above the transistor's $V_{CE}(Sat.)$, the initial assumption of the transistor being in the active mode is true.

3.2.5 A Brief Review

After gaining some experience with calculating the operating point of a transistor, let us briefly review what has been studied so far. We started with a transistor with three terminals, which was a new type of circuit element compared with the usual 2-terminal components we had been familiar with. As a result, two voltages and three currents were defined for the transistor to describe its electrical behavior. To relate these variables, a few equations were introduced expressing the relationship between the currents and voltages. To simplify dealing with these equations and variables, we tried to find the key variables, based on which all the other variables can be found. In this sense, the collector current was taken as the main current for the transistor from which the other two currents (i.e., base and emitter currents) can be determined. Then, we dealt with the voltages. According to the exponential characteristic equation for the collector current, it was clear that the base-emitter voltage plays the key role in establishing the collector current through the transistor. It was demonstrated that due to the very sharp slope of the exponential characteristic curve, very small changes in the base-emitter voltage results in large changes in the collector current. In other words, we observed that within a wide range, different collector currents were corresponding to base-emitter voltages that were so close to each other that they could all be approximated to a single value, $V_{BE}(ON)$. This made our analysis much easier, because the base-emitter voltage is now a known variable, helping us to write KVL equations associated with the loops passing through the transistor via the base-emitter. An important point experienced in the examples was that the KVL equation passing through the base-emitter could lead us to the collector current. It was also demonstrated that the collector-emitter voltage could be obtained from another KVL equation written for a loop that passed through the collector-emitter nodes. It can be concluded that *the collector-emitter voltage of a transistor is determined by the other elements connected to the transistor.* The transistor, in fact, does not contribute to the collector-emitter voltage at all; it is this voltage that decides whether the transistor should be in the active or saturation mode (provided that enough base-emitter voltage has been applied).

Example 3.4

For the bias circuit in Figure Ex. 3.4(a) with: $\beta = 100$, $V_{BE}(ON) = 0.7$ V, $V_{CE}(Sat.) = 0.1$ V, $V_{CC} = 5$ V, and $R_B = 200\ k\Omega$,

(a) Find the transistor's collector current, I_C.

(b) Find the range for R_C over which the transistor stays in the active mode.

Solution:

The collector current can be obtained from a KVL equation written for the path starting from V_{CC}, going through R_B, and passing the transistor through base-emitter, as previously seen in Example 3.1:

$$\text{KVL 1: } V_{CC} = R_B I_B + V_{BE}(ON)$$

which gives

$$I_B = \frac{V_{CC} - V_{BE}(ON)}{R_B}$$

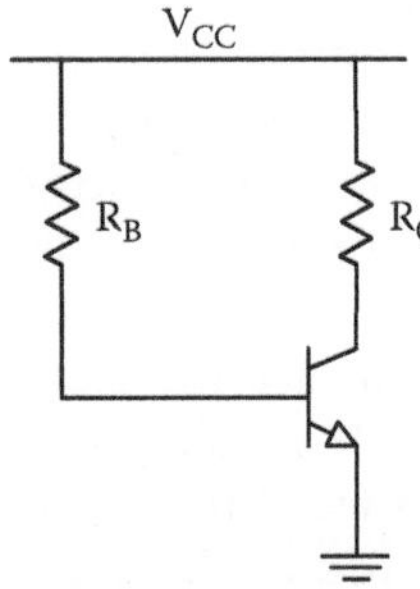

FIGURE EX. 3.4(a)

Without writing a KVL equation, we could also think of the base current as the current flowing through R_B. This way, the base current could be found by simply dividing the voltage across R_B that is $V_{CC} - V_{BE}(ON)$ by its resistance, which would result in the same equation as given previously.

In either case, replacing the known parameters with their numeric values we have:

$$I_B = \frac{5^V - 0.7^V}{200^{k\Omega}} = 21.5^{\mu A},$$

and the collector current is found as

$$I_C = \beta I_B = 100 \times 21.5^{\mu A} = 2.15^{mA}.$$

Of course, one of the ways to find the collector-emitter voltage is to solve the KVL equation written for a loop that passes through the transistor via the collector-emitter, as previously seen in Example 3.1. A simpler interpretation, illustrated in Figure Ex. 3.4(b), can be as follows: The upper end of resistor R_C is at V_{CC}, and the current flowing through this resistor causes a voltage drop of $R_C I_C$ across it. Thus, the voltage at the other end of the resistor, which is the collector voltage as well, will be:

$$V_C = V_{CC} - V_{RC} = V_{CC} - R_C I_C.$$

Because the emitter is grounded, the collector voltage is also the collector-emitter voltage:

$$V_{CE} = V_{CC} - R_C I_C.$$

To keep the transistor in the active mode, collector-emitter voltage should be larger than $V_{CE}(Sat.)$, and thus:

$$V_{CC} - R_C I_C > V_{CE}(Sat.).$$

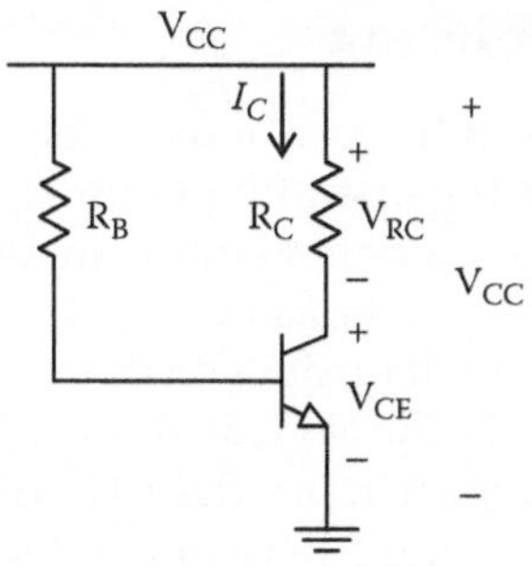

FIGURE EX. 3.4(b)

Rearranging this inequality for R_C gives:

$$R_C < (V_{CC} - V_{CE}(Sat.))/I_C,$$

and by replacing the parameters with their values, the final result will be achieved:

$$R_C < (5^V - 0.1^V)/2.15^{mA}$$

$$R_C < 2.28^{k\Omega}.$$

3.2.6 Bipolar Transistor in Saturation

The previous example indicates that with a certain amount of collector current, a transistor can be either in active mode or in saturation mode, depending on how the other elements connected to the transistor provide it with the collector-emitter voltage.

Because this book has been focused only on linear amplifiers, we do not deal with the analysis of the circuits in which the transistor is saturated, but one point is worth noting for this case. To solve a bias network, the transistor is initially assumed to be in the active mode as the whole biasing concept is all about putting the transistor in this mode. This is why we are allowed to use all the current and voltage relationships before we make sure that the transistor is really in the active mode. Once the circuit is analyzed and the collector-emitter voltage is achieved, the initial assumption should be checked. If it is confirmed that the transistor is in the active mode, all the results are valid and can be used for further analyses. Otherwise, if it is concluded that the transistor is saturated, no achieved current or voltage value is valid. The only result that is valid is the fact that the transistor is not active, and the circuit should be reanalyzed with proper methods depending on the resulted transistor's mode of operation.

3.3 Voltage-Divider Biasing

In the examples presented in the previous section, one of the drawbacks of the biasing network is that the operating point significantly depends on the value of the transistor's β parameter. Unfortunately, this parameter has relatively large variance around its typical value. This makes us think of biasing methods that are less dependent on or even independent of transistor parameters. One such approach is the so-called *voltage-divider biasing*, which is presented in Figure 3.6. In this circuit, the DC voltage required for the base is prepared by a voltage divider network, which consists of the voltage source, V_{CC}, and the resistors, R_{B1} and R_{B2}.

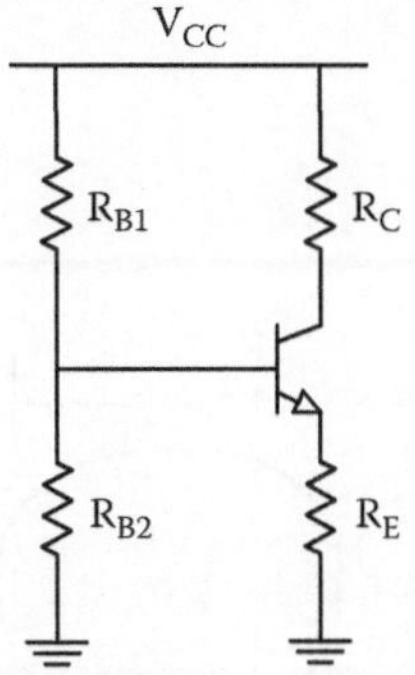

FIGURE 3.6
Voltage-divider biasing.

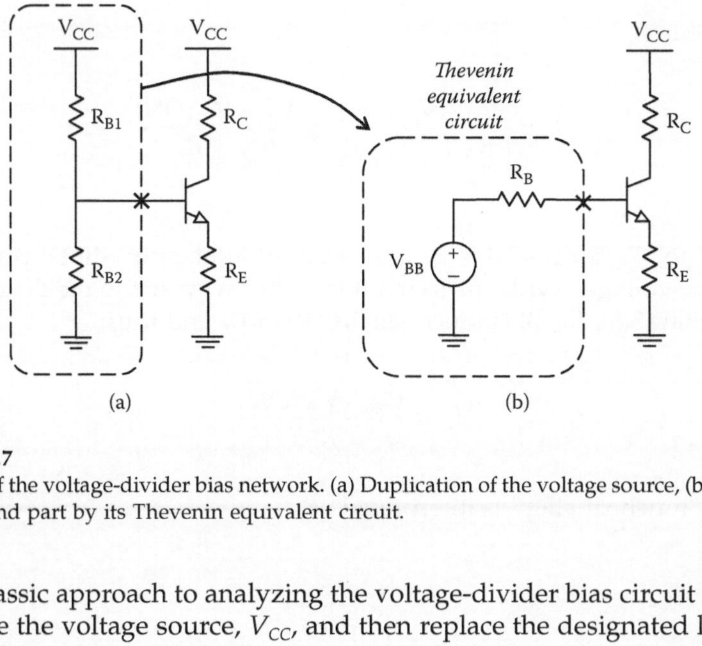

FIGURE 3.7
Analysis of the voltage-divider bias network. (a) Duplication of the voltage source, (b) replacing the left-hand part by its Thevenin equivalent circuit.

The classic approach to analyzing the voltage-divider bias circuit is to first duplicate the voltage source, V_{CC}, and then replace the designated left-hand part of the circuit with its Thevenin equivalent circuit (see Figure 3.7), in which:

$$V_{BB} = \frac{R_{B2}}{R_{B1} + R_{B2}} V_{CC} \text{ and } R_B = R_{B1} \| R_{B2}. \tag{3.4}$$

Now, it is easy to find the base current using the KVL equation written for the loop depicted in Figure 3.8:

$$\text{KVL: } V_{BB} = R_B I_B + V_{BE}(ON) + R_E I_E,$$

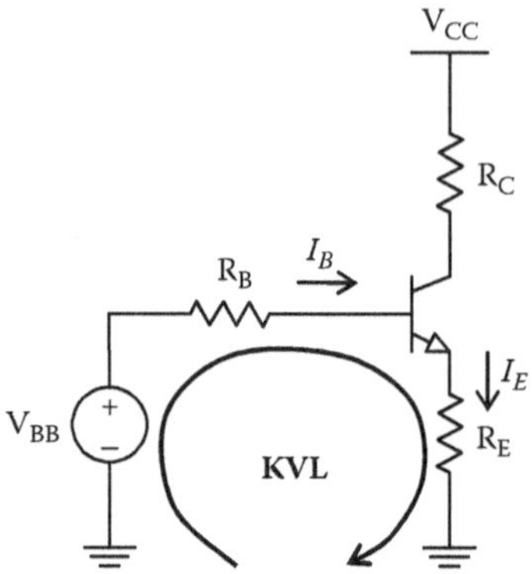

FIGURE 3.8
Finding collector current in the voltage-divider bias network.

which returns:

$$I_C = \frac{V_{BB} - V_{BE}(ON)}{\left(\dfrac{\beta+1}{\beta}.R_E\right) + \dfrac{R_B}{\beta}} \cong \frac{V_{BB} - V_{BE}(ON)}{R_E + \dfrac{R_B}{\beta}} \tag{3.5}$$

if solved for I_C. Because β is usually a large number, it will be practical to design the voltage-divider bias circuit in such a way that in the denominator of Equation (3.5), R_B/β, is much smaller than R_E and thus:

$$I_C \cong \frac{V_{BB} - V_{BE}(ON)}{R_E} \tag{3.6(a)}$$

Being "much smaller" is interpreted as

$$\frac{R_B}{\beta} \le \frac{R_E}{10} \tag{3.6(b)}$$

based on the "±10% approximation rule" previously described. In such cases, the collector current will be independent of the transistor's β parameter.

Example 3.5

A transistor with β = 100, $V_{BE}(ON)$ = 0.7 V, and $V_{CE}(Sat.)$ = 0.1 V is used in the voltage-divider bias circuit, which is depicted in Figure Ex. 3.5(a). Find the bias point (I_C, V_{CE}) for the transistor.

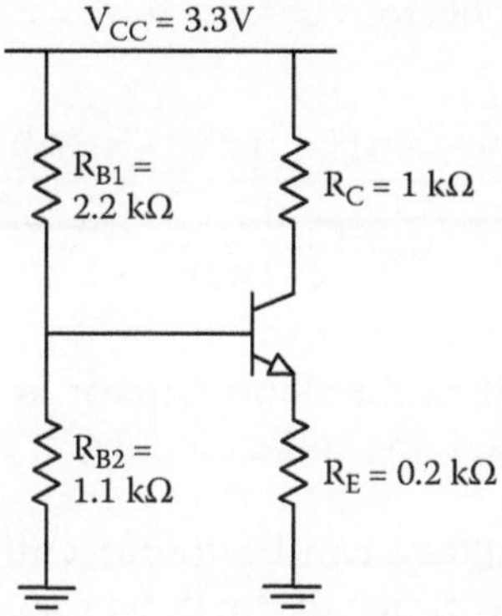

FIGURE EX. 3.5(a)

Solution:
First of all, the left-hand branch of the circuit is replaced with its Thevenin equivalent (see Figure Ex. 3.5(b)), where

$$V_{BB} = \frac{R_{B2}}{R_{B1} + R_{B2}} . V_{CC} = \frac{1.1}{1.1 + 2.2} \times 3.3^V = 1.1^V$$

$$R_B = R_{B1} \| R_{B2} = 1.1^{k\Omega} \| 2.2^{k\Omega} = 0.73^{k\Omega}$$

Then, KVL equation is written for the loop that passes through the transistor via base-emitter:

$$\text{KVL: } V_{BB} = R_B I_B + V_{BE}(ON) + R_E I_E,$$

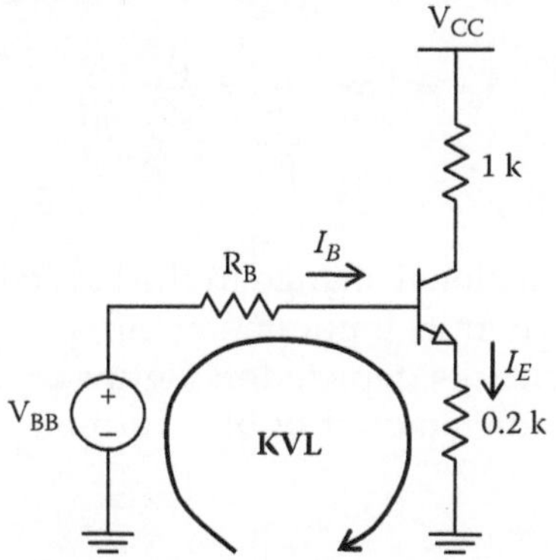

FIGURE EX. 3.5(b)

which, if solved for the collector current, returns:

$$I_C = \frac{V_{BB} - V_{BE}(ON)}{R_E + \dfrac{R_B}{\beta}} = \frac{1.1^V - 0.7^V}{0.2^{k\Omega} + \dfrac{0.73^{k\Omega}}{100}} = \frac{0.4^V}{0.2073^{k\Omega}} = 1.93^{mA} \cong 2^{mA}$$

Notice that the term R_B/β in the denominator is much smaller than R_E. If this term had been neglected, the collector current would have been obtained 2 mA.

The collector-emitter voltage can be found without a KVL equation. All we need is to calculate how much of the power-supply voltage, V_{CC}, appears across the resistors that are in series with collector-emitter (i.e., R_C and R_E) and then subtract it from V_{CC}:

$$V_{CE} = V_{CC} - (I_C R_C + I_E R_E) \cong V_{CC} - I_C(R_C + R_E)$$

Replacing the parameters with their values, we have:

$$V_{CE} = 3.3^V - 2^{mA}(1^{k\Omega} + 0.2^{k\Omega}) = 0.9^V,$$

which is larger than $V_{CE}(Sat.)$, and thus the transistor is in the active mode.

Example 3.6

When variation of a parameter, x, causes a circuit variable, y, to vary accordingly, *sensitivity* is the measure that is defined to express this relationship. Sensitivity of a transistor bias current, I_C, to the β parameter is defined as:

$$S_\beta^{I_C} = \lim_{\Delta\beta \to 0} \frac{\dfrac{\Delta I_C}{I_C}}{\dfrac{\Delta\beta}{\beta}} = \frac{\beta}{I_C} \cdot \frac{\partial I_C}{\partial \beta}$$

in which $\Delta I_C/I_C$ is the fractional change in the bias current that results from a given fractional change in its β parameter, $\Delta\beta/\beta$.

Find $S_\beta^{I_C}$ for the bias circuits depicted in Figure Ex. 3.6, and compare their β dependency for the same amount of bias current.

Solution:

It has been previously demonstrated that the bias current for Circuit I is written as:

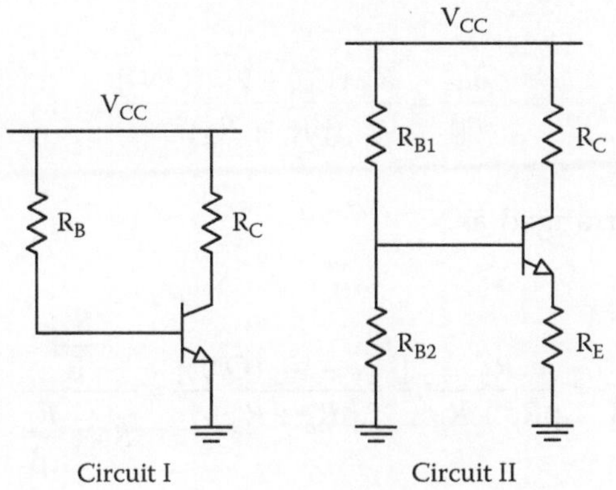

FIGURE EX. 3.6

$$I_C = \beta . \frac{V_{CC} - V_{BE}(ON)}{R_B}.$$

Therefore,

$$\frac{\partial I_C}{\partial \beta} = \frac{V_{CC} - V_{BE}(ON)}{R_B} = \frac{I_C}{\beta},$$

and the sensitivity of the bias current to β for Circuit I is

$$S_\beta^{I_C} = \frac{\beta}{I_C} . \frac{\partial I_C}{\partial \beta} = \frac{\beta}{I_C} . \frac{I_C}{\beta} = 1.$$

This means that if some fractional variation happens for β, the bias current will be subject to the same fractional changes. This is not acceptable in many applications as, for example, 20% variation in the β parameter around its typical value will cause 20% bias current variation around its nominal value.

For Circuit II, which is a voltage-divider bias circuit, the collector current has been already obtained as:

$$I_C = \frac{V_{BB} - V_{BE}(ON)}{R_E + \dfrac{R_B}{\beta}}$$

where

$$V_{BB} = \frac{R_{B2}}{R_{B1} + R_{B2}} V_{CC} \text{ and } R_B = R_{B1} \| R_{B2}.$$

Therefore,

$$\frac{\partial I_C}{\partial \beta} = \frac{R_B.[V_{BB} - V_{BE}(ON)]}{(\beta R_E + R_B)^2},$$

which can be rearranged as

$$\frac{\partial I_C}{\partial \beta} = \frac{R_B}{\beta R_E + R_B} \cdot \frac{[V_{BB} - V_{BE}(ON)]}{\beta R_E + R_B} = \frac{\frac{R_B}{\beta}}{R_E + \frac{R_B}{\beta}} \cdot \frac{I_C}{\beta}$$

and the sensitivity of the bias current to β for Circuit II is

$$S_\beta^{I_C} = \frac{\beta}{I_C} \cdot \frac{\partial I_C}{\partial \beta} = \frac{\beta}{I_C} \cdot \frac{\frac{R_B}{\beta}}{R_E + \frac{R_B}{\beta}} \cdot \frac{I_C}{\beta} = \frac{\frac{R_B}{\beta}}{R_E + \frac{R_B}{\beta}}.$$

which is always less than unity.

Based on the sensitivities obtained for the preceding circuits, it is obvious that Circuit II is always less sensitive to β variations than Circuit I. The sensitivity obtained for the voltage-divider bias circuit also confirms the previously discussed condition under which the bias current will be independent of β, that is:

$$\frac{R_B}{\beta} << R_E \quad \Rightarrow \quad S_\beta^{I_C} \cong 0.$$

3.3.1 Voltage-Divider Biasing Circuit: A Different View

The approach introduced to analyze the voltage-divider bias circuit is only an analysis method that gives us the collector current. To better understand the circuit, it will help if the functions of the key parts are understood. The voltage-divider bias circuit can be simply imagined as a two-part circuit (see Figure 3.9).

The right-hand part contains the transistor which if is provided with a proper base voltage will have the desired operating point. The left-hand part is a simple resistive voltage divider network, preparing a suitable base voltage, V_B, for the transistor given by:

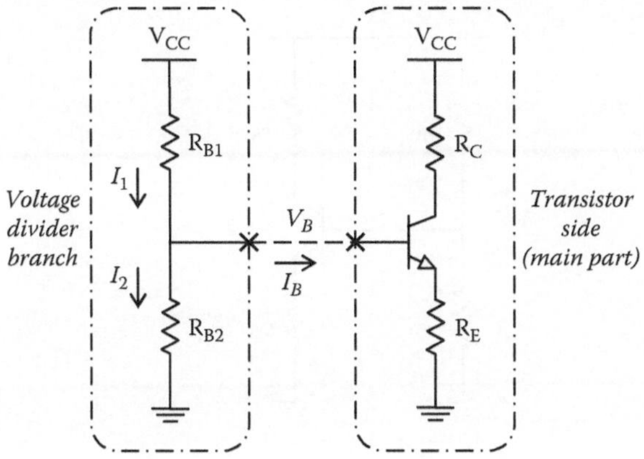

FIGURE 3.9
Voltage-divider bias network imagined as a two-part circuit.

$$V_B = \frac{R_{B2}}{R_{B1} + R_{B2}}.V_{CC}$$

which was already seen in Equation (3.4). The only concern about resistive voltage dividers of the type used in the voltage-divider bias circuit is that the well-known voltage division formula for their output voltage will not be precise enough if a nonnegligible current (compared with the current flowing through the voltage divider resistors) is drawn from their output node. Therefore, the base current, I_B, should be much smaller than I_1 and I_2, depicted in Figure 3.9. It can be shown that this condition for the base current is equivalent to the β-independence condition presented in Equation (3.6(b)).

Thinking this way, gives a good understanding about the voltage-divider bias circuit, which can be a great help when analyzing or designing the circuit. For instance, thinking of R_{B1} and R_{B2} forming a voltage divider when designing the voltage-divider bias circuit gives us a degree of freedom to change the resistance of both R_{B1} and R_{B2} while keeping their ratio constant to easily satisfy the β-independence condition for the circuit.

Example 3.7

The transistor used in the voltage-divider bias circuit of Figure Ex. 3.7(a) has the following specifications:

$$\beta = 120,\ V_{BE}(ON) = 0.7\ \text{V, and } V_{CE}(Sat.) = 0.2\ \text{V}.$$

Determine proper values for R_{B1} and R_{B2} to achieve a β-independent collector current of 1 mA.

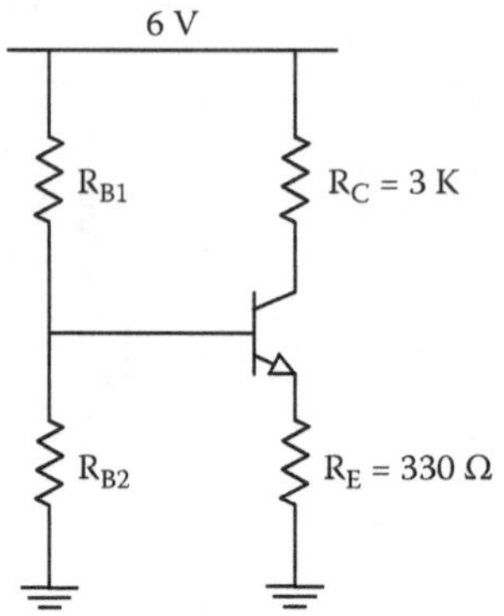

FIGURE EX. 3.7(a)

Solution:

As depicted in Figure Ex. 3.7(b), having a collector current of 1 mA requires a base voltage of:

$$V_B = V_{RE} + V_{BE}(ON) = I_E R_E + V_{BE}(ON) \cong I_C R_E + V_{BE}(ON)$$

$$= (1^{mA} \times 0.33^{k\Omega}) + 0.7^V = 1.03^V \cong 1^V.$$

In addition to generating a 1-V base voltage for the transistor, the voltage divider network should be designed so that the collector current of the transistor is β-independent. The former is achieved by properly ratioing R_{B1} and R_{B2}:

$$\frac{R_{B2}}{R_{B1} + R_{B2}} V_{CC} = \frac{R_{B2}}{R_{B1} + R_{B2}} \times 6^V = 1^V \Rightarrow R_{B1} = 5R_{B2}$$

and the latter is made possible by satisfying the following inequality:

$$\frac{R_{B1} \| R_{B2}}{\beta} \le \frac{R_E}{10}.$$

Using the specified ratio between R_{B1} and R_{B2}, this inequality can be simplified to:

$$R_{B1} \le \frac{6\beta R_E}{10}.$$

Replacing the known parameters with their numeric values, the final relationship is as follows:

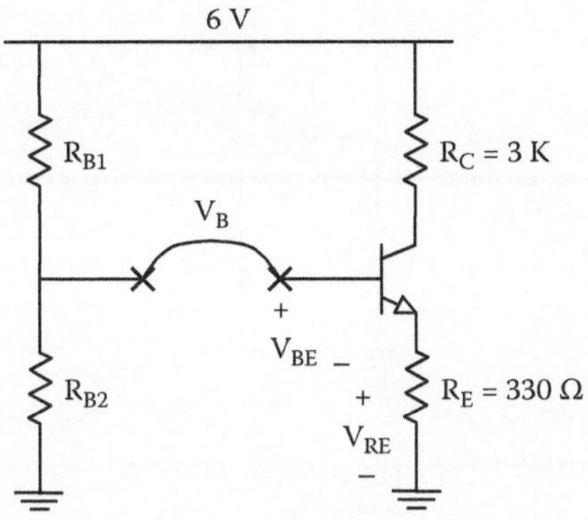

FIGURE EX. 3.7(b)

$$R_{B1} \le \frac{6 \times 120 \times 0.33^{k\Omega}}{10} \Rightarrow R_{B1} \le 23.76^{k\Omega}.$$

One of the possible designs for this circuit can be the choice of $R_{B1} = 20^{k\Omega}$ which gives:

$$R_{B2} = R_{B1}/5 = 20^{k\Omega}/5 = 4^{k\Omega}.$$

3.4 Biasing PNP Transistors

After exercising enough with biasing NPN transistors, now it is time to apply the same techniques to the circuits containing PNP transistors. Basic concepts of operating point calculations for PNP transistors are identical to those of NPN transistors; however, one needs to get used to the aforementioned differences between the currents and voltages of NPN and PNP transistors, as well as a circuit drawing difference. It is usually preferred, when drawing linear transistor circuits, that the nodes with larger DC voltages are placed higher than the ones with lower DC voltages. As a result, unlike NPN transistors where the collectors are usually drawn higher than their emitters, emitters of PNP transistors are often drawn higher than their collectors.

Example 3.8

A PNP transistor with $\beta = 70$, $V_{EB}(ON) = 0.7$ V, and $V_{EC}(Sat.) = 0.2$ V is used in the circuit in Figure Ex. 3.8(a). Assuming that $V_{CC} = 5$ V, $R_B = 330\ k\Omega$, and $R_C = 3\ k\Omega$, find the transistor's operating point (I_C, V_{EC}).

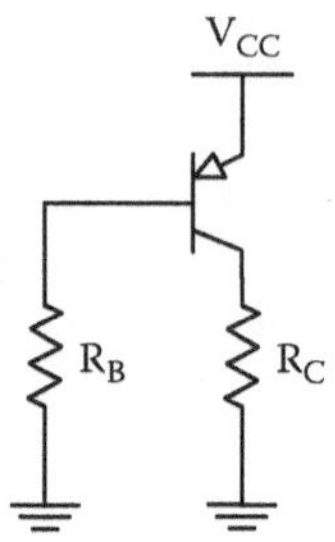

FIGURE EX. 3.8(a)

Solution:

Writing the KVL equation for the loop that passes through the transistor via emitter-base, depicted in Figure Ex. 3.8(b), gives:

$$\text{KVL 1: } V_{CC} = V_{EB}(ON) + V_{RB}$$

or

$$V_{CC} = V_{EB}(ON) + R_B I_B.$$

Replacing the known parameters with their numeric values we have:

$$5\text{ V} = 0.7\text{ V} + (330^{k\Omega} I_B),$$

which, if solved for I_B, gives:

$$I_B = \frac{5^V - 0.7^V}{330^{k\Omega}} = \frac{4.3^V}{330^{k\Omega}} = 13\ \mu A.$$

Finally, the collector current is found as

$$I_C = \beta I_B = 70 \times 13\ \mu A = 0.91\ mA.$$

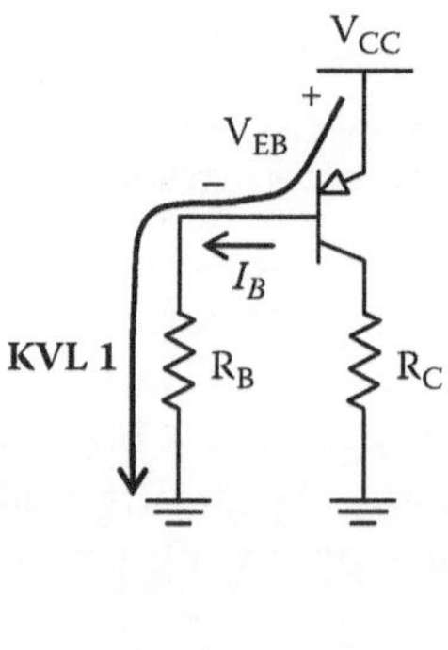

FIGURE EX. 3.8(b)

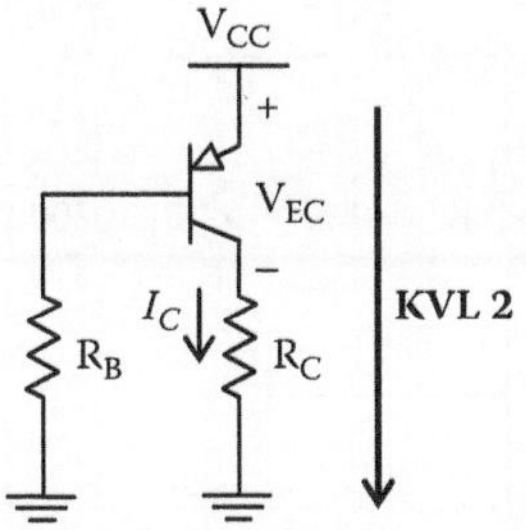

FIGURE EX. 3.8(c)

Now, to find the emitter-collector voltage, a second loop will be needed that passes both the emitter and collector. Such a loop is presented in Figure Ex. 3.8(c), which gives:

$$\text{KVL 2: } V_{CC} = V_{EC} + V_{RC}$$

or

$$V_{CC} = V_{EC} + R_C I_C.$$

Replacing the known parameters with their numeric values we have:

$$5\text{ V} = (3^{k\Omega} \times 0.91^{mA}) + V_{EC},$$

which, if solved for V_{EC}, gives:

$$V_{EC} = 5\text{ V} - (3^{k\Omega} \times 0.91^{mA}) = 2.27^{V}.$$

Because the emitter-collector voltage is well above the transistor's $V_{EC}(Sat.)$, it can be said that the initial assumption of the transistor being in the active mode is true.

Example 3.9

In the voltage-divider bias circuit depicted in Figure Ex. 3.9(a) a transistor with $\beta = 50$, $V_{EB}(ON) = 0.7V$, and $V_{EC}(Sat.) = 0.1\ V$ is used. Find the bias point (I_C, V_{EC}) for the transistor.

Solution:

First, the β-independence condition is checked to see if the simple voltage division analysis approach is applicable:

$$\frac{R_{B1} \| R_{B2}}{\beta} \le \frac{R_E}{10}.$$

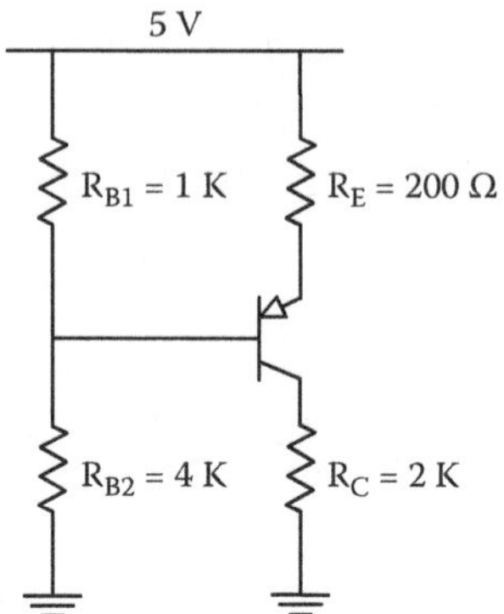

FIGURE EX. 3.9(a)

Replacing the numeric values:

$$\frac{1^{k\Omega} \| 4^{k\Omega}}{50} \le \frac{200^{\Omega}}{10}$$

and calculating the values of both sides:

$$16^{\Omega} \le 20^{\Omega}$$

confirms that one can go ahead and use the voltage division idea. With the straightforward resistances of R_{B1} and R_{B2}, it can be simply said that the 5-V supply voltage is distributed between them with the same ratio that they have: 1 V for the 1-$k\Omega$ R_{B1} and 4 V for the 4-$k\Omega$ R_{B2}. This means that the voltage divider branch provides the transistor with a base voltage of 4 V.

Now, to calculate the collector current, the analysis is concentrated on the part of the circuit in Figure Ex. 3.9(b).

A voltage difference of 1 V occurs between the upper end of R_E and the base of the transistor. This voltage appears across R_E and the emitter-base. The emitter-base takes its $V_{EB}(ON) = 0.7\ V$, and the rest remains across R_E, which determines the emitter current and consequently the collector current:

$$5^V - 4^V = V_{RE} + V_{EB}(ON),$$

and

$$1^V = (I_E \times 200^{\Omega}) + 0.7^V,$$

and finally:

$$I_C \cong I_E = 1.5^{mA}.$$

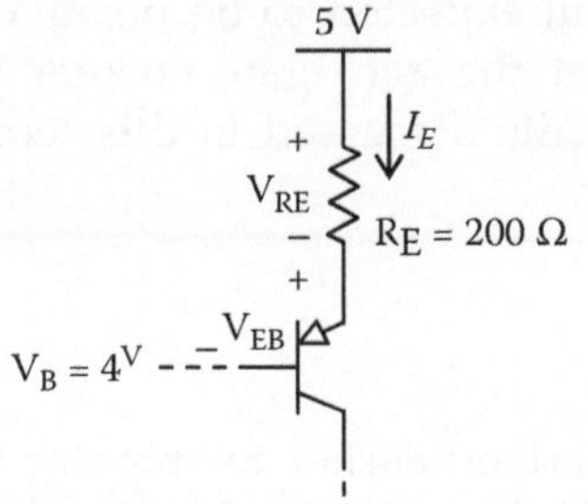

FIGURE EX. 3.9(b)

Emitter-collector voltage can be found by subtracting the voltages that appear across R_C and R_E from V_{CC}:

$$V_{EC} \cong V_{CC} - I_C(R_C + R_E) = 5^V - 1.5^{mA}(2^{k\Omega} + 0.2^{k\Omega}) = 1.7^V$$

which is larger than $V_{EC}(Sat.)$ meaning that the transistor is in the active mode.

3.5 Biasing MOS Transistors

As observed for bipolar transistors, no difference exists between the analysis approaches used for N-type and P-type transistors. Thus, to avoid repeating the same concepts once for negative-channel metal-oxide semiconductor (NMOS) transistors and once again for positive-channel metal-oxide semiconductor (PMOS) devices, let us only deal with NMOS transistors in this section. Keep in mind that for a PMOS transistor, the directions of currents are reversed, V_{SG} and V_{SD} are used instead of V_{GS} and V_{DS}, respectively, and the absolute value of the threshold voltage appears in the relationships.

It was studied earlier that for an NMOS transistor to be biased in the saturation mode, the following voltage conditions should be satisfied:

$$V_{GS} > V_{TH} \tag{3.7(a)}$$

and

$$V_{DS} > V_{DS}(Sat.), \tag{3.7(b)}$$

which imply applying proper voltages to the three terminals of the transistor. Under these conditions, the drain and source currents with the directions given in Figure 2.5 are all expected to be positive, and the gate current is zero. It will be seen that the zero gate current simplifies the bias point calculation for MOS circuits discussed in this section and the AC analysis that will be studied later.

3.5.1 Operating Point

Similar to what was described earlier for bipolar transistors, the operating point for an MOS transistor is represented by the drain current and the drain-source voltage (i.e., (I_D, V_{DS})).

The quadratic $i_D - v_{GS}$ characteristic curve of MOS transistors has a much slower slope compared with the exponential $i_C - v_{BE}$ characteristics of bipolar transistors. Therefore, the gate-source voltage of an MOS transistor cannot be treated almost as a constant when the transistor is biased in saturation mode. As seen in the previous chapter, V_{GS} is taken as a function of the drain current. Thus, in addition to the circuit equation relating the drain/source current and the gate-source voltage to each other, the quadratic $i_D - v_{GS}$ relationship is also used to calculate the bias point.

When a quadratic equation is solved, one usually ends up with two values as the answers, which are both valid mathematically. In the analysis of an MOS bias network, it is somehow different as the circuit can have only one operating point. Thus, only one of the drain current values is valid, but which one? One can simply use the answers for the drain current and calculate the gate-source voltage of the transistor. Only the value for the drain current that returns a reasonable gate-source voltage is valid. Similar to bipolar transistors, the operating point of an MOS transistor can be illustrated on both $i_D - v_{GS}$ and $i_D - v_{DS}$ characteristic curves. This is presented in Figure 3.10.

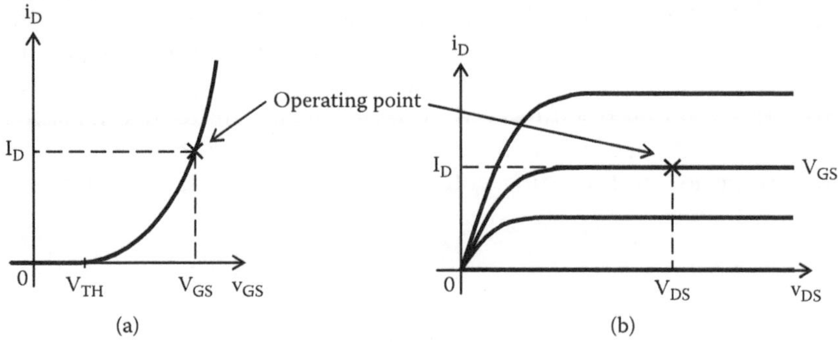

FIGURE 3.10
Operating points on (a) $i_D - v_{GS}$ characteristics (b) $i_D - v_{DS}$ characteristics.

Example 3.10

For the transistor used in the circuit in Figure Ex. 3.10(a) we have $k' = 100$ $\mu A/V^2$, W = 20 μm, L = 0.5 μm, and $V_{TH} = 0.8$ V. Calculate the drain current of the transistor.

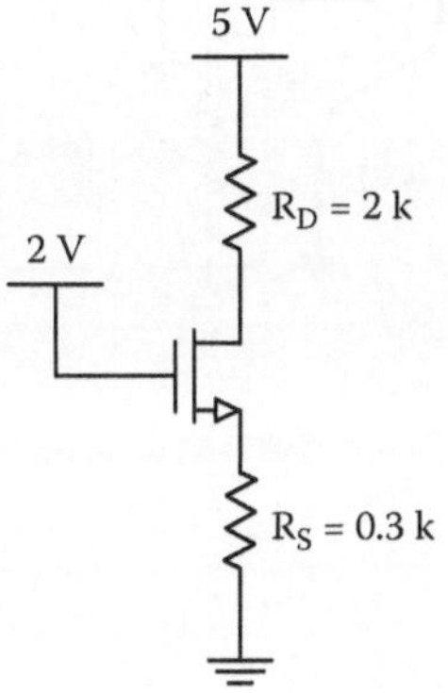

FIGURE EX. 3.10(a)

Solution:

The KVL equation that is needed is written for the path that starts from the 2-V voltage source connected to the gate, goes through the transistor via gate-source, and ends at the grounded end of R_S. For this path, illustrated in Figure Ex. 3.10(b), we have:

$$2^V = V_{GS} + V_{RS} = V_{GS} + I_D R_S,$$

which is one of the equations needed to find the drain current, and can be rewritten for V_{GS} as:

$$V_{GS} = 2^V - I_D R_S = 2^V - (0.3^{k\Omega} \times I_D).$$

The second equation is always the quadratic $i_D - v_{GS}$ characteristic equation:

$$i_D = \frac{\beta}{2}(v_{GS} - V_{TH})^2$$

in which

$$\beta = k'\frac{W}{L} = 100^{\mu A/V^2} \times \frac{20}{0.5} = 4^{mA/V^2}$$

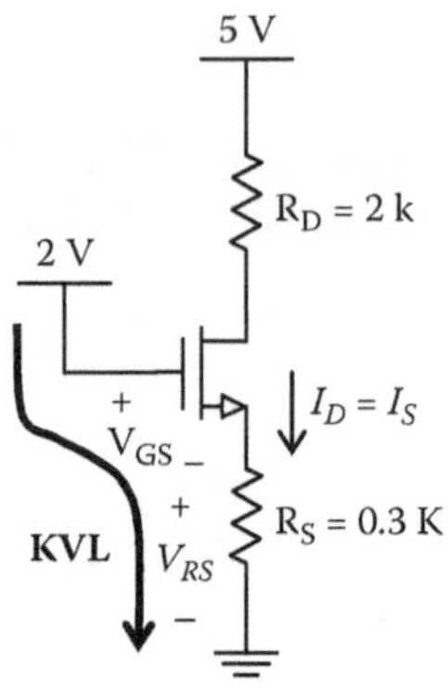

FIGURE EX. 3.10(b)

and $V_{TH} = 0.8^V$. Therefore,

$$\begin{cases} V_{GS} = 2 - 0.3 I_D \\ I_D = 2(V_{GS} - 0.8)^2 \end{cases},$$

and if V_{GS} in the quadratic equation is expressed in terms of the drain current, this gives:

$$I_D = 2(1.2 - 0.3 I_D)^2,$$

and

$$0.09 I_D^2 - 1.22 I_D + 1.44 = 0.$$

The two roots of this equation are $I_{D1} = 12.25^{mA}$ and $I_{D2} = 1.3^{mA}$, which if used to calculate V_{GS} give:

$$I_{D1} = 12.25^{mA} \Rightarrow V_{GS1} = -1.675^V,$$

and

$$I_{D2} = 1.3^{mA} \Rightarrow V_{GS2} = 1.61^V.$$

As we know, V_{GS} for an NMOS transistor consists of two terms, the threshold voltage and the overdrive voltage, which are both positive. Thus, I_{D1} is not acceptable, and $I_{D2} = 1.3^{mA}$ is chosen as the final answer.

Example 3.11

Figure Ex. 3.11 depicts the MOS version of the voltage-divider bias network. The transistor used in this circuit has the following specifications: $k' = 100\ \mu A/V^2$, W = 20 μm, L = 0.5 μm, and $V_{TH} = 0.8$ V.

Find the bias point (I_D, V_{DS}) for the transistor.

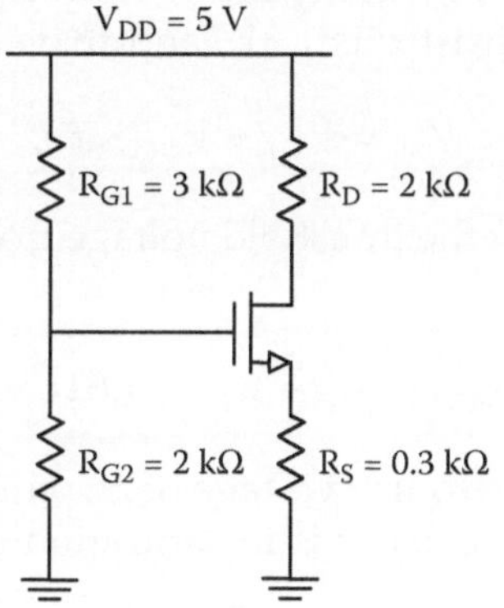

FIGURE EX. 3.11

Solution:

When using an NMOS transistor in voltage-divider bias configuration, no current is sunk from the voltage divider branch by the gate. Thus, the gate voltage is unconditionally written as:

$$V_G = \frac{R_{G2}}{R_{G1} + R_{G2}} V_{DD} = \frac{2}{2+3} \times 5^V = 2^V$$

Now, if the voltage divider network is represented by a voltage source connected to the gate, the circuit reduces to the simplified circuit analyzed in Example 3.10. Because the transistors used in both examples are identical, it can be said that the drain current is the same as before: $I_D = 1.3^{mA}$.

The drain-source voltage of an MOS transistor is determined similar to the collector-emitter voltage of a bipolar transistor. The transistor waits until all the external elements (usually in series with drain-source) take their voltages from the supply voltage, and then it gets the remainder of the voltage as its drain-source voltage. In this example, V_{DS} is found by subtracting the voltages that appear across R_D and R_S from V_{DD}:

$$V_{DS} = V_{DD} - I_D(R_D + R_S) = 5^V - 1.3^{mA}(2^{k\Omega} + 0.3^{k\Omega}) = 2^V.$$

To check whether the transistor is in saturation mode or not, we need to calculate the minimum drain-source voltage:

$$V_{DS}(Sat.) = \sqrt{\frac{2I_D}{\beta}} = \sqrt{\frac{2\times 1.3}{4}} = 0.81^V.$$

Alternatively, because we already know from Example 3.10 that the gate-source voltage of the transistor is 1.61^V, and that

$$V_{GS} = V_{TH} + V_{DS}(Sat.),$$

one could simply subtract the threshold voltage from the gate-source voltage to calculate $V_{DS}(Sat.)$:

$$V_{DS}(Sat.) = V_{OD} = V_{GS} - V_{TH} = 1.61^V - 0.8^V = 0.81^V.$$

Anyway, with the drain-source voltage of 2^V, which is larger than $V_{DS}(Sat.)$, it is confirmed that the transistor is in saturation mode.

Example 3.12

Find the bias point (I_D, V_{SD}) for the PMOS transistor depicted in Figure Ex. 3.12. The specifications of the transistor used in this circuit are: $k' = 70\ \mu A/V^2$, $W = 30\ \mu m$, $L = 0.5\ \mu m$, and $V_{TH} = -0.8$ V.

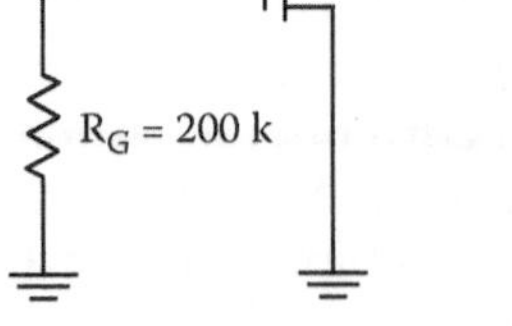

FIGURE EX. 3.12

Solution:

Because the gate current is zero, there will be no voltage drop across R_G, and thus, the gate will be at ground potential:

$$V_G = 0^V.$$

The source potential can also be found by subtracting the voltage drop across R_S from the supply voltage:

$$V_S = V_{DD} - R_S I_S,$$

and therefore:

$$V_{SG} = V_S - V_G = V_S = V_{DD} - R_S I_S$$

which becomes

$$V_{SG} = 5^V - 4^{k\Omega} I_S$$

after replacing the parameters with their numeric values. The other equation is always the quadratic equation for the drain current:

$$i_D = \frac{\beta}{2} . \left(v_{SG} - |V_{TH}|\right)^2$$

in which

$$\beta = k' . \frac{W}{L} = 70^{\mu A/V^2} \times \frac{30}{0.5} = 4.2^{mA/V^2}$$

and $|V_{TH}| = 0.8^V$. Therefore,

$$\begin{cases} V_{SG} = 5 - 4I_D \\ I_D = 2.1(V_{SG} - 0.8)^2 \end{cases}$$

and if V_{SG} in the quadratic equation is expressed in terms of the drain current, we have:

$$\frac{I_D}{2.1} = (4.2 - 4I_D)^2,$$

and

$$16I_D^2 - 34.1I_D + 17.64 = 0.$$

The two roots of this equation are $I_{D1} = 1.25^{mA}$ and $I_{D2} = 0.88^{mA}$, which if used to calculate V_{SG} give:

$$I_{D1} = 1.25^{mA} \Rightarrow V_{SG1} = 0^V,$$

and

$$I_{D2} = 0.88^{mA} \Rightarrow V_{SG2} = 1.48^V.$$

It is clear that $I_{D1} = 1.25^{mA}$ cannot be the answer because, as we know, V_{SG} for a PMOS transistor consists of two terms, the absolute value of the threshold voltage and the overdrive voltage, which are both positive. Thus, I_{D1} is not acceptable, and $I_{D2} = 0.88^{mA}$ is therefore chosen as the final answer.

The source-drain voltage of the transistor, V_{SD}, is determined by subtracting the voltage that appears across R_S from V_{DD}:

$$V_{SD} = V_{DD} - I_D R_S = 5^V - (0.88^{mA} \times 4^{k\Omega}) = 1.48^V.$$

At this point, the minimum drain-source voltage is needed to check whether the transistor is in saturation mode or not. As explained in the previous example, this voltage can be simply calculated as

$$V_{SD}(Sat.) = V_{OD} = V_{SG} - |V_{TH}| = 1.48^V - 0.8^V = 0.68^V.$$

With the source-drain voltage of 1.48^V, which is larger than $V_{SD}(Sat.)$, the initial assumption that the transistor is in saturation mode is confirmed.

3.6 Simulation Examples

Example 3.13

This example is intended to examine the effect of R_C in the operating mode of the circuit shown in Figure Ex. 3.13(a). Simulate the circuit, sweep the resistance R_C from 1 Ω to 5 kΩ, and discuss on the operating mode of the transistor. For the transistor, use the model of 2N3904 NPN transistor.

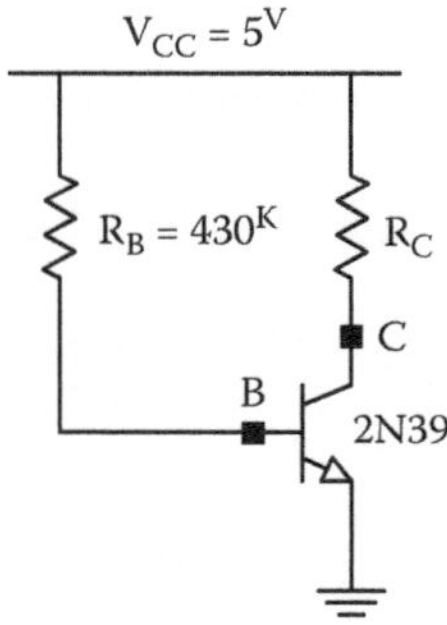

FIGURE EX. 3.13(a)

Solution:

```
VDC VCC 0 5V
RB VCC B 430k
RC VCC C {R}
Q1 C B 0 Q2N3904
.PARAM R 1k
.DC PARAM R 1 5k 10
.LIB nom.lib
.PROBE
.END
```

The collector current, I_C(Q1), the voltage across R_C, V_{CC} – $V2(R_C)$, and the collector-emitter voltage of the transistor, V(C), resulting from the simulation are presented in Figure Ex. 3.13(b). These traces clearly explain the following.

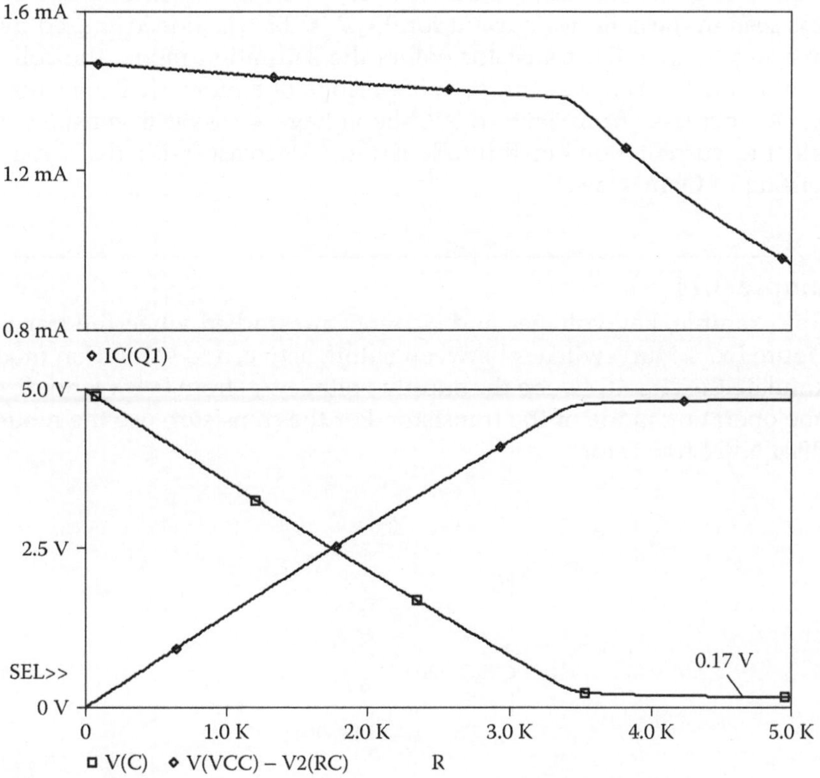

FIGURE EX. 3.13(b)

As long as the transistor is in the active mode, its collector-emitter voltage, which in this case is the same as the collector voltage, is related to the collector current and to R_C as:

$$V_{CE} = V_{CC} - I_C R_C.$$

This means that as the resistance R_C grows, assuming that the collector current is constant, V_{CE} decreases proportionally. This regime continues up to the point where the transistor reaches the edge of saturation mode (i.e., $V_{CE} = V_{CE}(Sat.)$). With a 5-V power supply, a collector current of around 1.5 mA, and a relatively small $V_{CE}(Sat.)$, the transistor is expected to be at the edge of saturation when R_C reaches around 3.5 $k\Omega$. From this point on, the transistor enters the saturation mode, V_{CE} stays at around $V_{CE}(Sat.)$, which is something between $0.15^V \sim 0.2^V$ in this example, and the behavior of the circuit changes accordingly. As seen, a "knee" point occurs in all the traces at around $R_C = 3.5\ k\Omega$.

Based on the equations that have been derived for this circuit, the collector current is supposed to be independent of R_C, but it is not. It was already explained that the collector current is also a weak function of the collector-emitter voltage in the active mode due to the Early effect. The decrement that is seen in the collector current for $0 < R_C < 3.5^{k\Omega}$ is indeed caused by the decreasing V_{CE}. As the transistor enters the saturation mode, the collector current tends to decrease with R_C with a slope of almost −1. The fact is that when R_C increases from $3.5^{k\Omega}$ to $5^{k\Omega}$, the voltage across it is constant. As a result, the current flowing through it must decrease with the same rate according to Ohm's law.

Example 3.14

In this example, key voltages and currents are studied when the transistor in Figure Ex. 3.14(a) switches between cutoff, active, and saturation modes.

Simulate the circuit, sweep the supply voltage V_{BB} from 0–5 V, and discuss on the operating mode of the transistor. For the transistor, use the model of 2N3904 NPN transistor.

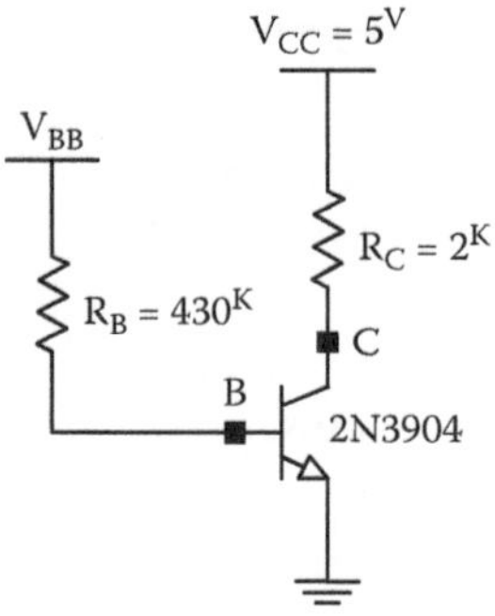

FIGURE EX. 3.14(a)

Solution:

```
VDC1 VCC 0 5V
VDC2 VBB 0 1V
RB VBB B 220k
RC VCC C 2k
Q C B 0 Q2N3904
.DC lin VDC2 0 5V 0.01V
.LIB nom.lib
.PROBE
.END
```

The base and collector currents, I_B(Q) and I_C(Q), base voltage, *V*(*B*), and collector-emitter voltage, *V*(*C*), resulting from the simulation are presented in Figure Ex. 3.14(b). The following explanation analyzes these traces:

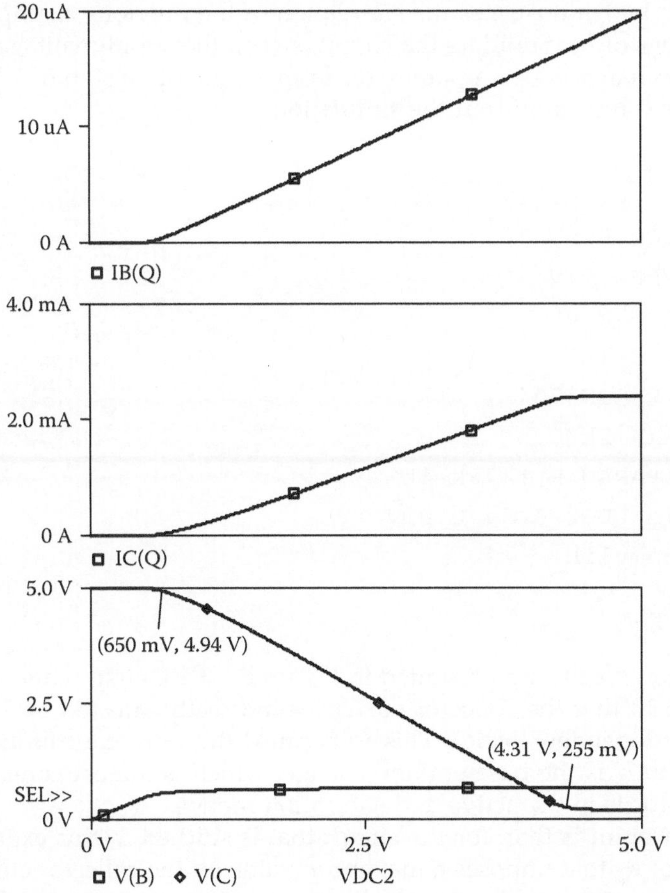

FIGURE EX. 3.14(b)

As long as V_{BB} is smaller than 0.65 V, the base-emitter is not supplied with enough voltage to exit the cutoff mode. Thus, both I_B and I_C are zero. With no current flowing through R_C, the voltage drop across it is zero, and the collector node will stay at $V_{CC} = 5^V$. When V_{BB} exceeds $V_{BE}(ON)$, the transistor enters the active mode and the base voltage remains almost constant at $V_{BE}(ON)$. In other words, the voltage across R_B increases at the same rate as V_{BB}, resulting in an increasing I_B, and consequently I_C, as indicated. The collector current increases up to the point where V_{CE} reaches V_{CE}(Sat.). From this point on, I_C does not follow I_B anymore even though the base current increases. This is because the voltage across R_C is almost constant in saturation mode, as explained in the previous example. Under these conditions in saturation mode, it can be said that the transistor's β parameter, which, by definition, is the collector current divided by the base current, drops.

Example 3.15

This example demonstrates the simulation of the previous example for different values of R_C. Simulate the circuit, sweep the supply voltage V_{BB} from 0–5 V, and change R_C, as a parameter from 1–5 k with a step of 1 k. Discuss on the traces resulting from the simulation.

Solution:

```
VDC1 VCC 0 5V
VDC2 VBB 0 1V
RB VBB B 220k
RC VCC C {R}
Q C B 0 Q2N3904
.PARAM R 2k
.STEP PARAM R 1k 5k 1k
.DC lin VDC2 0 5V 0.01V
.LIB nom.lib
.PROBE
.END
```

Simulation results are presented in Figure Ex. 3.15. First, while we saw in Example 3.13 that the collector current is indirectly affected by R_C, here I_B remains independent of R_C. This is because the only contribution of the transistor to I_B is the base-emitter voltage, which is almost constant while the transistor is in the active and saturation modes.

The next point is that, for the circuit that is studied in this example, it is the value of R_C that imposes a maximum value on the collector current. The larger the resistance of R_C, the smaller the maximum allowed collector current would be.

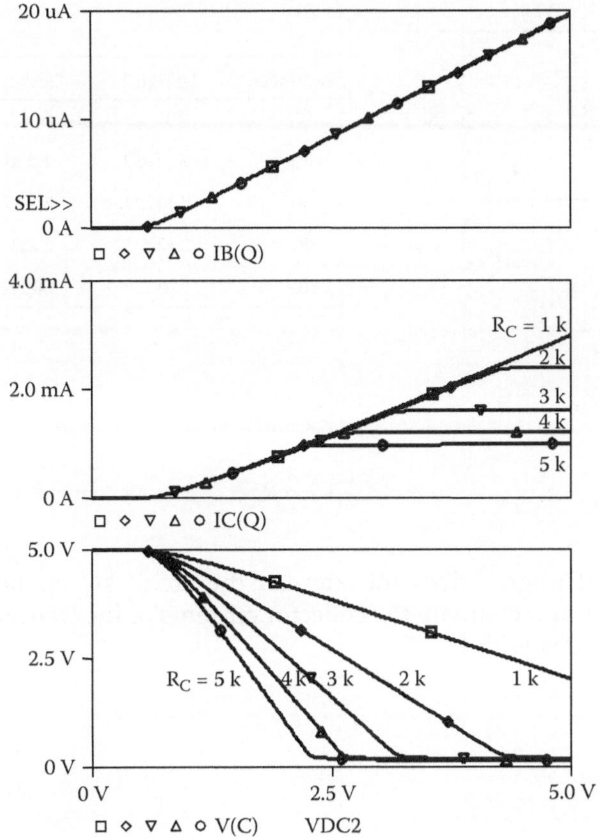

FIGURE EX. 3.15

The traces drawn in the bottom-most plot have different slopes, which vary by R_C. In other words, in the active mode V_{BB} variations are converted to variations in the collector voltage with different slopes. It is worth noting, for instance, that for $R_C = 5\ k\Omega$, the transition between the cutoff and saturation modes occurs with a slope greater than 2. Indeed, this behavior demonstrates the amplification capability of the circuit, which will be studied in the next chapter.

Example 3.16

A voltage-divider bias circuit is presented in Figure Ex. 3.16(a). For the NPN transistor we have $I_S = 1.5\ E-15\ A$ and $\beta = 120$. In this example, the circuit is simulated to observe the sensitivity of the bias current to the β parameter of the transistor. To do this, two sets of elements are used to achieve the same bias current, $I_C = 1.7\ mA$. One of the designs (Design 2) fulfills the β-independence requirement:

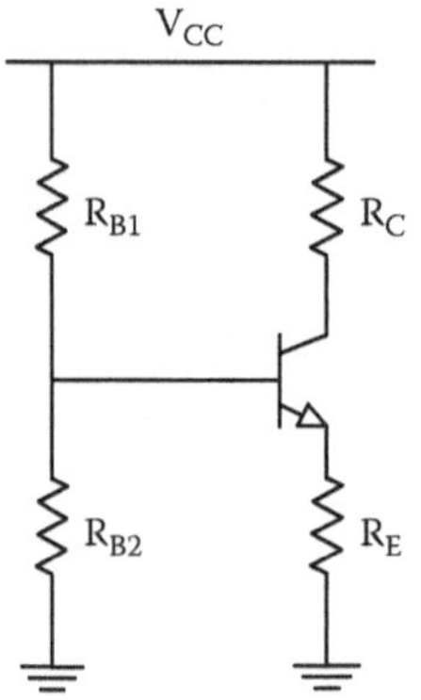

Parameter	Design 1	Design 2
V_{CC}	5 V	5 V
R_{B1}	220 kΩ	2.2 kΩ
R_{B2}	330 kΩ	3.3 kΩ
R_C	2 kΩ	2 kΩ
R_E	220 Ω	680 Ω

FIGURE EX. 3.16(a)

$$\frac{R_B}{\beta} << R_E ,$$

and the other (Design 1) does not. Simulate the circuit, sweep the β parameter from 50 to 150, and compare the collector currents of the transistors in terms of their variations with β.

Solution:

```
VDC VCC 0 5V
RB11 VCC B1 220k
RB12 B1 0 330k
RC1 VCC C1 2k
RE1 E1 0 220
Q1 C1 B1 E1 NPNmodel
RB21 VCC B2 2.2k
RB22 B2 0 3.3k
RC2 VCC C2 2k
RE2 E2 0 680
Q2 C2 B2 E2 NPNmodel
.MODEL NPNmodel NPN (IS=1.5E-15 BF={beta})
.PARAM beta 120
.DC param beta 50 150 1
.PROBE
.END
```

The traces, IC(Q1) and IC(Q2), depicted in Figure Ex. 3.16(b), are the collector currents for Design 1 and Design 2, respectively. As seen, both circuits are designed for a collector current of $I_C = 1.7^{mA}$ at $\beta = 120$. The ratio

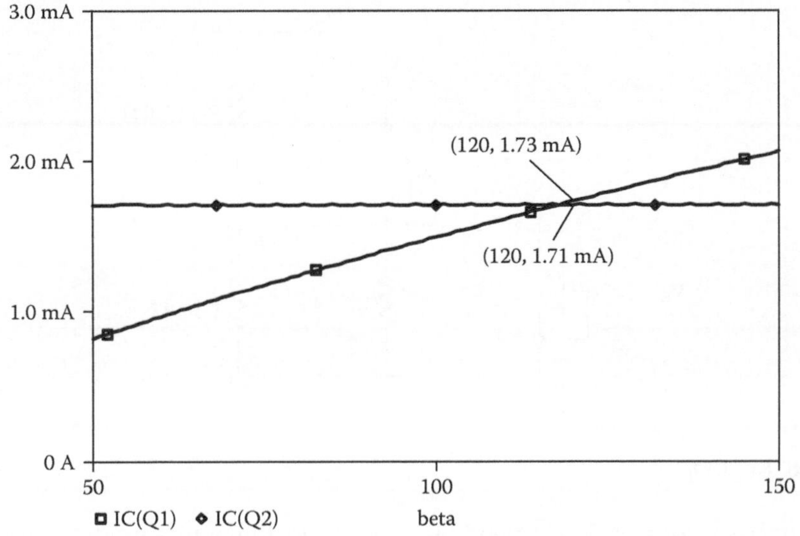

FIGURE EX. 3.16(b)

between R_{B1} and R_{B2} is the same in both designs, but the resistances in Design 2 are much smaller than the resistances in Design 1, which make the circuit β-independent. R_E is also intentionally chosen different to tune the bias currents at the same level.

Example 3.17

It was studied earlier that the base-emitter voltage of a bipolar transistor in the active mode can be approximated to an almost constant voltage level, $V_{BE}(ON)$. This property is useful in occasions where a constant voltage difference between two nodes is needed. A circuit that generates such a DC voltage difference is called a *DC level shifter.* Although this is what a voltage source is supposed to do, DC level shifters are widely used in analog electronics to avoid several voltage sources in a single circuit. The simplest implementation for a DC level shifter is the so-called *diode-connected transistor,* which is a bipolar transistor with its base and collector terminals connected together. (In the MOS counterpart of this implementation, the gate and the drain terminals are tied together.) The PNP realization of this idea, in which the voltage $V_{EB}(ON)$ appears across the two-terminal element, is illustrated in Figure Ex. 3.17(a). When the transistor is diode connected with enough applied voltage, it is guaranteed to be in the active mode if it is a bipolar transistor, and in the saturation mode if it is an MOS transistor.

In Figure Ex. 3.17(b), two diode-connected transistors are used in a bias circuit to generate a voltage difference of $2V_{EB}(ON)$ between the upper end of R_E and the base of Q_1. From the total voltage $2V_{EB}(ON)$, first Q_1 takes the

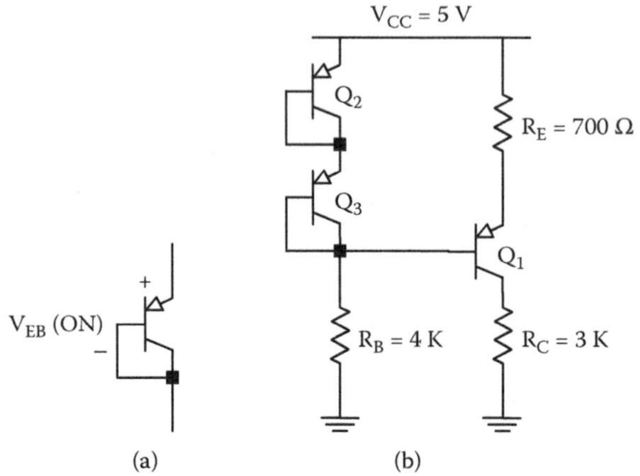

FIGURE EX. 3.17

emitter-base voltage that it needs to operate in the active mode. Then, the rest of the applied voltage ($2V_{EB}(ON) - V_{EB1}(ON)$) appears across R_E, which determines the emitter current and consequently the collector current:

$$I_C \cong I_E = [2V_{EB}(ON) - V_{EB1}(ON)]/R_E \cong V_{EB}(ON)/R_E.$$

With $V_{EB1}(ON) = 0.7^V$ and $R_E = 0.7^{k\Omega}$, collector current of Q_1 is expected to be around 1^{mA}. Assuming $I_S = 1.0E-15A$ and $\beta = 70$ for the transistors, simulate the circuit and find the key voltages and currents in the circuit.

Solution:

```
VDC VCC 0 5V
RB B 0 4k
RE VCC E 700
RC C 0 3k
Q1 C B E PNPmodel
Q2 C2 C2 VCC PNPmodel
Q3 B B C2 PNPmodel
.MODEL PNPmodel PNP (IS=1.0E-15 BF=70)
.OP
.END
```

All the DC variables of the circuit are reported in the output file that is outlined next. Although it can be observed that their collector currents are different by more than 10%, emitter-base voltages of the three transistors are almost the same at around 0.7^V. It is also interesting that the collector current of Q_1 closely agrees with our theoretical expectation mentioned previously ($1mA$). Finally, the fact that the three emitter-collector voltages are well above $V_{EC}(Sat.)$ indicates that the transistors are in the active mode.

```
****SMALL SIGNAL BIAS SOLUTION TEMPERATURE= 27.000 DEG C *****

NODE VOLTAGE    NODE VOLTAGE    NODE VOLTAGE    NODE VOLTAGE
(B)  3.5780     (C) 2.9891      (E)  4.2926     (C2) 4.2890

(VCC)5.0000

    VOLTAGE SOURCE CURRENTS
    NAME         CURRENT

    VDC         -1.891E-03

    TOTAL POWER DISSIPATION   9.45E-03  WATTS

** OPERATING POINT INFORMATION   TEMPERATURE = 27.000 DEG C **

**** BIPOLAR JUNCTION TRANSISTORS

NAME          Q1           Q2           Q3
MODEL         PNPmodel     PNPmodel     PNPmodel
IB           -1.42E-05    -1.24E-05    -1.24E-05
IC           -9.96E-04    -8.68E-04    -8.68E-04
VBE          -7.15E-01    -7.11E-01    -7.11E-01
VBC           5.89E-01     0.00E+00     0.00E+00
VCE          -1.30E+00    -7.11E-01    -7.11E-01
BETADC        7.00E+01     7.00E+01     7.00E+01
GM            3.85E-02     3.36E-02     3.36E-02
RPI           1.82E+03     2.09E+03     2.09E+03
RX            0.00E+00     0.00E+00     0.00E+00
RO            1.00E+12     9.63E+11     9.63E+11
CBE           0.00E+00     0.00E+00     0.00E+00
CBC           0.00E+00     0.00E+00     0.00E+00
CJS           0.00E+00     0.00E+00     0.00E+00
BETAAC        7.00E+01     7.00E+01     7.00E+01
CBX/CBX2      0.00E+00     0.00E+00     0.00E+00
FT/FT2        6.13E+17     5.34E+17     5.34E+17
        JOB CONCLUDED
        TOTAL JOB TIME              .02
```

Example 3.18

This example is considered for biasing NMOS transistors. In the circuit in Figure Ex. 3.18, M_1 is in diode-connected configuration to provide M_2 with the proper gate voltage required to be biased in saturation mode. For the transistors, assume:

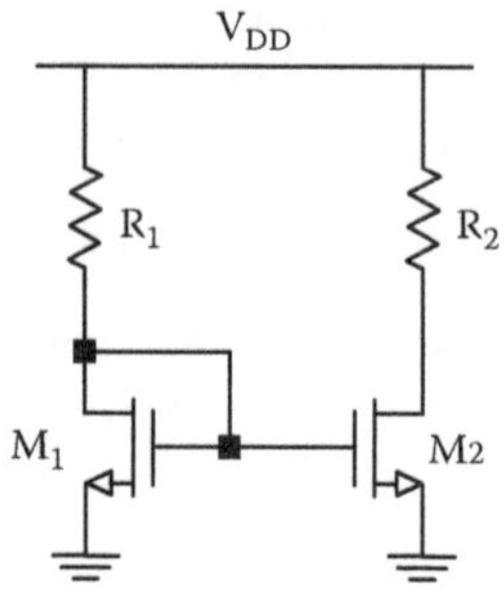

FIGURE EX. 3.18

$$k' = 100\ \mu A/V^2,\ V_{TH} = 0.8\ \text{V},\ (\text{W/L})_1 = 10\ \mu m/0.5\mu m,$$

and

$$(\text{W/L})_2 = 20\ \mu m/0.5\ \mu m.$$

Simulate the circuit with $V_{DD} = 3.3^V$, $R_1 = 50^{k\Omega}$, and $R_2 = 50^{k\Omega}$, and obtain the key voltages and currents for the transistors.

Solution:

```
VDC VDD 0 3.3V
R1 VDD D1 50k
R2 VDD D2 25k
M1 D1 D1 0 0 NMOSmodel W = 10u L = 0.5u
M2 D2 D1 0 0 NMOSmodel W = 20u L = 0.5u
.MODEL NMOSmodel NMOS (VTO = 0.7 kp = 100u)
.OP
.END
```

A text list of the simulation results is outlined next, as reported by the simulation output file. Both transistors have the same gate-source voltage, which is a result of their electrical connection. Because the W/L ratio for M_2 is twice that of M_1, its drain current is twice as big as M_1, as is theoretically expected to be. The voltage $V_{DS}(Sat.)$ is the same for both transistors (~ 0.22^V) as it is a function of I_D to (W/L) ratio. Comparing drain-source voltages of M_1 and M_2 with $V_{DS}(Sat.)$, it can be said that both transistors are in saturation.

```
*** SMALL SIGNAL BIAS SOLUTION  TEMPERATURE = 27.000 DEG C ***

NODE VOLTAGE    NODE VOLTAGE    NODE VOLTAGE

(D1)   .9183    (D2)    1.3946   (VDD) 3.3000
```

```
    VOLTAGE SOURCE CURRENTS
    NAME        CURRENT

    VDC         -1.429E-04

    TOTAL POWER DISSIPATION   4.72E-04  WATTS

*** OPERATING POINT INFORMATION  TEMPERATURE=27.000 DEG C ***

**** MOSFETS

NAME         M1                       M2
MODEL        NMOSmodel                NMOSmodel
ID            4.76E-05                 9.53E-05
VGS           9.18E-01                 9.18E-01
VDS           9.18E-01                 1.39E+00
VBS           0.00E+00                 0.00E+00
VTH           7.00E-01                 7.00E-01
VDSAT         2.18E-01                 2.18E-01
Lin0/Sat1    -1.00E+00                -1.00E+00
if           -1.00E+00                -1.00E+00
ir           -1.00E+00                -1.00E+00
TAU          -1.00E+00                -1.00E+00
GM            4.37E-04                 8.73E-04
GDS           0.00E+00                 0.00E+00
GMB           0.00E+00                 0.00E+00
CBD           0.00E+00                 0.00E+00
CBS           0.00E+00                 0.00E+00
CGSOV         0.00E+00                 0.00E+00
CGDOV         0.00E+00                 0.00E+00
CGBOV         0.00E+00                 0.00E+00
CGS           0.00E+00                 0.00E+00
CGD           0.00E+00                 0.00E+00
CGB           0.00E+00                 0.00E+00

        JOB CONCLUDED

        TOTAL JOB TIME              .02
```

3.7 Problems

3.1 Determine the mode of operation for the transistor in each of the circuits in Figure Prob. 3.1. Let $|V_{BE}(ON)| = 0.7$ V, $|V_{CE}(Sat.)| = 0.2$ V, and $\beta = 100$.

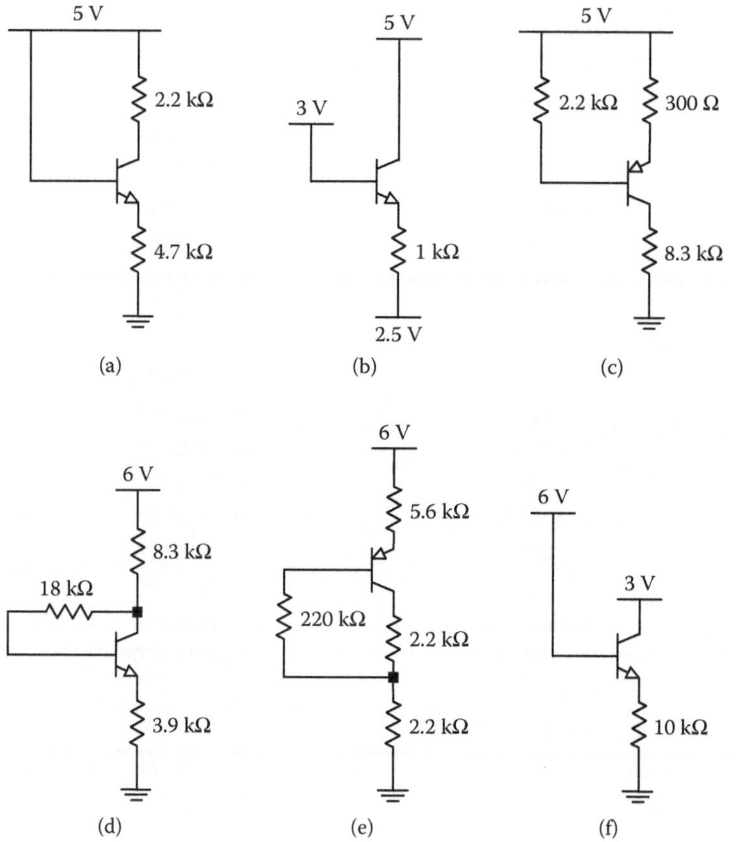

FIGURE PROB. 3.1

3.2 Assuming $|V_{BE}(ON)| = 0.7$ V, $|V_{CE}(Sat.)| = 0.1$ V, and $\beta = 100$, find the collector, base, and emitter voltages and currents for each of the circuits in Figure Prob. 3.2.

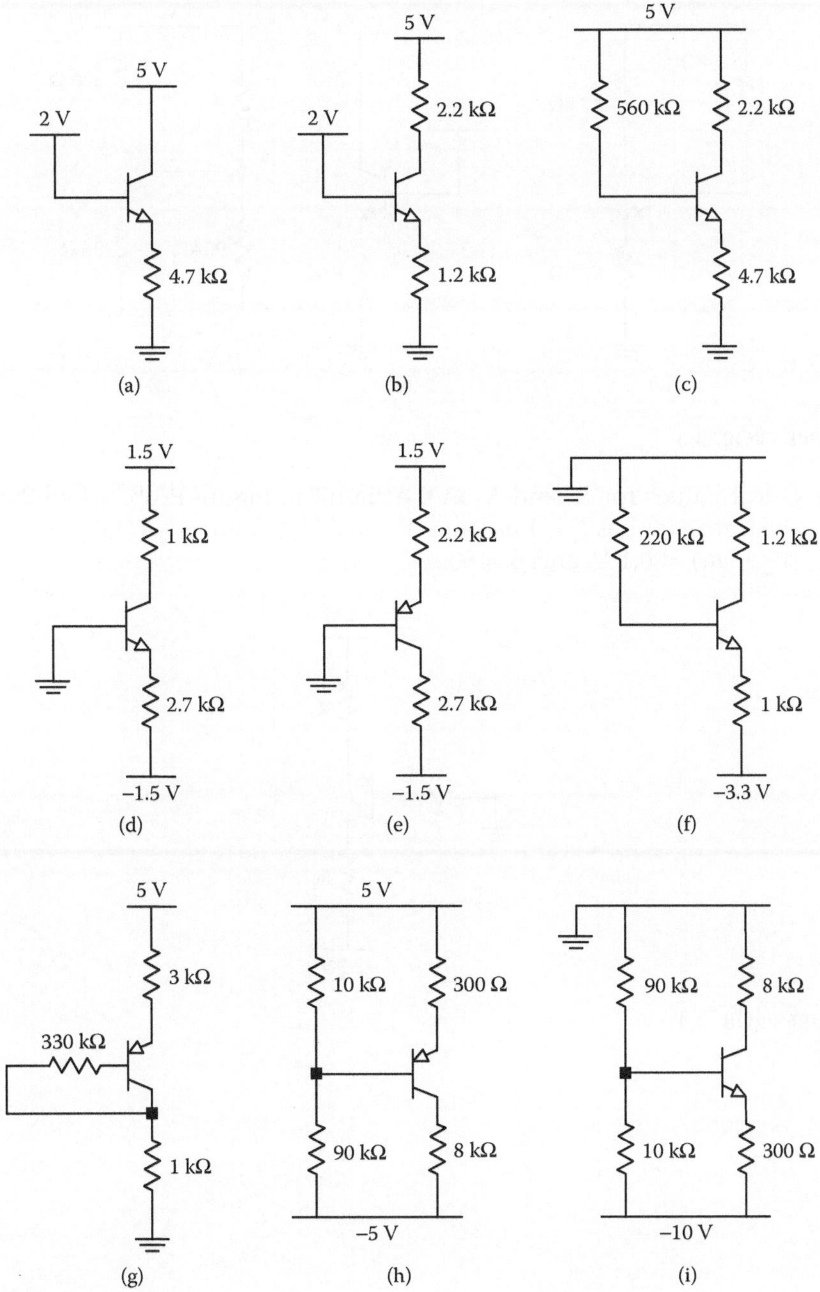

FIGURE PROB. 3.2

3.3 Assuming $|V_{BE}(ON)| = 0.7$ V, $|V_{CE}(Sat.)| = 0.2$ V, and $\beta = 100$, find a range for R in each of the circuits in Figure Prob. 3.3 to guarantee that the transistor is in the active mode.

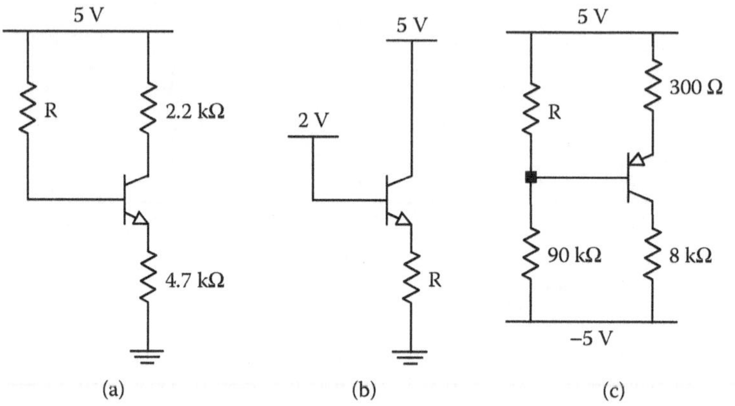

FIGURE PROB. 3.3

3.4 Select values for R_1 and R_2 in the circuit in Figure Prob. 3.4 so that the transistor has $I_C = 1\ mA$ and $|V_{CE}| = 2$ V. Assume $|V_{BE}(ON)| = 0.7\ V$, $|V_{CE}(Sat.)| = 0.1$ V, and $\beta = 80$.

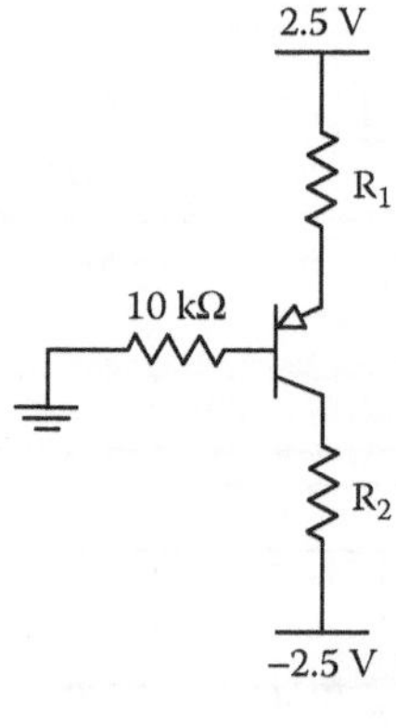

FIGURE PROB. 3.4

3.5 Assuming $V_{BE}(ON) = 0.7$ V, $V_{CE}(Sat.) = 0.1$ V, and $\beta = 120$ for the transistor used in the voltage-divider bias circuit in Figure Prob. 3.5,

(a) Find the operating point of the transistor (I_C, V_{CE}), and make sure that the transistor is biased in the active mode.

(b) What happens if a 1-kΩ resistor is then connected from the collector to ground?

(c) What happens if a 100-Ω resistor is connected from the base to ground?

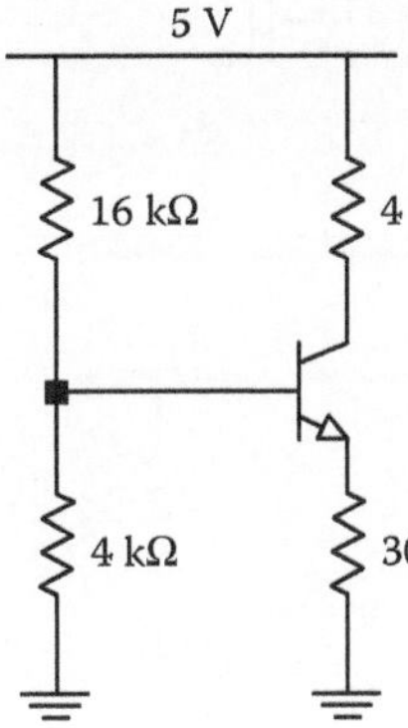

FIGURE PROB. 3.5

3.6 The circuit in Figure Prob. 3.6 is biased in active mode, and its collector, base, and emitter voltages are supposed to be 5 V, 2.5 V, and 1.8 V, respectively. After the circuit is made, it is observed that it is not functioning as expected. The collector, base, and emitter voltages are measured using a multimeter and are found to be 10 V, 0 V, and 0 V, respectively. It is suspected that one of the resistors is burned; it can become either open or a short circuit. Discuss all the possibilities that could have caused this failure.

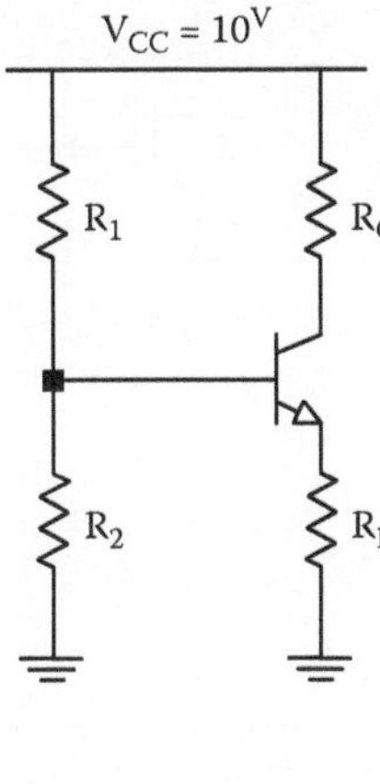

FIGURE PROB. 3.6

3.7 Assuming $|V_{TH}| = 0.7$ V and $\beta = 1\ mA/V^2$, find the drain current for each of the circuits in Figure Prob. 3.7.

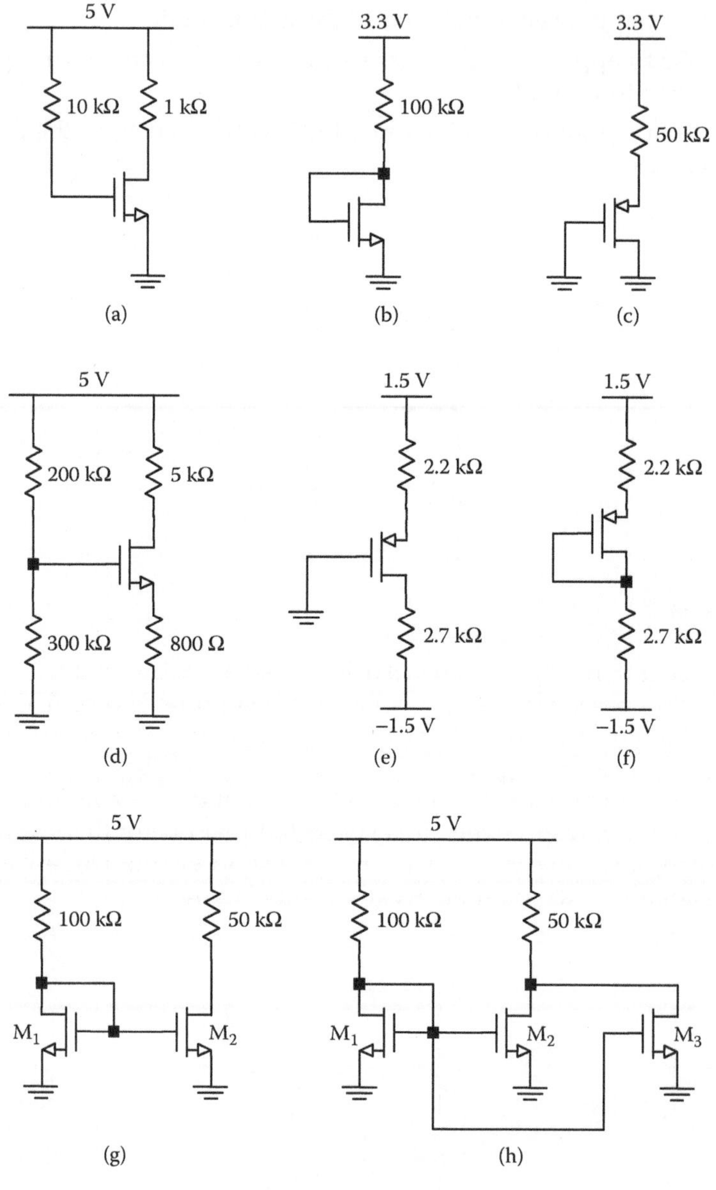

FIGURE PROB. 3.7

3.8 Assuming $V_{TH} = 0.7$ V and $k' = 100\ \mu A/V^2$, find values of W/L ratio for each of the transistors in the circuit in Figure Prob. 3.8 required to achieve $V_A = 2$ V and $V_B = 1$ V.

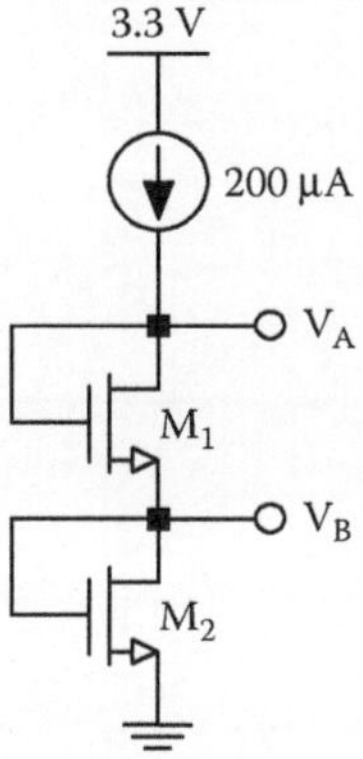

FIGURE PROB. 3.8

3.9 Assuming $V_{TH} = 0.7$ V and $\beta = 1\ mA/V^2$ for each of the transistors used in Figure Prob. 3.9,

(a) Find the operating point (I_D, V_{DS}) for M_1 in each of the circuits with $V_{DD} = 5$ V.

(b) Reduce the power-supply voltage by 10% and repeat part (a).

(c) Calculate sensitivity of I_{D1} to V_{DD} in each of the circuits.

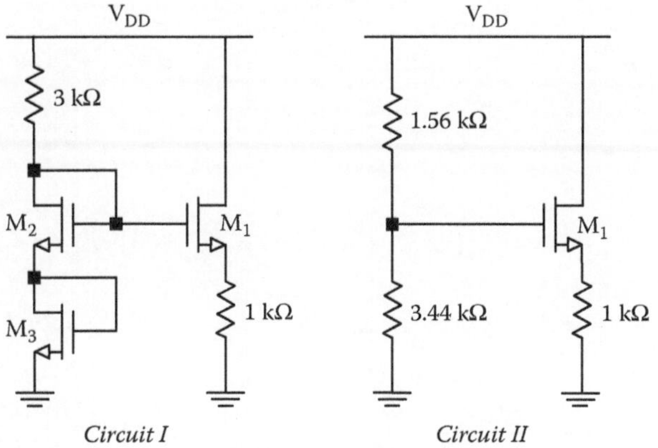

FIGURE PROB. 3.9

Chapter 4

Single-Stage Amplifiers

4.1 Introduction

The simplest amplifiers are made with a single transistor as the amplifying device and a few passive elements to establish the required bias conditions. These amplifiers are known as *single-stage amplifiers*. This chapter starts with how a transistor can perform small-signal amplification, then introduces different configurations for single-stage amplifiers, and finally presents a powerful technique to analyze such amplifiers.

To avoid possible confusion, the first half of this chapter deals with metal-oxide semiconductor (MOS) amplifiers only, whereas the second half discusses bipolar amplifiers.

4.2 The Transistor as an Amplifier

As mentioned in the previous chapter, assuming that a transistor is properly biased in the saturation mode, it is expected to linearly amplify small signals despite its nonlinear characteristics studied in Chapter 2. But how?

Let us start with the simplified parabolic (nonlinear) relationship between i_D and v_{GS} for an MOS transistor in the saturation mode, which is:

$$i_D = \frac{\beta}{2}(v_{GS} - V_{TH})^2. \tag{4.1}$$

Supposing that the transistor is biased at the operating point $Q(V_{GS}, I_D)$, also called the *quiescent point*, if the gate-source voltage, v_{GS}, is slightly changed by Δv_{GS}, then this change will be translated to a change in i_D (Δi_D) as dictated by Equation (4.1). This is illustrated in Figure 4.1.

If the variations in both i_D and v_{GS} are small compared with the associated bias values, that is,

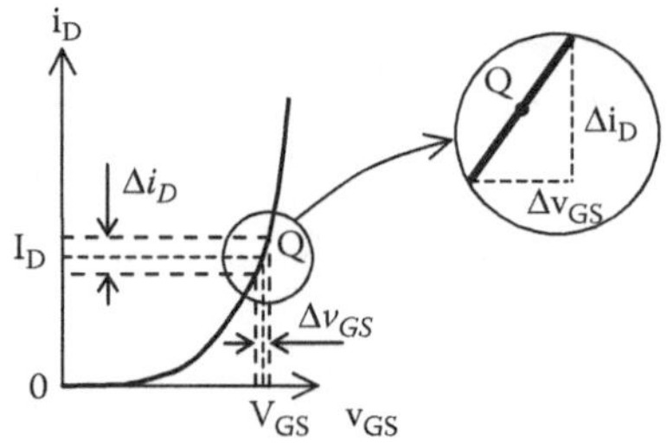

FIGURE 4.1
$i_D - v_{GS}$ characteristic curve for an NMOS transistor.

$$\frac{\Delta v_{GS}}{V_{GS}} \ll 1 \text{ and } \frac{\Delta i_D}{I_D} \ll 1, \tag{4.2}$$

then it can be said that the part of the curve that is spanned as the result of the variations in the gate-source voltage can be approximated by a straight line, with a slope that is determined by the curve's slope at point Q:

$$Slope@Q = \frac{\partial i_D}{\partial v_{GS}}\Big|_Q \tag{4.3}$$

Referring to the zoomed area in Figure 4.1, Δv_{GS} and Δi_D are related to each other by the same slope:

$$\Delta i_D = (Slope@Q) \times \Delta v_{GS}. \tag{4.4}$$

If it is assumed that both Δv_{GS} and Δi_D are periodic variations with zero mean value, then they are indeed AC components of v_{GS} and i_D. Thus, Equation (4.4) can be rewritten as:

$$i_d = (Slope@Q) \times v_{gs}. \tag{4.5}$$

Now that the curve's slope at the operating point relates the AC (signal-dependent) components of the drain current i_D and the gate-source voltage v_{GS}, it deserves to be properly named. Referring to Equation (4.3), it is obvious, by looking at the units that this slope is of conductance nature. Because this conductance converts a voltage component between two nodes to a current component that flows between two *different* nodes, it is called the transistor's *transconductance*, and is usually named as g_m. Therefore, the transconductance for an MOS transistor is defined as:

$$g_m = \frac{\partial i_D}{\partial v_{GS}}\Big|_Q \tag{4.6}$$

and differentiating Equation (4.1) with respect to v_{GS}, one can write:

$$g_m = \left.\frac{\partial\left[\frac{\beta}{2}(v_{GS}-V_{TH})^2\right]}{\partial v_{GS}}\right|_Q = \beta(v_{GS}-V_{TH})\big|_Q = \sqrt{2\beta i_D}\,\big|_Q, \tag{4.7}$$

and thus:

$$g_m = \sqrt{2\beta I_D}, \tag{4.8}$$

which is the formula for calculating a transistor's transconductance at a given bias current. Rewriting Equation (4.5) as:

$$i_d = g_m v_{gs}, \tag{4.9}$$

it can be easily said that the transistor amplifies the small variations in v_{GS} and provides amplified variations of i_D.

Remember that we started with a nonlinear relationship between i_D and v_{GS} (Equation (4.1)), and now *Equation (4.9) represents a linear relationship between i_d and v_{gs}, which is the basis of linear amplification by a transistor in small-signal mode,* discussed in the next section.

Example 4.1

For an NMOS transistor with $k' = 100\ \mu A/V^2$, W = 20 μm, and L = 1 μm biased at I_D = 1 mA, what will be the total drain current when a variation of 1^{mV}Sinωt is added to the gate-source voltage around its bias value?

Solution:

The first step is to find the transistor's transconductance at the given bias point:

$$\beta = k'W/L = 2\frac{mA}{V^2}$$

$$g_m = \sqrt{2\beta I_D} = \sqrt{2\times 2mA/V^2 \times 1mA} = 2\frac{mA}{V}$$

Then, using Equation (4.9) we have:

$$i_d = g_m v_{gs} = 2\ mA/V \times 1^{mV}\text{Sin}(\omega t) = 2^{\mu A}\text{Sin}(\omega t)$$

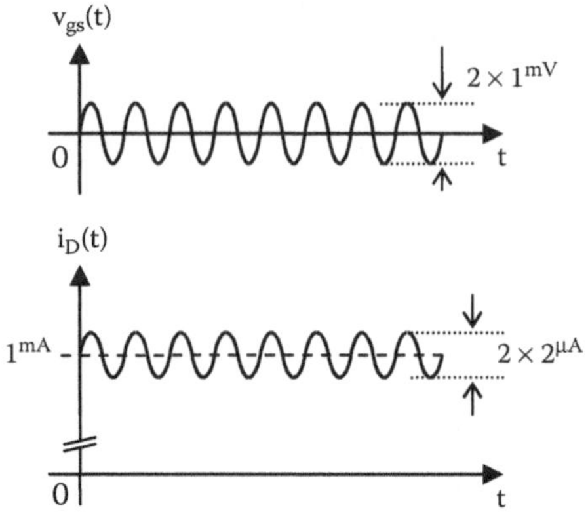

FIGURE EX. 4.1

and the total drain current is achieved as:

$$i_D = I_D + i_d = 1^{mA} + 2^{\mu A}\text{Sin}(\omega t).$$

Figure Ex. 4.1 illustrates the waveforms for both v_{gs} and i_D.

4.2.1 Small-Signal Model for a Transistor

As we have learned, the transistor can demonstrate a linear behavior in small-signal mode. For the circuits that contain one or more transistors operating in small-signal mode, the analysis will be significantly simplified if each transistor is modeled by simple circuit elements. It should be emphasized that such a model, called the *small-signal model,* is intended to describe the transistor's function only for small variations of transistor's voltages and currents referred to as the AC components. However, the results of the analysis methods that will be developed in this chapter and Chapters 7 and 8 can be extended to nonperiodic and sometimes even DC signals.

To form a small-signal model for an MOS transistor, let us start from the amplification represented by Equation (4.9). This equation states that the transistor senses the gate-source voltage and generates a current flowing from the drain to the source, which is proportional to the gate-source voltage of (all AC components), with the transistor's transconductance, g_m, as the proportionality factor. Based on this explanation, the simplest small-signal model for an MOS transistor is presented in Figure 4.2. It should be noted that no difference exists between the small-signal models for NMOS and PMOS transistors because they include only AC components of the currents/voltages, not the whole signals or their DC components.

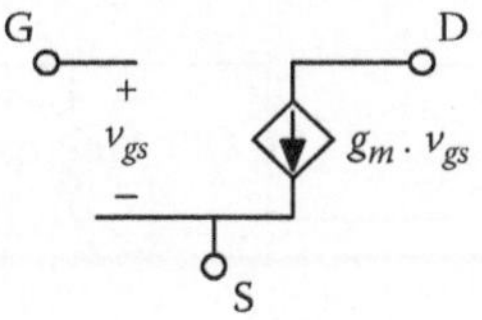

FIGURE 4.2
The simplest small-signal model for an MOS transistor.

Even this simple model is enough to achieve relatively accurate results about the operation of a transistor and the circuits that are built upon it, but in some cases we may need to add one more important phenomenon.

As we know, a more complete formula for the drain current of an MOS transistor is:

$$i_D = \frac{\beta}{2}(v_{GS} - V_{TH})^2(1 + \lambda v_{DS}), \tag{4.10}$$

which includes the effect of variations of drain-source voltage in the drain current due to channel-length modulation introduced in Chapter 2. Differentiating the above equation with respect to v_{DS}, one can write:

$$\frac{\partial i_D}{\partial v_{DS}}\Big|_Q = \lambda \frac{\beta}{2}(v_{GS} - V_{TH})^2\Big|_Q = \frac{\lambda . I_D}{1 + \lambda V_{DS}}, \tag{4.11}$$

which is, in fact, the slope of the $i_D - v_{DS}$ characteristic curve at the operating point $Q(I_D, V_{DS})$ (see Figure 4.3).

Again, the slope that relates variations in i_D and v_{DS} to each other, is a conductance, and defined as:

$$g_o = \frac{\partial i_D}{\partial v_{DS}}\Big|_Q . \tag{4.12}$$

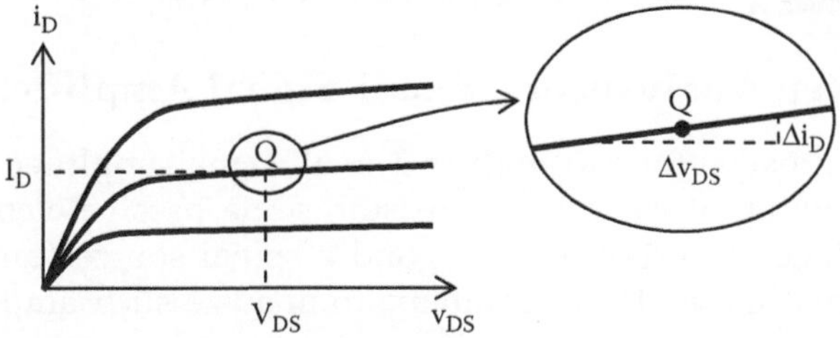

FIGURE 4.3
$i_D - v_{DS}$ characteristic curves for an MOS transistor.

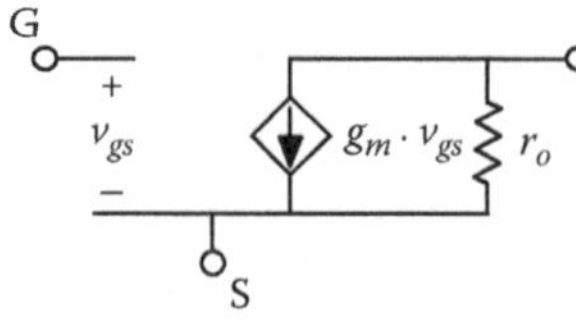

FIGURE 4.4
Small-signal model for an MOS transistor including the effect of channel-length modulation.

Because λ is typically a small number, using Equation (4.11) and Equation (4.12) g_o can be approximated to:

$$g_o \cong \lambda I_D \tag{4.13}$$

for a given bias current, I_D.

Unlike the case for g_m, the current that is caused by g_o (i.e., i_d) flows from the drain to the source, which is exactly where the voltage v_{ds} has been applied. Therefore, g_o is an equivalent conductance and not a transconductance. The element that will represent this phenomenon in the small-signal model of a transistor is a resistor with resistance:

$$r_o = \frac{1}{g_o} = \frac{1}{\lambda . I_D} \tag{4.14}$$

between the drain and the source. Figure 4.4 is the small-signal model that includes the resistance r_o representing channel length modulation.

Noting the fact that λ is a small number, r_o is usually a large resistance, typically on the order of couple of hundred $k\Omega$ to tens of MΩ.* Nevertheless, in the first few chapters that introduce the basic analysis methods, this resistance is sometimes not considered in the small-signal model of MOS transistors to keep the analysis methods as simple as possible.

4.3 Two-Step Analysis of a Small-Signal Amplifier

As we have learned before, to analyze a small-signal amplifier we are facing a circuit containing at least one transistor, some passive elements, one or more DC voltage or current sources, and a signal source (presumably AC signal). Different methods are available to analyze such amplifier circuits,

* For transistors with short channels (very small L), not studied in this book, this resistance can be much smaller.

but most of them are based on a two-step analysis method, which examines the DC and AC modes of a given circuit separately:

DC Analysis, in which all the AC sources are set to zero, and the circuit is treated exactly like a DC circuit. This means that the AC voltage and current sources are replaced with short circuits and open circuits, respectively, and the capacitors and inductors will be assumed as open circuits and short circuits, respectively.

AC Analysis, in which all the DC sources are set to zero. It should be noted that this is done with the assumption that all the active devices, including the amplifying transistor(s) will remain in the same operating point that has been set in DC analysis. In the AC analysis, the transistor(s) can be simply replaced with the equivalent small-signal model(s). The circuit can then be analyzed using well-known basic circuit analysis methods.

The final step, which resembles the *superposition* theorem studied in basic circuit theory for linear time-invariant circuits, combines the results of these two modes yielding a mathematical description of the real operation of the circuit and the currents and voltages of interest.

It is important to note that *if a small-signal amplifier circuit is only semilinear (i.e., under small-signal conditions, as previously described), it is expected to exhibit limited linear behavior.* In the previously mentioned AC mode, it is assumed that the circuit is operating under the small-signal conditions, and thus *the circuit obtained in the AC mode will be treated as a linear circuit and subject to all the rules that are defined and utilized for linear circuits.*

Example 4.2

For the circuit in Figure Ex. 4.2(a), assuming that $v_S(t) = 1^{mV}\text{Sin}(\omega t)$, find the total drain current. For the NMOS transistor we have: $V_{TH} = 0.5$ V, $k' = 100\ \mu A/V^2$, W = 20 μm, and L = 1 μm.

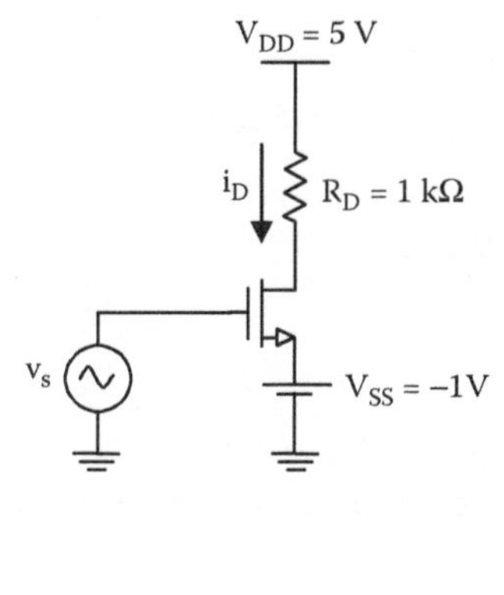

FIGURE EX. 4.2(a)

Solution:

The first step is the DC analysis. The schematic of the circuit in DC mode is depicted in Figure Ex. 4.2(b). As depicted, this circuit is used to calculate the DC component of the drain current, I_D:

$$\beta = k'.W/L = 2mA/V^2$$

$$I_D = \frac{\beta}{2}(V_{GS} - V_{TH})^2 = 1mA/V^2(1-0.5)^2 = 250\mu A.$$

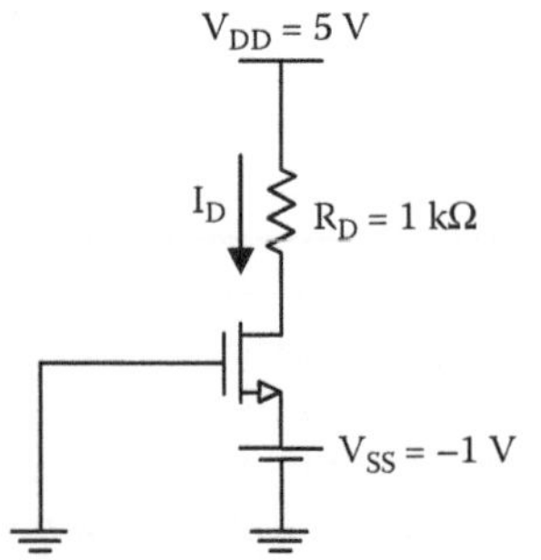

FIGURE EX. 4.2(b)

Proceeding to the second step, the circuit schematic in AC mode is obtained as presented in Figure Ex. 4.2(c). Remember that the symbol being used for power supplies like V_{DD}, represents an independent voltage source between the node it is connected to and the ground. Thus, setting V_{DD} to zero in the AC mode means that all the nodes that have been connected to V_{DD} in the original schematic will be shorted to the ground in the AC circuit schematic.

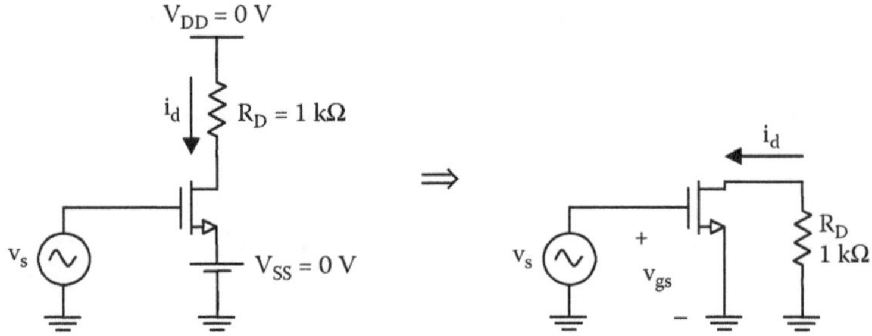

FIGURE EX. 4.2(c)

We know that the AC component of the drain current is determined by the AC component of the gate-source voltage and the transistor's transconductance:

$$g_m = \sqrt{2\beta I_D} = \sqrt{2 \times 2mA/V^2 \times 0.25mA} = 1mA/V,$$

$$i_d = g_m v_{gs} = 1\ mA/V \times 1^{mV}\text{Sin}\omega t = 1^{\mu A}\text{Sin}\omega t,$$

and the total drain current is achieved as:

$$i_D = I_D + i_d = 100^{\mu A} + 1^{\mu A}\text{Sin}\omega t.$$

Detailed examples of how this method is applied to amplifiers will be provided in the following sections where amplifiers are studied in more depth.

4.4 Coupling the Input/Output Signal to/from an Amplifier

We have so far learned how a transistor can amplify a signal in small-signal mode and how to analyze a circuit that contains one or more transistors. In this section, we will learn how a signal can be delivered to or from an amplifier.

Figure 4.5 presents some of the approaches used to couple a signal to the input port of an amplifier. The same techniques can be used to deliver the amplified signal from the amplifier to the load at the output.

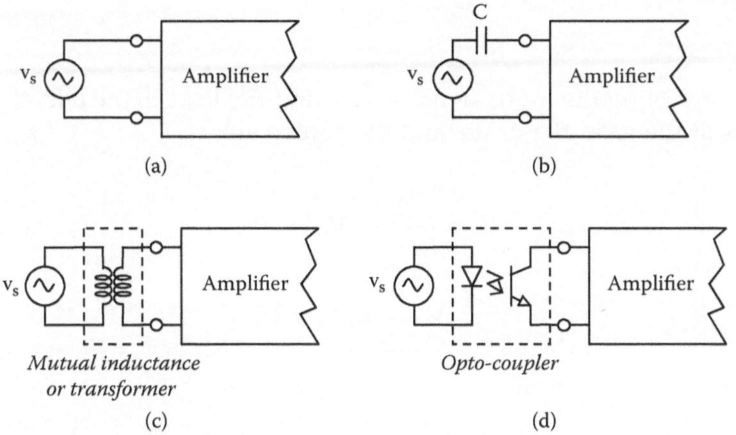

FIGURE 4.5
Different approaches to couple a signal to the input port of an amplifier: (a) direct, (b) capacitive, (c) inductive, and (d) optical.

Direct coupling provides a connection (short circuit) between the input signal source and the amplifier, present in both the DC and the AC schematics of the amplifier. As an example, the input signal source, v_S, in Example 4.2 is directly coupled to the amplifier, more specifically, to the gate of the transistor.

Example 4.3

Figure Ex. 4.3(a) includes a transistor with $\beta = 2\ mA/V^2$ biased at $I_D = 0.5\ mA$ in saturation mode, along with an input voltage signal source and a load, which are supposed to be coupled to the amplifier. What happens to the amplifier's bias conditions if the input signal source and the load are directly connected to the amplifier?

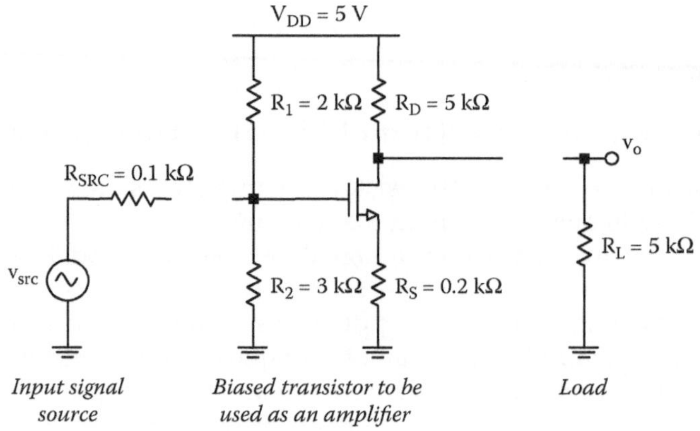

FIGURE EX. 4.3(a)

Solution:

Before connecting the input signal source and the load to the circuit, the DC voltages at the gate, the drain, and the source are:

$$V_G = \frac{R_2}{R_1 + R_2} . V_{DD} = 3V,$$

$$V_D = V_{DD} - R_D I_D = 2.5\ V,$$

and

$$V_S = R_S\, I_D = 0.1\ V.$$

Therefore,

$$V_{DS} = V_D - V_S = 2.4\ V,$$

and because the overdrive voltage for this transistor is:

$$V_{OD} = \sqrt{2I_D / \beta} = \sqrt{2 \times 0.5mA / (2mA / V^2)} \cong 0.7V,$$

the transistor is biased in saturation mode.

Now, let us connect the input signal source and the load to the biased transistor using simple wires (short circuits) and replace the input signal source with a short circuit for DC analysis (see Figure Ex. 4.3(b)).

Connecting R_S to the gate disturbs the gate DC voltage that has been already prepared by the bias network. The new gate DC voltage will be:

$$V_G = \frac{\left(R_2 \| R_{SRC}\right)}{R_1 + \left(R_2 \| R_{SRC}\right)} V_{DD} = 0.24V,$$

which is even smaller than a typical threshold voltage. This means that directly coupling the input signal source, in this example, disturbs the DC biasing and pushes the transistor into cutoff mode. This way, the circuit cannot be used as an amplifier anymore. Even if the transistor was still in saturation mode (obviously at a reduced drain current), the direct coupling between the amplifier and the load could cause another problem: The added load resistance (R_L) would draw current through R_D (in addition to the regular drain current that the transistor sinks) and as a result, the voltage drop across R_D would increase, reducing the drain DC voltage and thus V_{DS}. This may push the transistor from the saturation mode into the linear mode, depending on the circuit conditions.

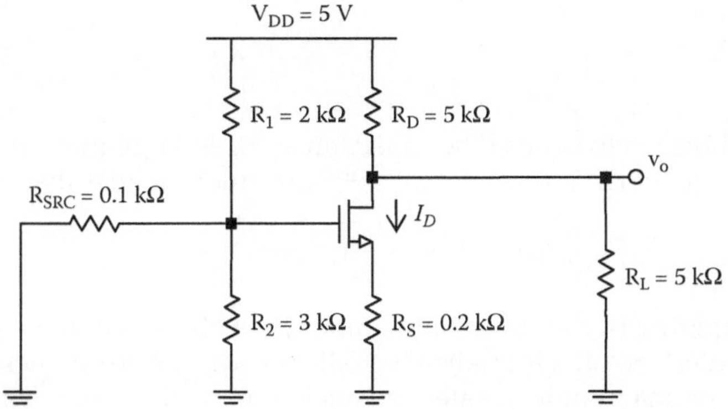

FIGURE EX. 4.3(b)

Ideally, it is desired to couple the input/output signal to/from the amplifier in the AC mode using short circuits, but Example 4.3 described a case where direct coupling will disturb the DC mode design. Using *capacitive coupling*, this problem can be overcome. Because the impedance of a capacitor is infinite at DC, it is virtually disconnected from the input in DC mode. If the capacitor is large enough, it still acts as a short circuit in AC mode, however, coupling the input signal to the amplifier. Note that when capacitively coupling, it is usually assumed that the magnitude of the coupling capacitor's impedance at the signal frequency is so small compared with either the input resistance of the amplifier or the other resistances in series with it that it can be approximated to a short circuit. To achieve such a small impedance for the coupling capacitor, its capacitance should be quite large, based on the minimum frequency component in the signal spectrum.

Example 4.4

The input signal source, $v_S(t) = 10^{mV}\text{Sin}(2\pi \times 10^{kHz}t)$, in Figure Ex. 4.4 is capacitively coupled to an amplifier with the input resistance of 100 $k\Omega$. Assuming that the internal resistance of the signal source, R_S, is 1 $k\Omega$, what range of capacitance would be appropriate for the coupling capacitor, C, so that the input signal is efficiently coupled to the amplifier?

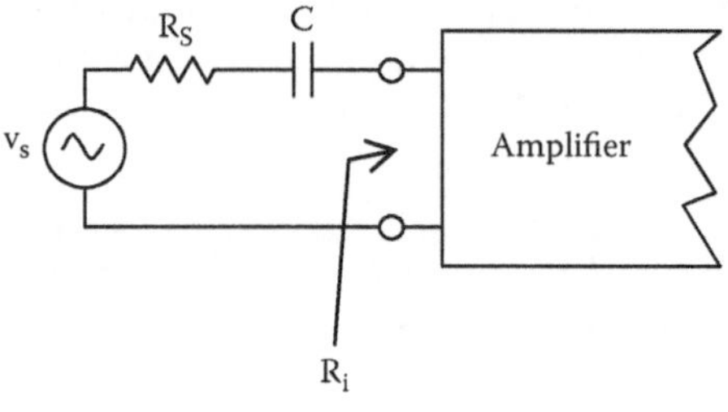

FIGURE EX. 4.4

Solution:

The best choice is to choose the capacitance, C, so large that the associated impedance at signal frequency (i.e., 10^{kHz}) is much smaller than R_S:

$$|Z_C(j2\pi \times 10^{kHz})| = 1/(2\pi.f.C) << R_S.$$

When comparing two numbers in the analysis of electronic circuits, the terms "much greater" or "much smaller" usually means that one of them is at least one order of magnitude greater or smaller than the other. Thus, in this example, the impedance of the coupling capacitor at the signal frequency should be at least 10 times smaller than R_S:

$$1/(2\pi \times 10^{\text{kHz}}C) \le R_S/10,$$

$$C \ge 10/(2\pi \times 10^4 \times 10^3),$$

$$C \ge 0.16\ \mu F.$$

The third coupling technique is *inductive coupling,* which enables us to transfer AC signal, but forces the input voltage to be zero in DC mode. This type of coupling is preferred in the cases where the transfer of a current signal is intended or the isolation between the signal ground and the amplifier's ground is desired. Finally, *optical coupling* is made possible by converting the input signal to light by a light-emitting diode (LED) and again converting the generated light to electric signal to be delivered to the amplifier. This method is mostly preferred when ground isolation is needed.

Because of their wider use in both discrete and integrated amplifiers, the first two signal coupling methods (i.e., direct and capacitive coupling) will be used throughout this book.

4.5 Basic Single-Stage Amplifier Configurations

As stated before, an MOS transistor has the potential of amplifying the input signal if properly applied to the transistor. An MOS transistor can be used as an amplifier with three possible configurations, which are illustrated in Figure 4.6. The input and output signals can be of either voltage or current type based on which one of the four basic types introduced in Chapter 1, is formed. As will be studied in the following sections, however, each configuration has its own properties that may make it suitable as a specific type of amplifier.

In the *common-source* (CS) configuration, the input is applied between the gate and the source and the output is taken between the drain and the source. In the *common-gate* (CG) configuration, the input signal is applied between

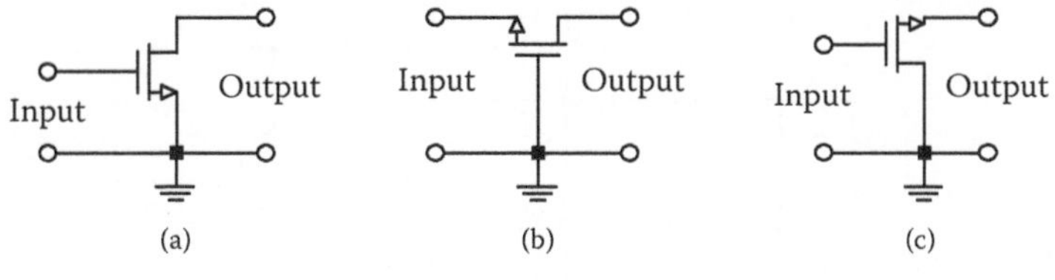

FIGURE 4.6
Basic configurations for a single-stage amplifier: (a) common source (CS), (b) common gate (CG), and (c) common drain (CD) or source follower (SF).

the source and the gate, and the output can be taken between the drain and the gate. The third configuration is the *common-drain* (CD) configuration or the *source follower* (SF) circuit in which the input is applied between the gate and the drain and the output is taken between the source and the drain. Despite the differences among these configurations, they share a common concept: The input signal changes the gate-source voltage, causing a proportional drain (or source) current variation. This current is either used directly as the output signal or is passed through a resistor at the output to make a voltage as the output signal.

It is worth noting that the three configurations illustrated in Figure 4.6 are drawn just to describe the basic concept of where the input signal is applied and where the output signal is taken from. As we know, to function as an amplifier, a transistor should be properly biased using a well-designed bias network.

4.5.1 Common-Source Configuration

Figure 4.7 is an example of the common-source configuration for a voltage amplifier. The transistor is biased using the voltage-divider biasing technique (presumably in the saturation mode) and capacitively coupled to the input signal source and the load. According to the two-step analysis technique previously described, the DC analysis is performed with the input AC signal source, v_{src}, shorted and the coupling capacitors opened. The resulting circuit will be a simple example of the voltage-divider biasing network studied in Chapter 3. To analyze the amplifier in the AC mode, as the second step, all the DC sources are set to zero and the coupling capacitors are shorted. The resulting AC schematic of the circuit is depicted in Figure 4.8, in which the small-signal voltage amplifier is designated using a dashed box. As explained before, this circuit can be treated as a linear amplifier. To analyze the circuit in AC mode, the transistor should be replaced with its small-signal

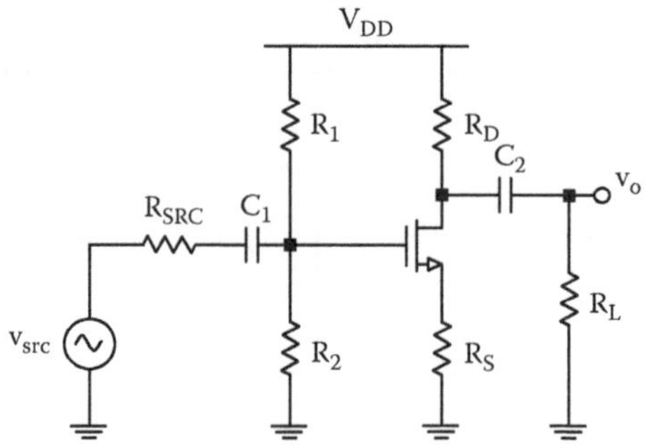

FIGURE 4.7
An example of the CS configuration for a voltage amplifier.

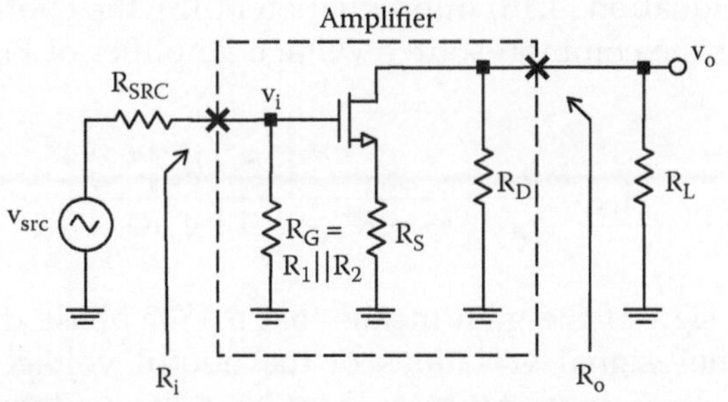

FIGURE 4.8
AC schematic for the circuit of Figure 4.7.

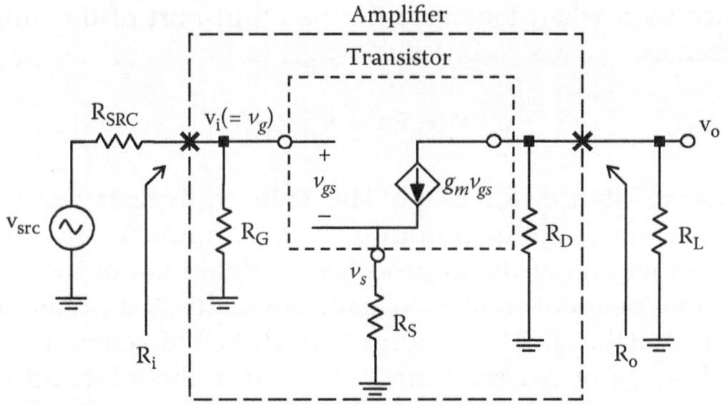

FIGURE 4.9
Replacing the transistor with its small-signal model.

model (see Figure 4.9). To keep the analysis of these circuits as simple as possible, the most basic small-signal model is going to be used in this section.

To determine the circuit's overall voltage gain, v_o/v_{src}, let us try to find the output voltage, v_o:

$$v_o = -(g_m v_{gs}).(R_D \| R_L). \tag{4.15}$$

Now, v_{gs} should be found as a function of the input source signal, v_{src}:

$$v_{gs} = v_g - v_s = \frac{R_G}{R_G + R_{SRC}} v_{src} - (g_m v_{gs}) R_S,$$

and:

$$v_{gs} = \frac{1}{1 + g_m R_S} \cdot \frac{R_G}{R_G + R_{SRC}} v_{src}. \tag{4.16}$$

Substituting Equation (4.16) into Equation (4.15), the overall voltage gain, $A_{VS} = v_o/v_{src}$, for the common-source voltage amplifier of Figure 4.7 is:

$$A_{VS} = \frac{v_o}{v_{src}} = \frac{R_G}{R_G + R_{SRC}} \cdot \frac{-g_m\left(R_D \| R_L\right)}{1 + g_m R_S}. \tag{4.17}$$

The negative sign of the gain means that a 180° phase difference occurs between the small-signal variations of the output voltage and the input signal. It is of great importance to note that the voltages and currents achieved using AC analysis are only the AC components, and need to be summed with the associated DC components to achieve the complete signals.

The input and output resistances of an amplifier can also be obtained through AC analysis. As presented in Figure 4.9, the input resistance, R_i, is the resistance seen when looking into the input port of the amplifier, and can be written as:

$$R_i = R_G = R_1 \| R_2 . \tag{4.18}$$

The output resistance, R_o, is the Thevenin equivalent resistance of the circuit when looking into the output port of the amplifier. Recall from basic circuit theory that one of the approaches to obtain the output resistance is to set all the independent sources to zero. Doing this, the circuit of Figure 4.9 will be reduced to the circuit in Figure 4.10. This circuit consists of two parts. Because no voltage or current source is found in the left-hand part, all the voltages and currents in this part, including v_g, are zero. On the other side, the current $g_m v_{gs}$ passes through R_S and generates the source voltage, v_s. Therefore, one can write:

$$v_s = R_S\,(g_m v_{gs}) = R_S\, g_m\,(0 - v_s) = -R_S\, g_m v_s \tag{4.19}$$

Because we know that both R_S and g_m are positive, this equation will not hold unless $v_s = 0$. As a result, the dependent current source becomes zero,

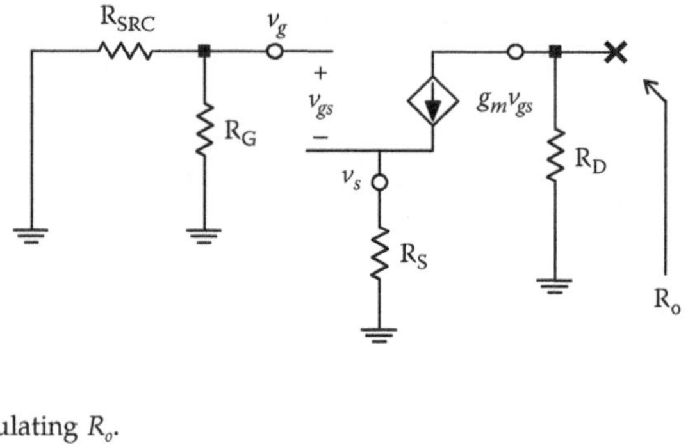

FIGURE 4.10
Schematic for calculating R_o.

and is electrically equivalent to an open circuit. This means that the output resistance that we are looking for is:

$$R_o = R_D. \tag{4.20}$$

The overall gain of the amplifier (Equation (4.17)) can be rewritten as:

$$A_{VS} = \frac{v_o}{v_{src}} = \frac{R_G}{R_G + R_{SRC}} \cdot \frac{-g_m}{1 + g_m R_S} \cdot \frac{R_D R_L}{R_L + R_D}, \tag{4.21}$$

and then

$$A_{VS} = \frac{v_o}{v_{src}} = \frac{R_i}{R_i + R_{SRC}} \cdot \frac{-g_m R_D}{1 + g_m R_S} \cdot \frac{R_L}{R_L + R_o} \tag{4.22}$$

to clearly describe the attenuations at the input and output, as well as the pure gain of the amplifier. The first and the third terms represent the attenuations at the input and output, respectively, and the second term is the pure voltage gain of the amplifier itself, which is in fact v_o/v_i of Figure 4.8.

Example 4.5

Given $\beta = 20\ mA/V^2$ for the transistor, and $R_1 = 2\ k\Omega$, $R_2 = 3\ k\Omega$, $R_D = 5\ k\Omega$, $R_S = 1\ k\Omega$, $R_L = 20\ k\Omega$, and $R_{SRC} = 0.1\ k\Omega$ for the amplifier in Figure 4.7, and supposing that the transistor is biased at 0.5 mA in saturation mode, (a) determine the overall gain, $A_{VS} = v_o/v_{src}$, and (b) discuss the strengths and weaknesses in the design of this amplifier.

Solution:

(a) For this amplifier we have:

$$g_m = \sqrt{2\beta I_D} = \sqrt{2 \times 20mA/V^2 \times 0.5mA} = 4.47mA/V$$

$$R_i = R_G = R_1 \| R_2 = 1.2\ k\Omega,$$

and

$$R_o = R_D = 5\ k\Omega,$$

and the overall gain given in Equation (4.22) is calculated as:

$$A_{VS} = \frac{v_o}{v_{src}} = \frac{R_i}{R_i + R_{SRC}} \cdot \frac{-g_m R_D}{1 + g_m R_S} \cdot \frac{R_L}{R_L + R_o},$$

$$= \frac{1.2k}{1.2k + 0.1k} \cdot \frac{-(4.47m \times 5k)}{1 + (4.47m \times 1k)} \cdot \frac{20k}{20k + 5k}$$

$$= 0.92 \times (-4.09) \times 0.8 = -3$$

(b) As the preceding equation says, 92% of the source signal (v_{src}) is delivered to the amplifier as the input voltage, v_i, which demonstrates a good design at the input. The amplifier's pure gain is only −4.09, which is lower than the typical gain expected from a well-designed single-stage amplifier. Finally, the attenuation at the output wastes 20% of the amplified signal and allows only 80% of the amplified signal to be delivered to the load, which may not be considered as a good design in some applications.

4.5.1.1 Bypass Capacitor

When designing voltage amplifiers, sometimes the overall gain is quite low, but is not due to attenuation at the input or output. The problem is in the design of the amplifier itself. The amplifier studied in Example 4.5 is an example of such cases in which the pure gain of the amplifier is only around −4. Let us take a closer look at this amplifier and figure out why its gain is so low. The amplification term in the overall gain calculated in the previous example has a relatively large numerator ($g_m R_D = 23.35$), but the not-so-small denominator ($1 + g_m R_S \approx 5.5$) results in a small value for this term (−4.09). To increase this term, we can either increase the numerator or decrease the denominator. Because g_m appears in both the numerator and the denominator, the only reasonable way to make the numerator larger is by increasing R_D. Notice that a larger R_D increases the voltage drop across it and so the drain voltage will, in turn, be lowered, placing a limit on R_D. This limit occurs because we would like to keep the transistor in saturation by providing a drain-source voltage greater than the overdrive voltage of the transistor. The interested reader can show that for this example, the maximum possible resistance for R_D to keep the transistor in saturation is approximately 8.6 $k\Omega$, which in turn increases the amplifier's pure gain to a maximum of −7. Now, let us consider decreasing the denominator. Again, reducing g_m is not an option, and we can only choose a smaller R_S, preferably zero to achieve the maximum possible gain. Unlike R_D, R_S contributes to the bias current of the transistor, and any change in R_S will disturb the biasing of the amplifier. To conclude, it seems that there is a conflict between the AC requirements and the DC design of the amplifier: On one hand, in the DC mode, we would like to keep R_S as it is, and on the other hand, in AC mode, the best design

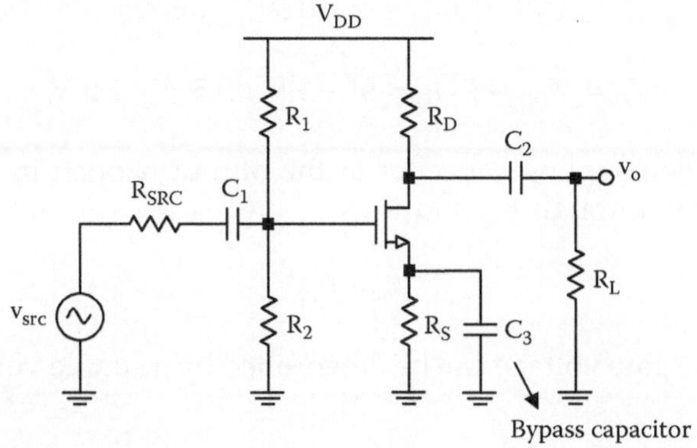

FIGURE 4.11
A CS amplifier with bypass capacitor.

would be to short this resistor. To overcome this problem, adding a capacitor, referred to as the *bypass* or *decoupling* capacitor, in parallel with R_S, will be the suggested solution (see Figure 4.11).

The function of the bypass capacitor can also be interpreted as follows: As illustrated in Figure 4.9, the output voltage is generated by the drain current, which is directly proportional to the gate-source voltage (all in the AC mode). Only a fraction of the input voltage is delivered to the transistor as v_{gs} for amplification, however, and the rest of the input voltage appears across R_S. Clearly, to increase the gain of the CS amplifier, we need to provide the entire input signal across the gate-source terminals. By shorting R_S in the AC mode, the bypass capacitor helps the transistor to receive and amplify the entire input voltage, and as a result, the gain increases.

Now that we have become familiar with the basic of operation of a CS amplifier using formulas and equations, let us study what happens in this amplifier in more detail. The following example is of crucial importance in understanding how the voltages and currents in a CS amplifier vary when the input signal source is connected to the amplifier.

Example 4.6

For the amplifier of Example 4.5, assuming that the source signal is $v_{src}(t) = 1^{mV}\text{Sin}(\omega t)$, (a) obtain the total voltages at the gate, source, and drain of the transistor, as well as the total drain current and output voltage; then sketch these signals (b), add a bypass capacitor, and find the current and voltages mentioned in (a).

Solution:

Let us start with the **DC analysis** for both (a) and (b). Knowing that $I_D = 0.5\ mA$, we have:

$$V_S = R_S I_D = 1^{k\Omega} \times 0.5^{mA} = 0.5 \text{ V},$$

$$V_D = V_{DD} - R_D I_D = 5^V - 5^{k\Omega} \times 0.5^{mA} = 2.5 \text{ V},$$

and because the coupling capacitor at the output is open in DC mode, no current can flow through R_L, and thus:

$$V_O = 0 \text{ V}.$$

In addition, the gate voltage can be determined by a simple voltage division:

$$V_G = \frac{R_2}{R_1 + R_2}.V_{DD} = \frac{3k}{2k + 3k} \times 5V = 3V.$$

(a) From the **AC analysis** of the circuit before adding the bypass capacitor we have:

$$v_g = \frac{R_G}{R_G + R_{SRC}}.v_{src}$$

$$= \frac{1.2k}{1.2k + 0.1k} \times 1^{mV} Sin(\omega t)$$

$$= 0.92^{mV} Sin(\omega t).$$

To calculate the drain current in the AC mode as $i_d = g_m v_{gs}$, only the gate-source voltage is needed, which has already been defined for this circuit in Equation (4.16) as:

$$v_{gs} = \frac{1}{1 + g_m R_S} \cdot \frac{R_G}{R_G + R_{SRC}}.v_{src} = \frac{1}{1 + g_m R_S}.v_g$$

$$= \frac{1}{1 + (4.47m \times 1k)} \times 0.92^{mV} Sin(\omega t)$$

and thus:

$$v_{gs} = 0.168^{mV} Sin(\omega t).$$

As can be seen, less than 20% of the voltage that is delivered to the amplifier (at the gate) appears across the gate-source to generate the drain current:

$$i_d = g_m v_{gs} = 4.47^{mA/V} \times 0.168^{\mathrm{mV}}\, Sin(\omega t) = 0.75^{\mu\mathrm{A}}\, Sin(\omega t).$$

Taking a quick look at the AC schematic of the circuit illustrated in Figure 4.9, the voltages at the drain and the source can easily be obtained as:

$$v_s = R_S i_d = 1^{k\Omega} \times 0.75^{\mu\mathrm{A}}\, Sin(\omega t) = 0.75^{\mathrm{mV}}\, Sin(\omega t),$$

and

$$v_d = v_o = -\left(R_D \| R_L\right) i_d = -4^{k\Omega} \times 0.75^{\mu\mathrm{A}}\, Sin(\omega t) = -3^{\mathrm{mV}}\, Sin(\omega t),$$

which is in complete agreement with the overall gain of –3 achieved in Example 4.5. Now, the entire signals can be obtained by superposing the results from DC and AC modes:

$$i_D = I_D + i_d = 0.5^{mA} + 0.75^{\mu A}\mathrm{Sin}(\omega t),$$

$$v_G = V_G + v_g = 3^V + 0.92^{mV}\mathrm{Sin}(\omega t),$$

$$v_S = V_S + v_s = 0.5^V + 0.75^{mV}\mathrm{Sin}(\omega t),$$

$$v_D = V_D + v_d = 2.5^{\mathrm{V}} - 3^{mV}\mathrm{Sin}(\omega t),$$

and

$$v_O = V_O + v_o = 0^{\mathrm{V}} - 3^{mV}\mathrm{Sin}(\omega t) = -3^{mV}\mathrm{Sin}(\omega t).$$

Notice the difference between v_D and v_O, which highlights the role of the coupling capacitor at the output. The AC component of the drain voltage is coupled to the load, whereas the DC component is rejected.

The signals obtained previously are sketched in Figure Ex. 4.6(a). Please note the 180° phase difference between the input and output voltages.

(b) Adding the bypass capacitor, the AC schematic in Figure Ex. 4.6(b) is achieved. Performing the **AC analysis** for this circuit we obtain:

$$v_g = v_{gs} = \frac{R_G}{R_G + R_{SRC}} . v_{src} = \frac{1.2k}{1.2k + 0.1k} \times 1^{\mathrm{mV}}\, Sin(\omega t) = 0.92^{\mathrm{mV}}\, Sin(\omega t),$$

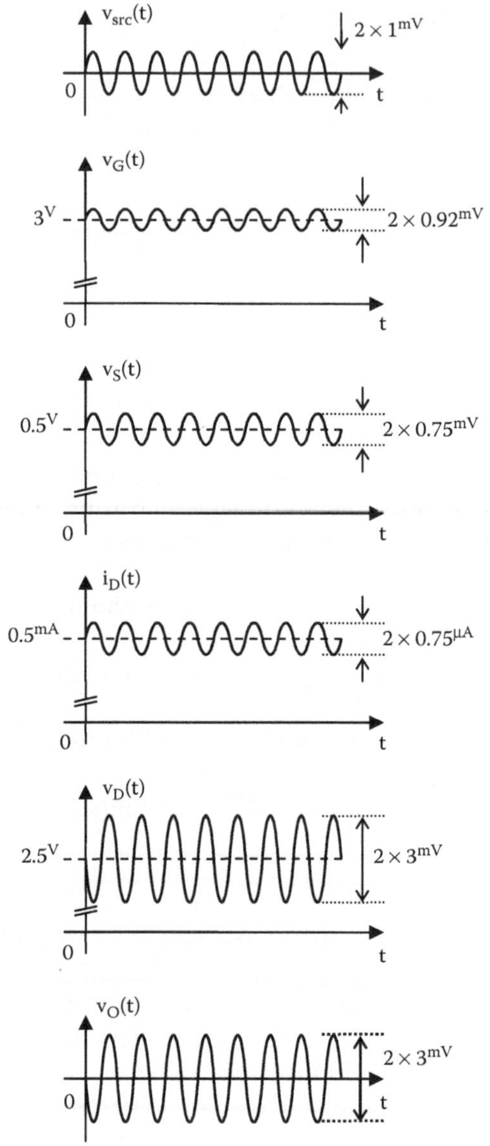

FIGURE EX. 4.6(a)

and

$$i_d = g_m v_{gs} = 4.47^{mA/V} \times 0.92^{\mathrm{mV}} Sin(\omega t) = 4.11^{\mu\mathrm{A}} Sin(\omega t),$$

and thus:

$$v_d = v_o = -\left(R_D \| R_L\right) i_d = -4^{k\Omega} \times 4.11^{\mu\mathrm{A}} Sin(\omega t) = -16.44^{\mathrm{mV}} Sin(\omega t).$$

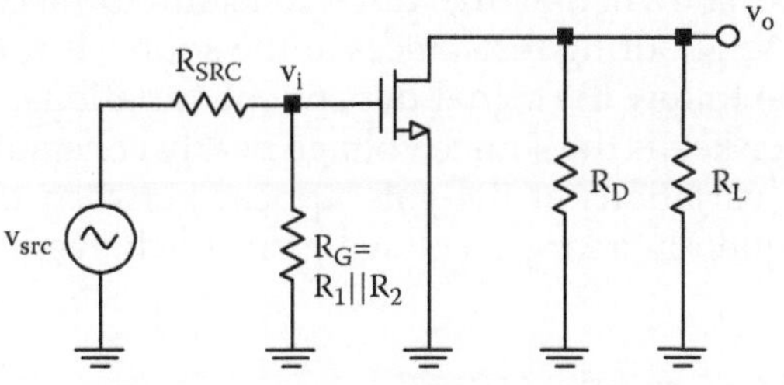

AC schematic after adding the bypass capacitor

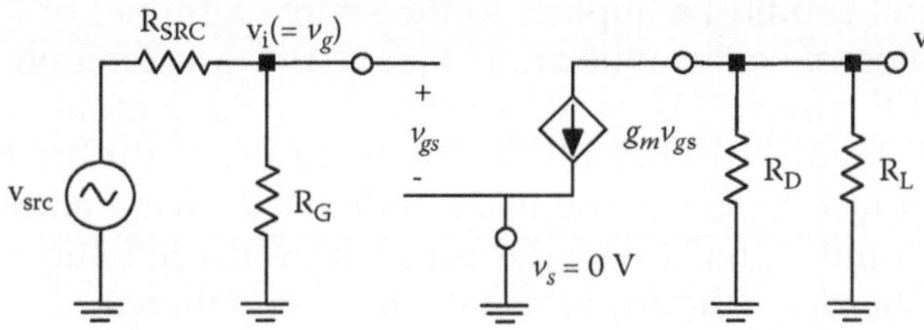

Replacing the transistor with its small-signal model

FIGURE EX. 4.6(b)

Remember that because of the short circuit provided by the bypass capacitor in the AC mode, the AC component of the source voltage is zero. Finally, the signals of interest are obtained as:

$$i_D = I_D + i_d = 0.5^{mA} + 4.11^{\mu A}\text{Sin}(\omega t),$$

$$v_G = V_G + v_g = 3^V + 0.92^{mV}\text{Sin}(\omega t),$$

$$v_S = V_S + v_s = 0.5^V + 0 = 0.5^V,$$

$$v_D = V_D + v_d = 2.5^V - 16.44^{mV}\text{Sin}(\omega t),$$

and

$$v_O = V_O + v_o = 0^V - 16.44^{mV}\text{Sin}(\omega t) = -16.44^{mV}\text{Sin}(\omega t).$$

A third explanation can be provided regarding how a bypass capacitor helps the CS amplifier to have a higher voltage gain. Suppose that the bypass capacitor is connected between the source and the ground before the input signal is applied to the amplifier. Because the transistor has already been biased and a 0.5-V DC voltage is established at the source, the bypass capacitor is charged to the same voltage level. Now, if the input signal source is connected to the circuit, signal-dependent voltage variations are coupled to the gate. It can be shown that the time constant formed by the bypass capacitor and the neighboring resistances at the source is too large to let the voltage at this node follow the signal-dependent variations. This means that the bypass capacitor keeps the source voltage nearly constant and as a result, the entire voltage variations at the gate appear between the gate and the source, and consequently a larger voltage gain is achieved.

4.5.2 Common-Gate Configuration

As explained earlier, to design a voltage amplifier with CG configuration, the input signal should be applied to the source terminal of the transistor and the output is taken from the drain. Figure 4.12 is a common-gate voltage amplifier, and Figure 4.13 is its AC schematic. Looking at the AC schematic in Figure 4.13, it may be difficult to see how it relates to the original circuit in Figure 4.12. In fact, to satisfy the usual style for drawing amplifier circuits in which the input signal enters the circuit from the left and goes through the circuit to the right, the AC schematic has been altered.

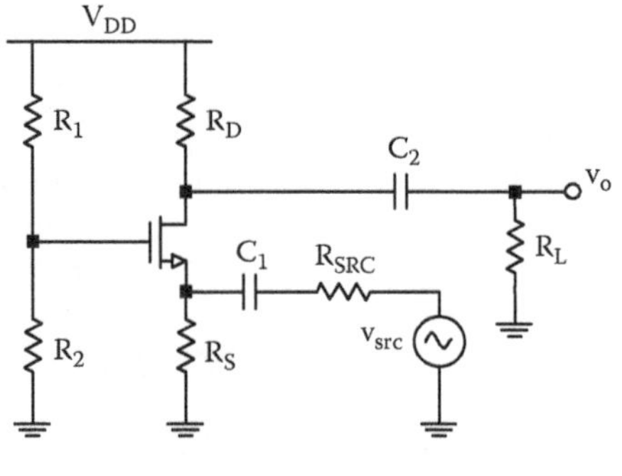

FIGURE 4.12
Circuit schematic of a common-gate voltage amplifier.

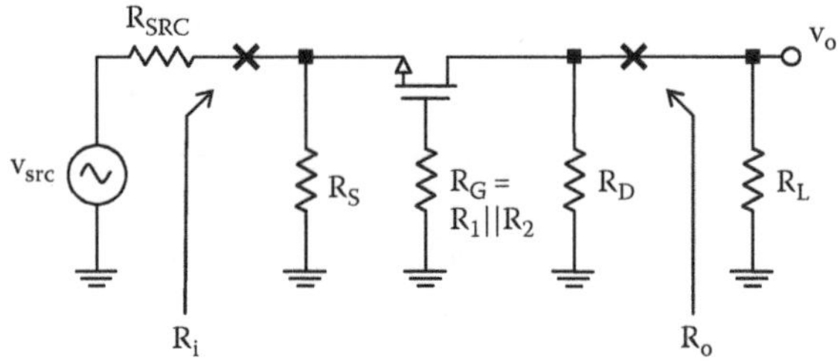

FIGURE 4.13
AC schematic of the amplifier of Figure 4.12.

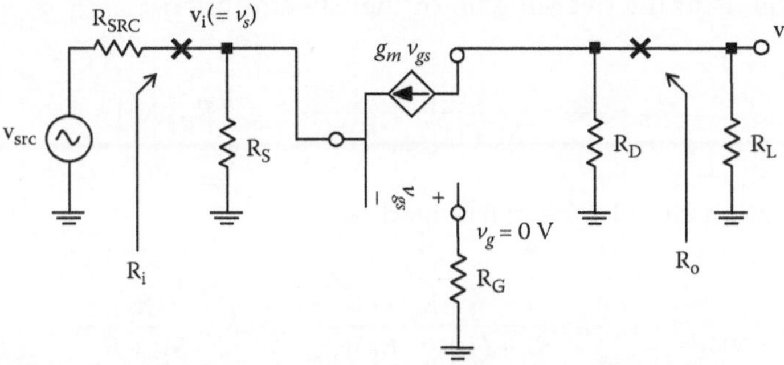

FIGURE 4.14
AC schematic of the amplifier of Figure 4.12, with the transistor replaced with its small-signal model.

Figure 4.14 depicts the resulting circuit schematic when the transistor in Figure 4.13 is replaced by its small-signal equivalent circuit. Now, we can obtain the amplifier's overall gain and input and output resistances. As before, the output voltage is generated when the drain current, $g_m v_{gs}$, flows through $R_D \| R_L$:

$$v_o = -(g_m v_{gs})(R_D \| R_L),$$

and because no current is flowing through R_G, it is obvious that the gate voltage is zero, and thus $v_{gs} = v_g - v_s = -v_s$. It can be said that v_s is, indeed, generated by both the current coming from v_{src} and $g_m v_{gs}$:

$$v_s = [(v_{src} - v_s)/R_{SRC} + g_m v_{gs}].R_S$$

Replacing v_{gs} with $-v_s$ and rearranging the preceding equation, we have:

$$v_s = \frac{R_S}{R_S + (1 + g_m R_S) R_{SRC}} . v_{src},$$

and, therefore, the output voltage can be written as a function of the source signal as:

$$\begin{aligned} v_o &= -g_m v_{gs} \left(R_D \| R_L\right) \\ &= g_m v_s \left(R_D \| R_L\right) \\ &= \frac{R_S}{R_S + (1 + g_m R_S) R_{SRC}} g_m \left(R_D \| R_L\right) v_{src} \end{aligned}$$

which leads us to the overall gain of the CG amplifier:

$$A_{VS} = \frac{v_o}{v_{src}} = \frac{R_S}{R_S + (1 + g_m R_S) R_{SRC}} \cdot g_m \left(R_D \| R_L \right) \tag{4.23}$$

This equation can also be rearranged as:

$$A_{VS} = \frac{v_o}{v_{src}} = \frac{R_S}{R_S + (1 + g_m R_S) R_{SRC}} \cdot g_m R_D \cdot \frac{R_L}{R_D + R_L}. \tag{4.24}$$

Again, it can be concluded that the overall gain contains three terms: an attenuation at the input (the first term), a pure gain provided by the amplifier itself (the second term), and an attenuation at the output exactly as was observed for the CS amplifier (the third term). Due to the zero gate current in AC analysis, in this configuration the gate is always at ground potential, and the entire voltage that is delivered to the amplifier at the source of the transistor appears across the gate-source. As a result, the pure gain of the CG amplifier is always as large as the maximum pure gain that could be expected from a CS amplifier after adding a bypass capacitor.

Let us use some basic circuit analysis rules to determine the input resistance. Looking at the detailed AC schematic in Figure 4.15, the two resistors R_D and R_L can be absorbed by the current source $g_m v_{gs}$. In addition, keep in mind that $v_g = 0$. Thus, $v_{gs} = -v_s$, and it is not necessary to keep R_G in the circuit anymore because it is separated from the rest of the circuit. The resulting simplified circuit is depicted in Figure 4.16 in which v_{gs} has been replaced by $-v_s$. Considering the change in the direction of the current source, we end up with a dependent current source, $g_m v_s$, in parallel with R_S, where the current is a function of the voltage across it. As we know, based on Ohm's law, when the current flowing through an element is proportional to the

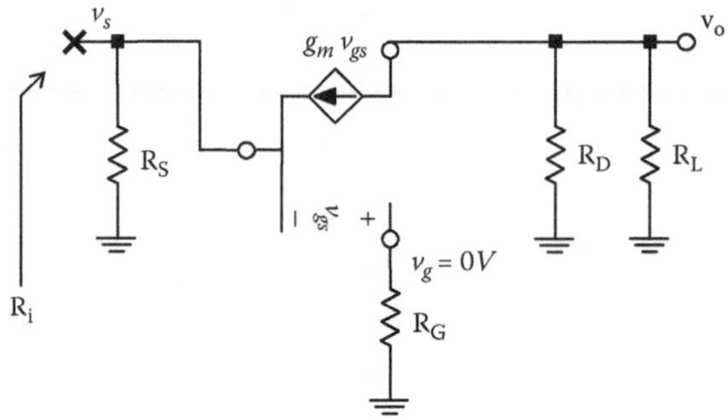

FIGURE 4.15
The detailed AC schematic of the amplifier of Figure 4.12, redrawn to determine the input resistance.

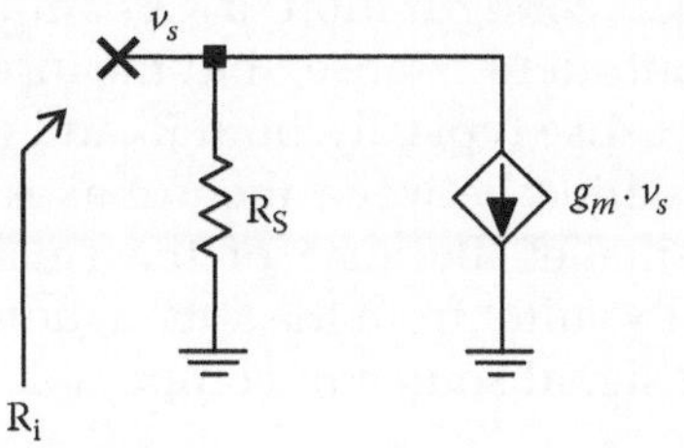

FIGURE 4.16
Simplified form of the circuit of Figure 4.15.

voltage across it, that element can be modeled by a resistor, and the resistance is given as:

$$v/i = v_s/g_m v_s = g_m^{-1}, \tag{4.25}$$

Thus, the input resistance will be the parallel combination of R_S and g_m^{-1}:

$$R_i = R_S \| g_m^{-1}. \tag{4.26}$$

As mentioned earlier, to determine the output resistance, all the independent current sources should be zeroed; thus, the circuit can be reduced to the circuit depicted in Figure 4.17. The only current flowing through $R_S \| R_{SRC}$ is $g_m v_s$ and the voltage v_s is written as:

$$v_s = -g_m v_s (R_S \| R_{SRC}). \tag{4.27}$$

As seen earlier in Equation (4.19), this equation can hold under one condition only, and that is when $v_s = 0$. If so, the current source is opened and we simply have:

$$R_o = R_D. \tag{4.28}$$

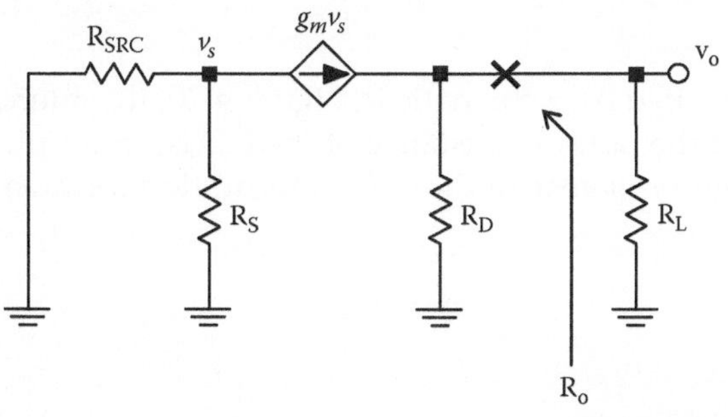

FIGURE 4.17
AC schematic of the amplifier of Figure 4.12, simplified to obtain the output resistance.

Compared with the CS configuration, a CG amplifier exhibits the same pure voltage gain and output resistance,* but the input resistance is typically much smaller. This is because typically, both R_S and g_m^{-1} are small resistances and the parallel combination between them makes the input resistance of Equation (4.26) even smaller. Because of the small input resistance, CG voltage amplifiers usually suffer from the attenuation at the input if the input signal is provided by a signal source of comparable internal resistance.

4.5.3 Common-Drain or Source-Follower Configuration

For the previous two configurations, it was demonstrated that a high input resistance is seen when the input signal is applied to the gate terminal, as well as a low resistance when looking into the source terminal. These two observations lead us to the third configuration for a voltage amplifier, which in a good design should ideally have a high input resistance and a low output resistance. The CD configuration for a voltage amplifier, also called *source follower*, is depicted in Figure 4.18. As indicated, the input signal is coupled to the gate where a high input resistance is present and the output is taken from the source where a low output resistance is expected. Unlike the previous amplifier configurations, the drain is directly connected to V_{DD}. The drain current still has signal-dependent variations as before, but there will be no R_D needed at the drain to convert those current variations to voltage. From the DC standpoint, we know that R_D does not contribute to the drain current, and designing the bias network without it keeps the drain voltage always at V_{DD}. Thus, the transistor will have an increased drain-source voltage that allows for higher voltage swings at the output.

To determine the AC specifications of the CD amplifier in Figure 4.18 (i.e., its overall gain, input, and output resistances), the AC schematic should be drawn as illustrated in Figure 4.19. The transistor is then replaced with its small-signal model as seen in Figure 4.20. The input resistance for this circuit is exactly the same as that of the CS configuration formulated in Equation (4.18):

$$R_i = R_G, \tag{4.29}$$

and from the detailed AC schematic in Figure 4.20, the interested reader can easily show that the output resistance of the CD configuration is exactly the same as the input resistance of the CG configuration formulated in Equation (4.26):

$$R_o = R_S \| g_m^{-1}. \tag{4.30}$$

* This is true as long as the output resistance of the transistor, r_o, can be ignored.

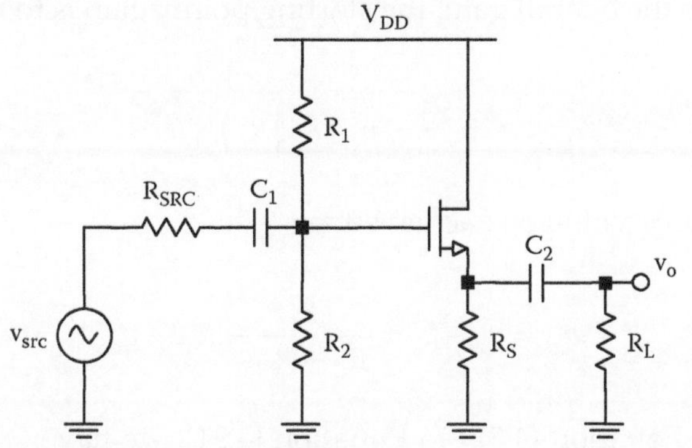

FIGURE 4.18
A voltage amplifier with CD configuration.

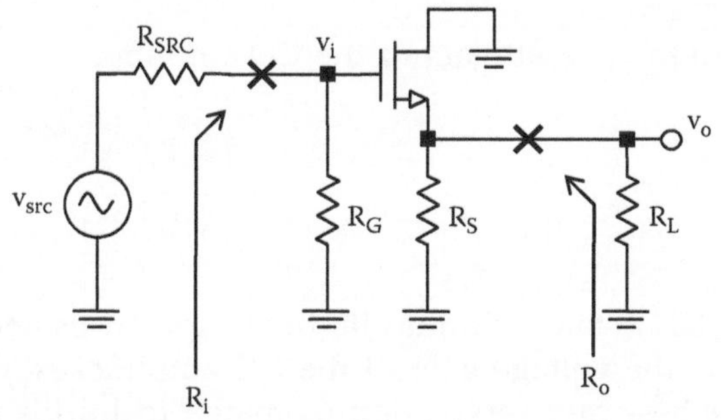

FIGURE 4.19
AC schematic of the CD amplifier in Figure 4.18.

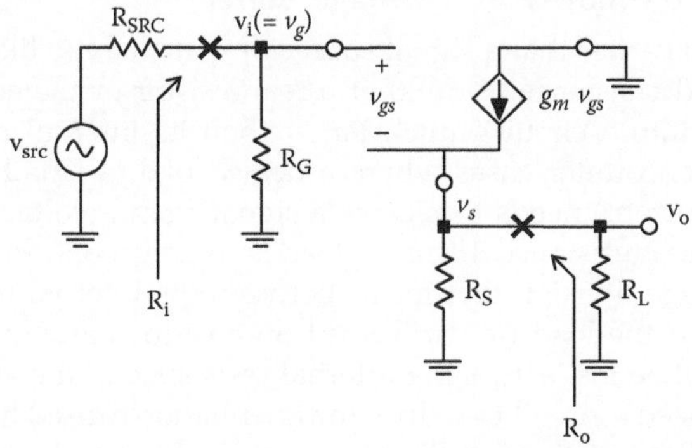

FIGURE 4.20
The AC schematic of the CD amplifier after replacing the transistor with its small-signal equivalent circuit.

To determine the overall gain, the starting point again is to find the output voltage:

$$v_o = g_m v_{gs}\left(R_S \| R_L\right). \tag{4.31}$$

The gate-source voltage is achieved as:

$$v_{gs} = v_g - v_s = \frac{R_G}{R_G + R_{SRC}}.v_{src} - v_o. \tag{4.32}$$

Substituting Equation (4.32) in Equation (4.31), we have:

$$v_o = \frac{R_G}{R_G + R_{SRC}}.\frac{g_m\left(R_S \| R_L\right)}{1 + g_m\left(R_S \| R_L\right)}.v_{src}, \tag{4.33}$$

which leads us to the overall gain of the CD amplifier:

$$A_{VS} = \frac{v_o}{v_{src}} = \frac{R_G}{R_G + R_{SRC}}.\frac{g_m\left(R_S \| R_L\right)}{1 + g_m\left(R_S \| R_L\right)}. \tag{4.34}$$

As Equation (4.34) states, despite its high input resistance and the low output resistance, the voltage gain of the CD amplifier is always less than unity, and in the best case can be approximated to 1. This means that this amplifier cannot be expected to amplify the input voltage at all; however, the good news is that one unique function that the CD voltage amplifier, or source follower (SF), can perform is *voltage buffering*.

4.5.3.1 Source Follower as a Voltage Buffer

It was explained earlier that a signal source of voltage type likes to deliver its voltage to a load resistance, R_L, (either a real resistor or the equivalent input resistance seen into a circuit) much larger than its internal resistance, R_{SRC}. Sometimes we encounter cases where a heavy load (a small resistance that draws a large current) needs to receive a signal from a voltage signal source with large internal resistance. If such a load is directly connected to the signal source, the voltage division that forms between the internal resistance of the source (R_{SRC}) and the load (R_L) will result in a serious attenuation and most of the signal will be lost across the internal resistance of the signal source. In such cases, we need a circuit to reduce the attenuation caused by the mismatch between what the signal source likes to see as the load and what the real load is. The ideal interface circuit is one that exhibits a large input resistance (much larger than the internal resistance of the signal source) and a small output resistance (much smaller than the load resistance) and a unity voltage gain.

Such circuits are called *voltage buffers* and as indicated, the source follower can be a good voltage buffer if properly designed.

Example 4.7

A voltage signal provided by a source with 2 $k\Omega$ internal resistance is intended to be delivered to a 1-kΩ load. Use the source follower of Figure 4.18 with $V_{DD} = 5$ V, $R_1 = 60\ k\Omega$, $R_2 = 90\ k\Omega$, $R_S = 1\ k\Omega$, $V_{TH} = 0.5$ V, and $\beta = 20\ mA/V^2$ to buffer the signal source and compare it with the case where the load is directly connected to the signal source.

Solution:

Starting with **DC analysis** we have:

$$V_G = \frac{R_2}{R_1 + R_2}.V_{DD} = \frac{90k}{60k + 90k} \times 5V = 3V,$$

and the source voltage is:

$$V_S = R_S I_D = (1k)\ I_D.$$

So, replacing the gate-source voltage:

$$V_{GS} = V_G - V_S = 3\ V - (1k)\ I_D$$

in the quadratic equation for the drain current, we have:

$$I_D = \frac{\beta}{2}(V_{GS} - V_{TH})^2 = \frac{20mA / V^2}{2}(3V - (1k)I_D - 0.5)^2,$$

and

$$I_D^2 - 5.1I_D + 6.25 = 0$$

is the equation that should be solved for I_D. The two possible roots of this quadratic equation are 2 mA and 3 mA; however, 3 mA is not acceptable because it produces 3 V across R_S, which results in $V_{GS} = 0$ V, and the transistor would not be in saturation. Thus, the drain current can only be 2 mA, the voltage across R_S (i.e., $V_S = R_S I_D$) is 2 V, and

$$V_{DS} = V_D - V_S = 5\ V - 2\ V = 3\ V.$$

Because the drain-source voltage is greater than the overdrive voltage:

$$V_{OD} = \sqrt{2I_D / \beta} = \sqrt{2 \times 2mA / (20mA / V^2)} \cong 0.45V,$$

the transistor is biased in saturation mode.

Now, after calculating the transistor's transconductance as:

$$g_m = \sqrt{2\beta.I_D} = \sqrt{2 \times 20mA/V^2 \times 2mA} = 8.94mA/V$$

and knowing that

$$R_G = R_1 \| R_2 = 60\ k\Omega \| 90\ k\Omega = 36\ k\Omega,$$

it is time to perform the AC analysis and calculate the amplifier's overall gain and input and output resistances formulated in Equation (4.34), Equation (4.29), and Equation (4.30), respectively:

$$A_{VS} = \frac{v_o}{v_{src}} = \frac{R_G}{R_G + R_{SRC}} \cdot \frac{g_m (R_S \| R_L)}{1 + g_m (R_S \| R_L)}$$

$$= \frac{36k}{36k + 2k} \cdot \frac{8.94m(1k \| 1k)}{1 + 8.94m(1k \| 1k)} = 0.95 \times 0.82 = 0.78,$$

$$R_i = R_G = 36\ k\Omega,$$

and

$$R_o = R_S \| g_m^{-1} = 1\ k \| 112\ \Omega = 101\ \Omega.$$

As stated by the overall gain found previously, 78% of the source signal is delivered to the load, whereas if the load were to be directly connected to the signal source, we would have:

$$\frac{v_o}{v_{src}} = \frac{R_L}{R_L + R_{SRC}} = \frac{1k}{1k + 2k} = 0.33,$$

indicating that the load would receive only 33% of the signal.

4.6 Analysis by Inspection

It is assumed that the reader is familiar with some circuit analysis methods and rules that allow one to analyze linear electric circuits, without being bogged down in solving several equations with too many variables. Equivalent resistance, capacitance, and self-inductance, as well as voltage division, current division, node, and mesh analysis techniques are examples of such simplifying circuit analysis techniques and rules.

Unlike the well-known elements, such as resistors, capacitors, and self-inductors with straightforward linear relationships between the voltages across them and the currents flowing through them, a transistor is a nonlinear element. Even dividing the analysis of a small-signal transistor amplifier into DC and AC analyses does not change the fact that a transistor is still a three-terminal element. Studying the role of transistor, especially in AC mode, will not be as simple as dealing with the previously mentioned two-terminal elements. This section introduces a method that drastically facilitates the analysis of a small-signal transistor circuit in the AC mode. In this method, called *analysis by inspection*, under certain conditions, the transistor can be replaced with a simple resistor, which makes the analysis much easier.

4.6.1 Virtual Resistances Seen into a Transistor

If a small-signal linear circuit operates in its midband frequency range* and does not include any kind of feedback loop, then *in the AC analysis* the transistor can sometimes be replaced with a simple resistor connected from the terminal of interest to the ground. The so-called "virtual resistances" associated with each of the transistor's terminals (i.e., virtually connected from that terminal to the ground) are as follows.

4.6.1.1 The Virtual Resistance Seen into the Gate

The resistance seen into the gate of a transistor, illustrated in Figure 4.21 is almost infinite:

$$R_{Seen,G} = \infty. \tag{4.35}$$

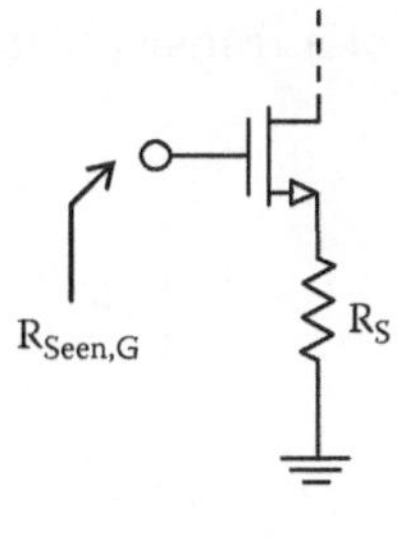

FIGURE 4.21
The virtual resistance seen into the gate of a transistor.

* Based on the assumptions so far discussed in AC analysis, when a small-signal amplifier is analyzed in the AC mode, it is indeed operating in its midband frequencies. A more accurate definition for the midband frequency range will be introduced in Chapter 7.

This is simply because of the insulator (silicon dioxide (SiO_2)) that exists between the gate and transistor's channel.*

Example 4.8

In the AC schematic in Figure Ex. 4.8(a) find $v_i(t)$ as a function of $v_{src}(t)$.

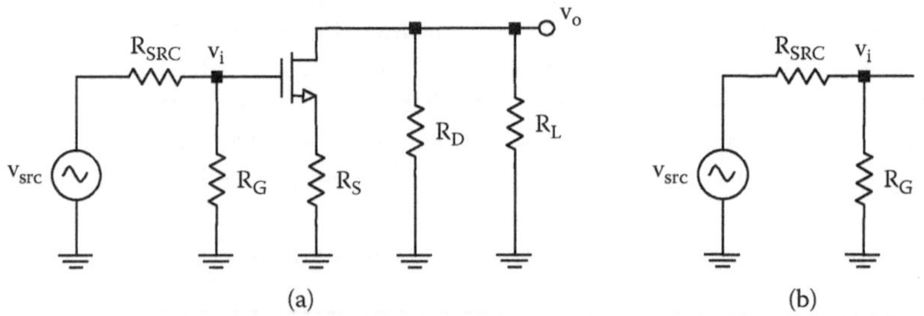

FIGURE EX. 4.8

Solution:

As presented in Figure Ex. 4.8(b), if the transistor is replaced with the virtual resistance seen into the transistor from its gate viewpoint (i.e., from the gate to the ground when looking from the gate), $v_i(t)$ will be a fraction of $v_{src}(t)$ determined by a simple voltage division:

$$v_i(t) = \frac{R_G}{R_G + R_{SRC}} . v_{src}(t)$$

4.6.1.2 *The Virtual Resistance Seen into the Source*

When looking into the source of a transistor (see Figure 4.22), the resistance seen into the source is**:

$$R_{Seen,S} = \frac{1}{g_m} . \left(1 + \frac{R_D}{r_o}\right) \tag{4.36}$$

For the cases where R_D is much smaller than r_o, the resistance seen reduces to $1/g_m$.

* It is again emphasized that these rules hold only for midband frequencies.

** This will be the resistance seen, assuming that the source is not grounded.

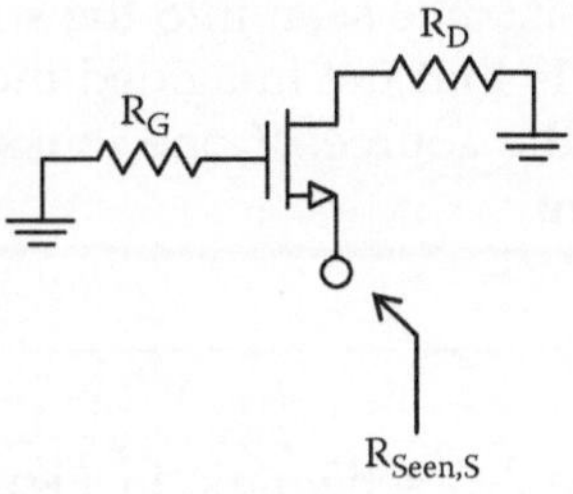

FIGURE 4.22
The virtual resistance seen into the source of a transistor.

To better understand why the transistor in Figure 4.22 is modeled approximately by $1/g_m$ when looking into the source, let us assume the transistor as a black box as illustrated in Figure 4.23. It is observed that a current proportional to the voltage applied across gate-source comes out of the source:

$$i_d = g_m v_{gs}.$$

According to this relationship, it makes sense to imagine that a resistance of $1/g_m$ is connected between the source and the gate that makes the current i_d when a voltage v_{gs} appears across it. It should be noted, however, that because we know that no current flows through R_G, the gate is at the ground

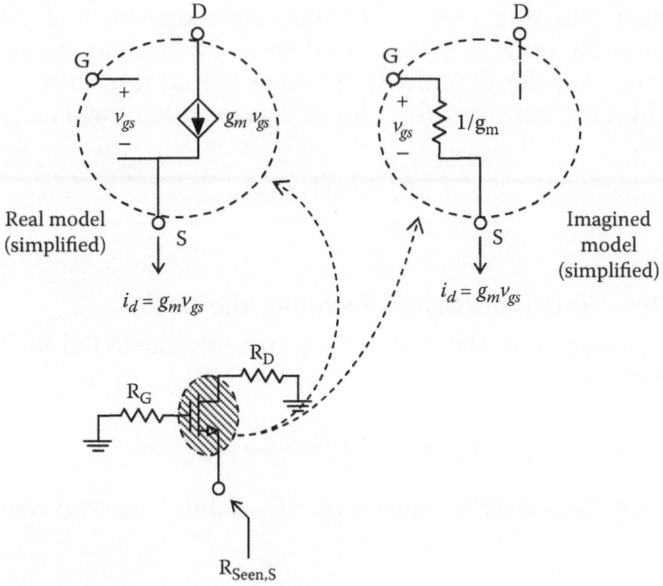

FIGURE 4.23
Imagination of the resistance seen into the source.

potential, and thus the resistance seen into the source is only $1/g_m$ and not $1/g_m + R_G$. Another point is that the imagined model in Figure 4.23 is only valid when looking into the source. If one looks into the gate, an infinite resistance will still be seen.

Example 4.9

For the amplifier with the AC schematic in Figure Ex. 4.9, find the input resistance, R_i.

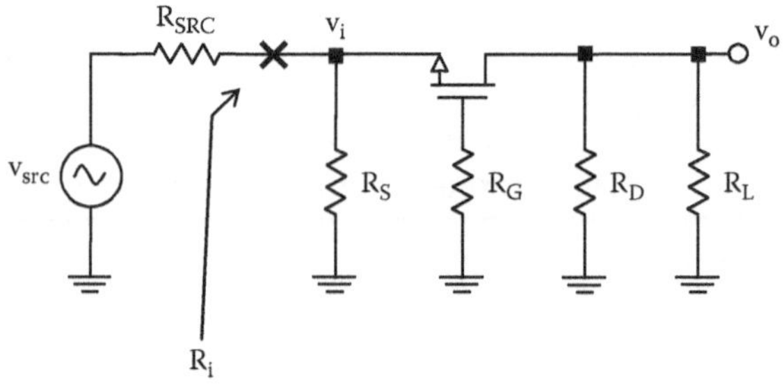

FIGURE EX. 4.9

Solution:

As depicted in Figure Ex. 4.9, R_i is the parallel combination of R_S and the resistance seen into the source of the transistor. Supposing that the resistance in the drain of the transistor (connected from the drain to the ground) (i.e., $R_D \| R_L$) is much smaller than the transistor's output resistance, r_o, the resistance seen into the transistor from the source viewpoint will be $1/g_m$. Thus:

$$R_i = R_S \| (1/g_m)$$

4.6.1.3 *The Virtual Resistance Seen into the Drain*

The resistance seen into the drain of a transistor, illustrated in Figure 4.24, is determined as:

$$R_{Seen,D} = r_o(1 + g_m R_S) \tag{4.37}$$

and because $g_m R_S$ is usually much larger than unity, one can write:

$$R_{Seen,D} \approx g_m r_o R_S \tag{4.38}$$

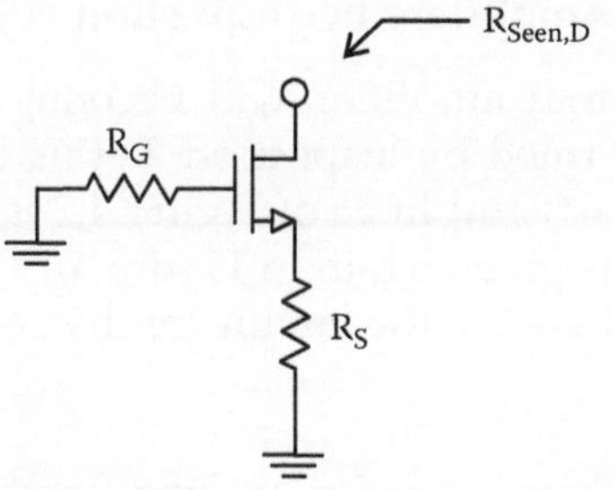

FIGURE 4.24
The virtual resistance seen into the drain of a transistor.

Example 4.10

It was discussed earlier that adding a bypass capacitor significantly increases the voltage gain of a CS amplifier. Discuss the effect that such a capacitor can have on the amplifier's output resistance.

Solution:
Studying the CS amplifier in Figure Ex. 4.10, the output resistance of the amplifier without a bypass capacitor is

$$R_o = R_D \| r_o(1 + g_m R_S)$$

and will reduce to

$$R_o = R_D \| r_o$$

when C_S is added. As long as R_D is much smaller than r_o, however, this decrease in the output resistance does not affect the amplifier's voltage gain.

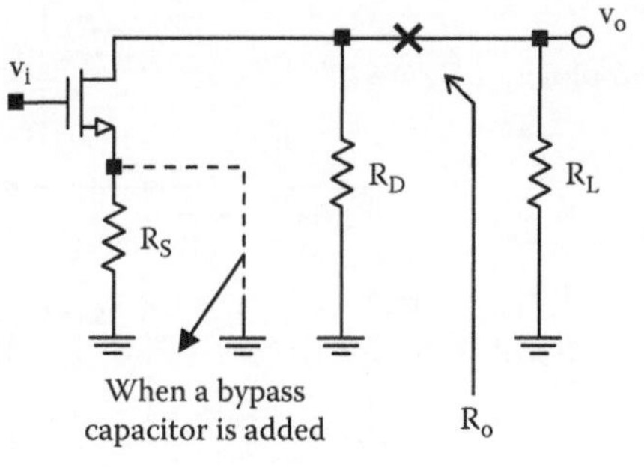

FIGURE EX. 4.10

4.6.2 AC Analysis of Amplifiers by Inspection

Analysis of a small-signal amplifier will become much easier and more straightforward if performed by inspection. Let us describe this method by using an example. The CS amplifier of Figure 4.7 is our example. AC schematic of this amplifier is given again in Figure 4.25 for convenience.

As illustrated in Figure 4.26, the circuit can be conceptually divided into three parts:

1. The **input part**, which provides an *input signal* to the amplifier (v_i) as a fraction of the *source signal*, v_{src} (Figure 4.26(a))
2. The **amplifier part**, which provides a drain/source current (i_d) as the result of amplifying the input signal (Figure 4.26(b)), and
3. The **output part**, which converts the amplified current, i_d, into the *output signal*, v_o (Figure 4.26(c)).

Before starting to find the overall gain of the amplifier, the reader should notice that two main points make the three parts in Figure 4.26 different from just three simple pieces of the circuit in Figure 4.25: The first point is that R_i in Figure 4.26(a) represents both R_G and the resistance seen into the gate of the transistor. As discussed earlier, however, the latter is assumed to

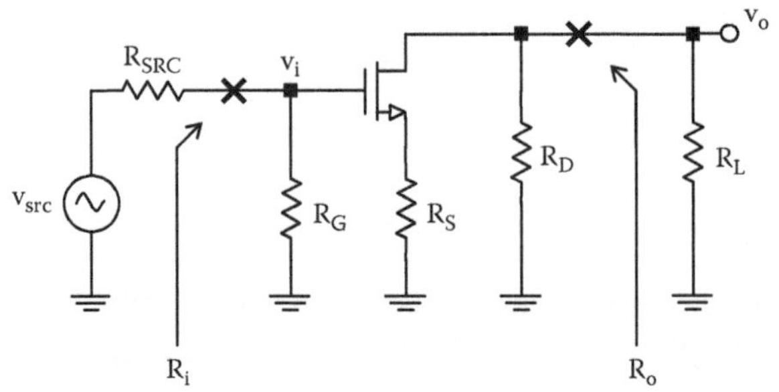

FIGURE 4.25
AC schematic of the CS amplifier of Figure 4.7.

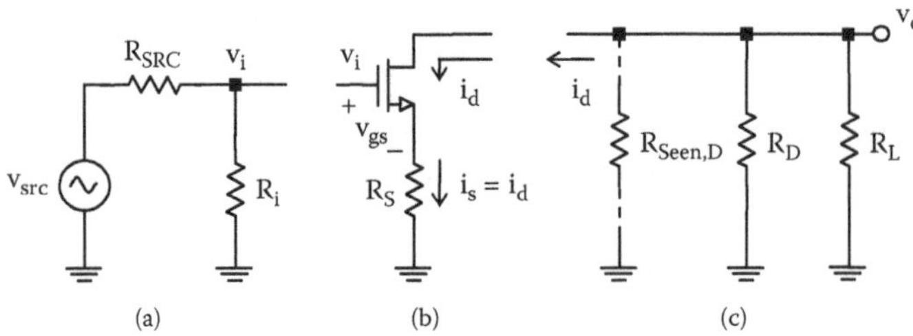

FIGURE 4.26
Three main parts of the amplifier from Figure 4.25: (a) the input part, (b) the amplifier part, and (c) the output part.

be infinite, and thus $R_i = R_G$. The second issue is the third resistor connected from the output node to the ground, by dashed lines, in Figure 4.26(c). This is the resistance seen from the drain of the transistor to the ground when looking into the drain, which is:

$$R_{Seen,D} = r_o(1 + g_m R_S) \approx g_m r_o R_S.$$

To find the overall voltage gain of the amplifier of Figure 4.25, one should determine the contribution of each of the previously mentioned parts. The role of the input part is to deliver the source signal, v_{src}, to the amplifier. As indicated in Example 4.8, only a fraction of the source signal can be delivered to the amplifier part, calculated using simple voltage division:

$$v_i(t) = \frac{R_G}{R_G + R_{SRC}} . v_{src}(t) \tag{4.39}$$

For the second step, because we are interested in figuring out how i_d is generated by the amplifier part, let us look at the amplifier part, from the source viewpoint, in Figure 4.27. As indicated, the resistance seen from the source to the gate is assumed to be $1/g_m$, which is in series with R_S, and the current we are interested in, i_d, can be determined simply by Ohm's law:

$$i_d = v_i \bigg/ \left(R_S + \frac{1}{g_m} \right). \tag{4.40}$$

We found the AC drain current using the circuit in Figure 4.27(b), but it should be noted that this circuit has been drawn from the source viewpoint and cannot be used to find other voltages or currents from the other terminal viewpoints. For example, it cannot be said that in Figure 4.27(a) a current equal to i_d is drawn from v_i, or a resistance of $(1/g_m)+R_{src}$ is seen when looking into the gate of the transistor.

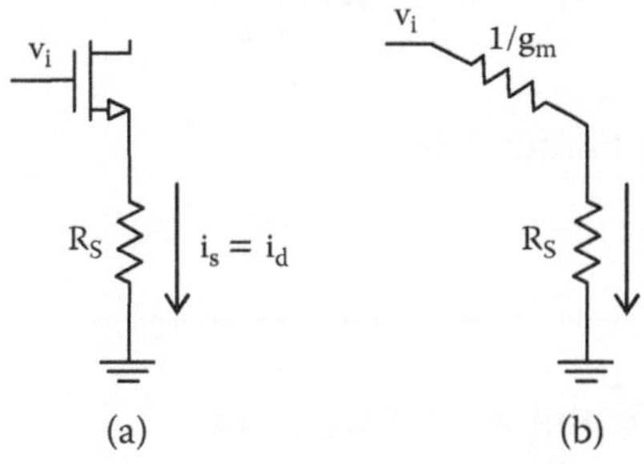

FIGURE 4.27
(a) The amplifier part (b) its equivalent circuit from the source viewpoint.

The third step in finding the gain of the amplifier is to figure out how the output voltage is generated. Taking a quick look at Figure 4.26(c), one can easily say that the output voltage, v_o, is created by passing the current i_d through the parallel combination of $R_{Seen,D}$, R_D, and R_L, and thus one can write:

$$v_o = -i_d(R_{Seen,D}\|R_D\|R_L). \tag{4.41}$$

To find the overall voltage gain for the amplifier, Equation (4.39)–Equation (4.41) are combined to give:

$$\frac{v_o}{v_{src}} = \frac{R_G}{R_G + R_S} \cdot \frac{-\left(R_{Seen,D}\,\|\,R_D\,\|\,R_L\right)}{R_{Src} + 1/g_m} \tag{4.42}$$

The overall gain achieved consists of two terms: the attenuation at the input and the amplification, which represents both the amplification by the amplifier and the attenuation at the output, discussed earlier.

As a rule of thumb to obtain a regular CS amplifier's voltage gain (excluding the attenuation at the input), one can write:

$$\frac{v_o}{v_i} = \frac{-(\text{Total resistance seen from the output node to the ground})}{\begin{array}{c}\text{Resistance from the input node, through gate-source to the}\\ \text{ground seen from the source viewpoint}\end{array}} \tag{4.43}$$

Equation (4.43) is written in such a way that its absolute value can also be used to obtain the voltage gain of a CG amplifier.

Finding the voltage gain of a CD amplifier by inspection is much simpler. In this configuration, the input voltage is applied to the gate and the output is taken from the source. Thus, looking at the circuit from the source viewpoint, the output is simply determined by a voltage division between $1/g_m$ and the resistance seen from the source to the ground, that is,

$$\frac{v_o}{v_i} = \frac{R_S}{R_S + 1/g_m} \tag{4.44}$$

This is illustrated in Figure 4.28.

4.7 Other Basic Types of Amplifiers

After studying voltage amplifiers and gaining enough experience in the analysis of amplifiers, it is time to deal with the other basic types of amplifiers:

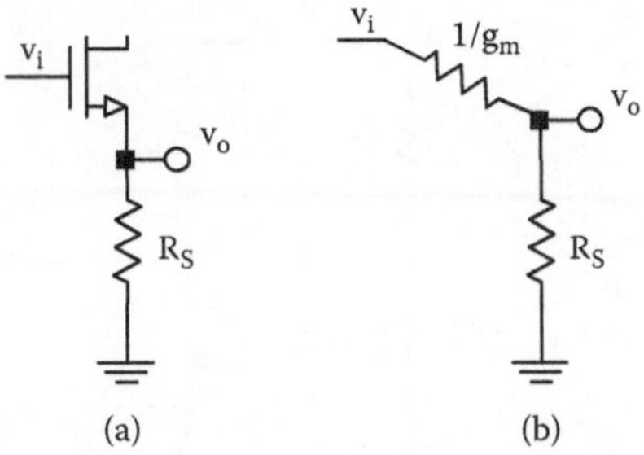

FIGURE 4.28
(a) A CD amplifier and (b) its equivalent circuit seen from the source viewpoint.

current amplifiers, transconductance amplifiers, and transresistance amplifiers. Most of the analysis methods used for voltage amplifiers can be used to analyze any other kind of small-signal circuit: the analysis is split into DC and AC analyses. DC analysis is performed exactly as mentioned earlier, whereas for the AC analysis, either one of the previously mentioned approaches (i.e., "replacement of transistor with the small-signal model" or the "analysis by inspection") can be used. As we know, the former calls for analysis using the well-known circuit analysis rules and solving a simultaneous equation system, and in the latter, the virtual resistance concept will be employed. The amplifier's input and output resistances will be determined again exactly as before. The only difference is that the gain should now be determined based on the type of the amplifier.

As mentioned before, any one of the three configurations studied thus far (i.e., common-source, common-drain, and common-gate) can be used to realize a single-stage amplifier of the three types mentioned previously in addition to the voltage amplifiers extensively studied before.

The following are examples of finding the gain for a transconductance amplifier and a current amplifier.

Example 4.11

A common-source transconductance amplifier is depicted in Figure Ex. 4.11(a). Having $V_{DD} = 5$ V, $R_{SRC} = 2\ k\Omega$, $R_L = 1$ kΩ, $R_1 = 60\ k\Omega$, $R_2 = 90\ k\Omega$, $R_S = 1\ k\Omega$, $R_D = 1k\Omega$, $V_{TH} = 0.5$V, $\beta = 20\ mA/V^2$, and $\lambda = 0.01\ \text{V}^{-1}$, find the overall transconductance, $G_{MS} = i_o/v_{src}$, and the input and output resistances of the amplifier.

Solution:

Because the transistor and the bias network are exactly what we had in Example 4.7, without redoing the DC analysis we have $I_D = 2\ mA$, $g_m = 8.94\ mA/V$, and $r_o = (\lambda I_D)^{-1} = 50\ k\Omega$. The amplifier's AC schematic will be as depicted in Figure Ex. 4.11(b).

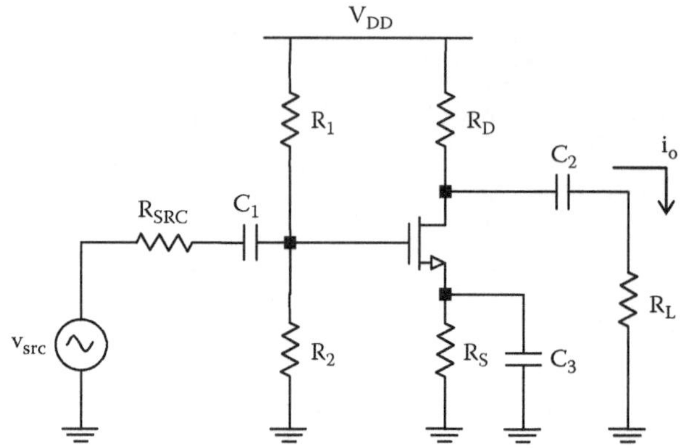

FIGURE EX. 4.11(a)

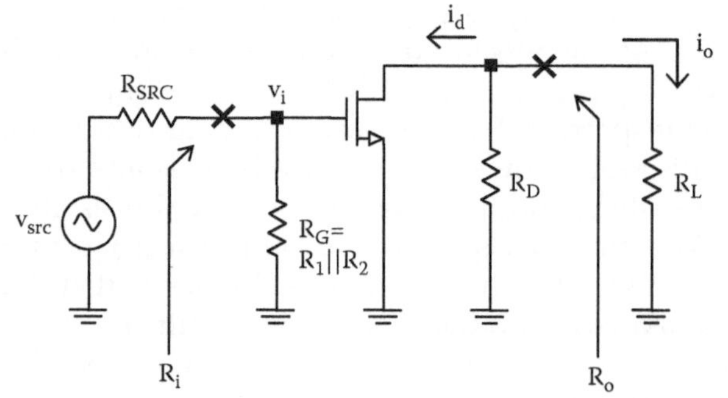

FIGURE EX. 4.11(b)

The input and output resistances are easily achieved by inspection as

$$R_i = R_G = 60\ k\|90\ k = 36\ k\Omega,$$

and

$$R_o = R_D\|r_o$$

$$= 1\ k\|50\ k$$

$$= 1\ k\Omega.$$

The overall transconductance can be written as the product of some partial terms indicating what happens to the signal when passing through the amplifier. The source signal, v_{src}, faces a voltage divider network formed by R_{SRC} and the input resistance of the amplifier $R_i = R_G$. The resulting voltage that is applied to the gate, v_i, is then converted to the drain current, i_d, by the transistor ($i_d = g_m v_{gs}$), and finally the output current is determined by a current division between R_D (neglecting the resistance seen into the drain of the transistor) and the load:

$$G_{MS} = \frac{i_o}{v_{src}} = \frac{v_i}{v_{src}} \cdot \frac{i_d}{v_i} \cdot \frac{i_o}{i_d} = \frac{R_G}{R_G + R_{SRC}} \cdot g_m \cdot \frac{-R_D}{R_D + R_L}$$

and

$$\begin{aligned} G_{MS} &= \frac{36k}{36k+2k} \times 8.94mA/V \times \frac{-1k}{1k+1k}, \\ &= 0.95 \times 8.94mA/V \times (-0.5) \\ &= -4.25mA/V. \end{aligned}$$

The overall transconductance achieved means that this is a good design at the input, which delivers 95% of the source signal to the amplifier. Furthermore, the amplifier is employing all the transconductance that the transistor is providing; however, because of the relatively low output resistance, which is comparable to the load resistance, half of the pure amplifier's transconductance is lost.

Example 4.12

As discussed already, a good current amplifier should have a small input resistance and a large output resistance compared with the signal source's internal resistance and the load resistance, respectively. From this standpoint, the common-gate configuration appears to be the best choice. Apply a current signal to the source of a voltage-divider biased transistor and taking the output from the drain, find the overall current gain that this configuration can provide.

Solution:

The circuit under study and its AC schematic are provided in Figure Ex. 4.12(a) and Figure Ex. 4.12(b), respectively.

While passing through this circuit, the input signal source current, i_{src}, first faces a current division between $R_{SRC} \| R_S$ and the resistance seen into the source of the transistor (i.e., $1/g_m$). Then, the transistor converts the source current to the drain current with a current gain of 1, and finally, through a current division between R_D and R_L the output current is determined:

$$A_{IS} = \frac{i_o}{i_{src}} = \frac{i_s}{i_{src}} \cdot \frac{i_d}{i_s} \cdot \frac{i_o}{i_d} = \frac{-R_S \| R_{SRC}}{(R_S \| R_{SRC}) + 1/g_m} \times 1 \times \frac{-R_D}{R_D + R_L}.$$

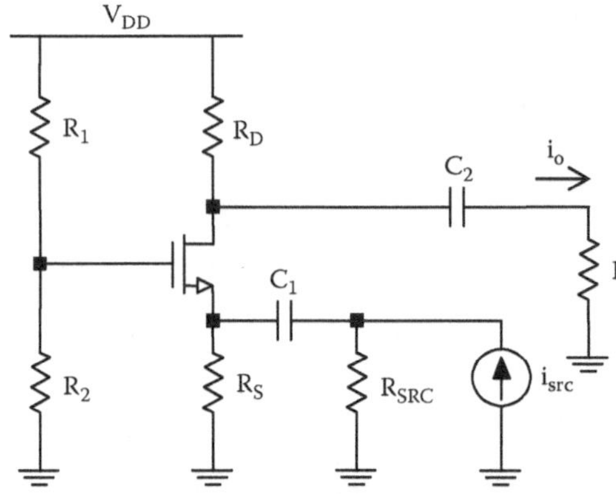

FIGURE EX. 4.12(a)

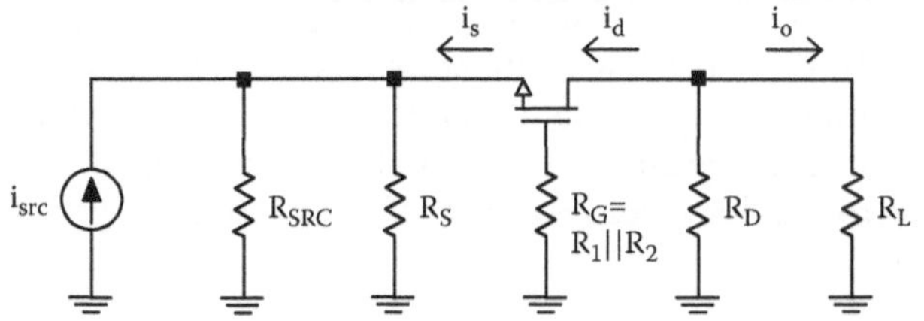

FIGURE EX. 4.12(b)

As seen, the contribution of the transistor to the overall current gain of this circuit is only a factor of 1. Taking the two current divisions at the input and output into account, and despite its good input and output resistances, the common-gate configuration's current gain does not exceed unity. From what was studied about the common-drain configuration for voltage amplifiers, it can be guessed that the common-gate configuration can play the role of a good *current buffer* wherever a signal source of current type has difficulty in driving a relatively large load resistance compared with its internal resistance.

4.8 Bipolar Amplifiers

After studying MOS single-stage amplifiers, it is time to deal with the amplifiers designed based on bipolar transistors. Bipolar transistors were introduced in Chapter 2 and biased in Chapter 3. In this section, we will learn how to use a bipolar transistor as an amplifier, and then how to analyze a bipolar single-stage amplifier.

4.8.1 Bipolar Transistor as an Amplifier

The exponential $i_C - v_{BE}$ characteristics of a bipolar transistor

$$i_C = I_S e^{v_{BE}/V_T} \tag{4.45}$$

is obviously nonlinear. Thus, an approach similar to what was performed for the MOS transistor can be taken. If the transistor is biased in the active mode and the variations in i_C and v_{BE} are kept small enough, then the previously mentioned exponential $i_C - v_{BE}$ characteristic curve around the operating point, $Q(I_C, V_{BE})$, can be approximated to a line segment as illustrated in Figure 4.29.

Based on what was explained for an MOS transistor, assuming that the variations in both i_C and v_{BE} are small compared with the associated bias values, that is,

$$\frac{\Delta v_{BE}}{V_{BE}} \ll 1 \text{ and } \frac{\Delta i_C}{I_C} \ll 1, \tag{4.46}$$

it can be said that the slope of the characteristic curve of Figure 4.29 at the operating point, $Q(I_C, V_{BE})$,

$$Slope@Q = \frac{\partial i_C}{\partial v_{BE}}\Big|_Q \tag{4.47}$$

is, in fact, the transistor's transconductance, being defined as:

$$g_m = \frac{\partial i_C}{\partial v_{BE}}\Big|_Q \tag{4.48}$$

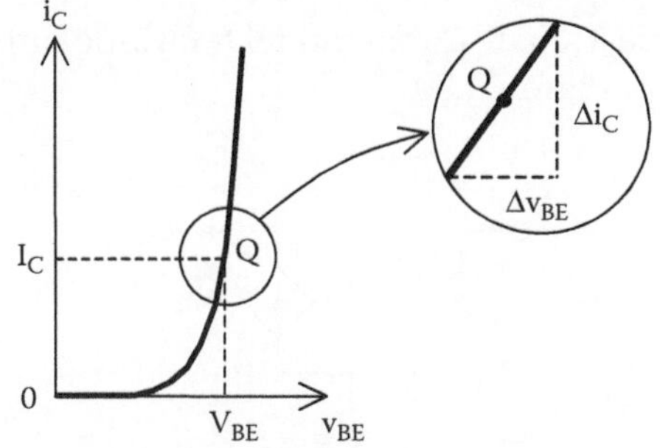

FIGURE 4.29
$i_C - v_{BE}$ characteristic curve for a bipolar transistor.

Differentiating Equation (4.45) with respect to v_{BE}, one can write:

$$g_m = \frac{\partial(I_S e^{v_{BE}/V_T})}{\partial v_{BE}}\bigg|_Q = \frac{I_S e^{v_{BE}/V_T}}{V_T}\bigg|_Q = \frac{i_C}{V_T}\bigg|_Q, \tag{4.49}$$

and thus:

$$g_m = \frac{I_C}{V_T}, \tag{4.50}$$

which is the formula for calculating the transconductance of a bipolar transistor for a given bias current.

For a bipolar transistor, the AC component of the collector current is proportional to that of the base-emitter voltage with the transistor's transconductance as the proportionality factor:

$$i_c = g_m v_{be}. \tag{4.51}$$

A bipolar transistor's small-signal model can be introduced based on what Equation (4.51) states: the current that enters the collector is proportional to the voltage across base-emitter. Unlike an MOS transistor where the gate oxide is an insulator and modeled by an open circuit between the gate and the source, for a bipolar transistor a resistance, called r_π, is assumed between the base and the emitter. This resistance relates the AC components of the base-emitter voltage and the base current, and is sometimes referred to as the *base-emitter dynamic resistance*. The resistance r_π is determined by dividing the thermal voltage, V_T, by the DC component of the base current:

$$r_\pi = \frac{V_T}{I_B} = \frac{V_T}{I_C/\beta} = \beta\frac{V_T}{I_C} = \beta/g_m = \beta r_e, \tag{4.52}$$

where $r_e = 1/g_m = V_T/I_C$.

This way, the simplest small-signal model for a bipolar transistor is formed, which is presented in Figure 4.30.

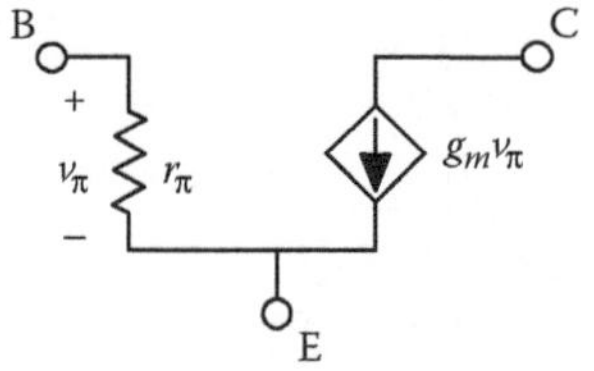

FIGURE 4.30
Simplified small-signal model for a bipolar transistor.

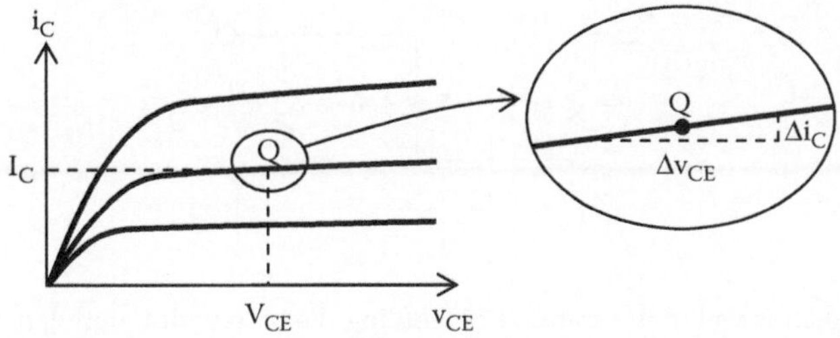

FIGURE 4.31
$i_C - v_{CE}$ characteristic curves for a bipolar transistor.

As expected, the emitter current is the summation of the collector and the base currents (all AC components):

$$i_e = i_c + i_b. \tag{4.53}$$

A more advanced small-signal model for a bipolar transistor includes the output resistance of the transistor. As was introduced in Chapter 2, similar to the channel-length modulation effect in MOS transistors, the base-width modulation effect in bipolar transistors makes the collector current vary slightly as the collector-emitter voltage varies. This behavior is observed in the $i_C - v_{CE}$ characteristics of a transistor as the curves' slopes in the active region (see Figure 4.31).

The more complete formula for the collector current of a bipolar transistor is:

$$i_C = I_S e^{v_{BE}/V_T}\left(1+\frac{v_{CE}}{V_A}\right), \tag{4.54}$$

which includes the base-width modulation effect. As illustrated in Figure 4.31, when biased at $Q(I_C, V_{CE})$, a variation of Δv_{CE} in the collector-emitter voltage causes a proportional variation of Δi_C in the collector current related to each other by the slope of the curve at the operating point. Differentiating the preceding equation with respect to v_{CE}, one can find the slope of the curve at the operating point, $Q(I_C, V_{CE})$, which is a conductance, as:

$$Slope@Q = g_o = \frac{\partial i_C}{\partial v_{CE}}\Big|_Q = \frac{1}{V_A}.I_S e^{v_{BE}/V_T}\Big|_Q = \frac{1}{V_A}.I_C\Big/\left(1+\frac{V_{CE}}{V_A}\right). \tag{4.55}$$

Because V_A is usually much greater than V_{CE}, Equation (4.55) reduces to:

$$Slope@Q = g_o \cong \frac{I_C}{V_A}, \tag{4.56}$$

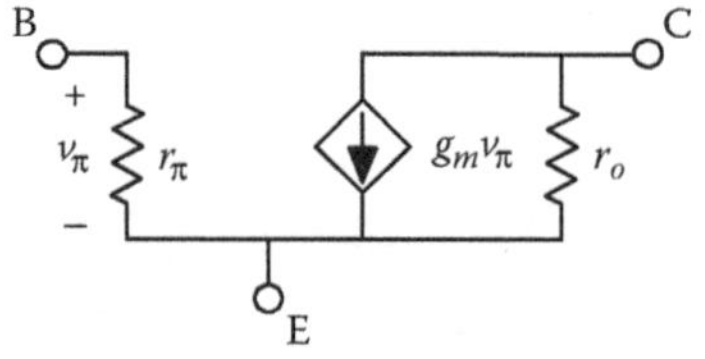

FIGURE 4.32
Small-signal model for a bipolar transistor including the base-width modulation effect.

The element that will represent this phenomenon in the small-signal model is a resistance of:

$$r_o = g_o^{-1} = \frac{V_A}{I_C} \tag{4.57}$$

between the collector and the emitter (see Figure 4.32).

Knowing the fact that V_A is a large voltage, r_o is usually a large resistance typically in the order of couple of hundred $k\Omega$ to tens of $M\Omega$.

4.8.1.1 Virtual Resistances Seen into a Bipolar Transistor

To be able to analyze a bipolar amplifier by inspection, we need to learn about the virtual resistances seen into a bipolar transistor. Now that we are familiar enough with the basics of AC analysis, let us derive the virtual resistances for a bipolar transistor by reasoning. Suppose that the voltage v_{be} is applied across the base-emitter of a transistor (see Figure 4.33). From Figure 4.30, we know that the current sunk into the base is:

$$i_b = v_{be}/r_\pi.$$

In other words, *from the base viewpoint, a resistance of* $r_\pi = \beta r_e$ *is seen between the base and the emitter.*

Now, let us add a resistor from the emitter to the ground as depicted in Figure 4.34. The voltage at the base can be written as:

$$v_b = v_{be} + v_e = i_b r_\pi + i_e R_E \approx i_b(\beta r_e) + (i_b\beta)R_E = i_b\beta(r_e + R_E)$$

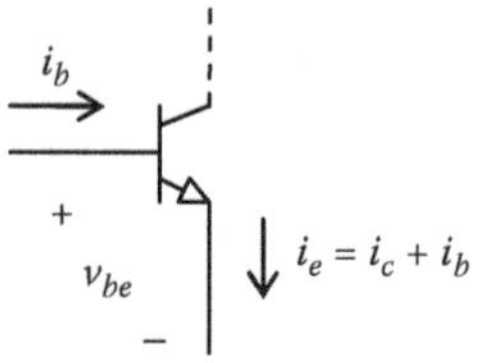

FIGURE 4.33
Base and emitter currents when a voltage is applied across base-emitter.

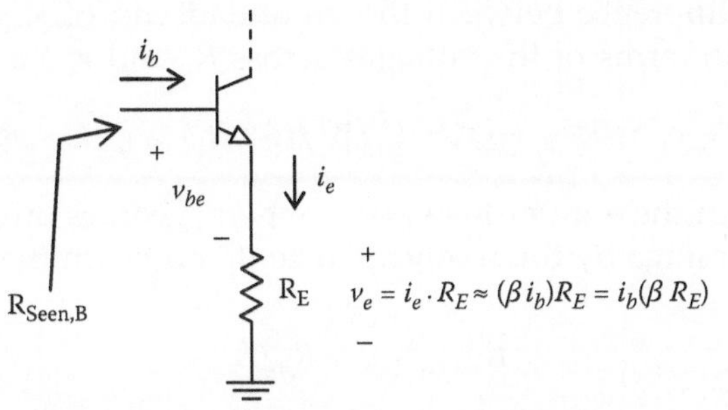

FIGURE 4.34
The virtual resistance seen into the base.

In other words, the virtual resistance *from the base viewpoint* connected from the base to the ground in Figure 4.34 is:

$$R_{Seen,B} = \beta(r_e + R_E). \tag{4.58}$$

It also can be said that from the base viewpoint, the resistance in the emitter (R_E) is first multiplied by β, and then considered in series with $r_\pi = \beta r_e$.

With similar reasoning, the virtual resistance seen into the emitter of a bipolar transistor can be derived. Once the voltage v_{be} is applied across the base-emitter of a transistor (see Figure 4.33), the current that is received from the emitter is:

$$i_e = i_c + i_b = (\beta+1)i_b = (\beta + 1)v_{be}/r_\pi = v_{be}/r_e,$$

which is obviously proportional to the applied voltage. Therefore, it can be *imagined* that *from the emitter viewpoint, a resistance of r_e is seen between the base and the emitter,* which is different from what was obtained from the base viewpoint. Now, let us connect the base to the ground using a resistor as presented in Figure 4.35.

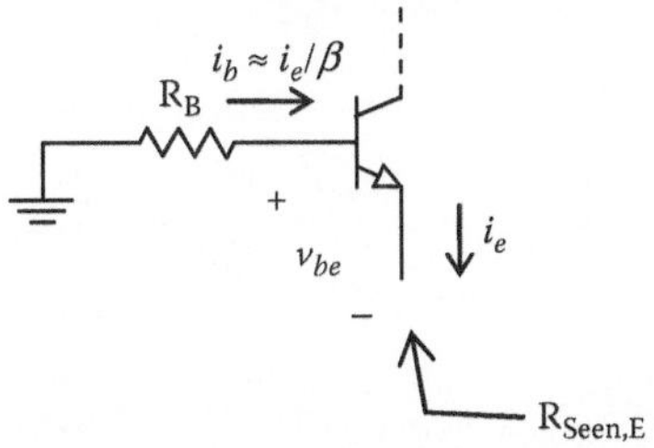

FIGURE 4.35
The virtual resistance seen into the emitter.

The voltage difference between the grounded end of R_B and the emitter can be written in terms of the voltages across R_B and r_π as:

$$i_b R_B + v_{be} = i_b R_B + i_b r_\pi \approx (i_e/\beta)\,(r_\pi + R_B) = i_e\,(r_e + R_E/\beta)$$

Therefore, from the emitter viewpoint, where i_e comes out, one can divide the preceding voltage by the received current and claim that a virtual resistance of:

$$R_{Seen,E} = r_e + R_B/\beta. \tag{4.59}$$

can be seen into the emitter. In other words, when looking into the emitter the resistances in the base are seen in series with r_e, but they appear smaller by a factor of β; however, we know that the majority of the emitter current is supplied by the dependent current source. The fact that this current is proportional to the current that really comes from the base (and apparently much larger) helps us to imagine a much smaller resistance from the emitter viewpoint in the circuit of Figure 4.35.

To prevent any confusion between the resistances seen from the emitter and the base viewpoints, the following two *assumptions* about the transistor, illustrated in Figure 4.36, can help:

- A resistance r_e connects the emitter to the base, which is assumed to be at the emitter side.
- It can be assumed that a monocle is at the base, with a magnification factor of β. We know that using this kind of monocle when looking through the eyepiece, the objects are seen larger, and looking in the reverse direction causes the objects to appear smaller by the same factor.

Now, it is time to deal with the virtual resistance seen into the collector of a bipolar transistor. The most general case is the case where resistors are connecting both the emitter and the base to the ground (see Figure 4.37).

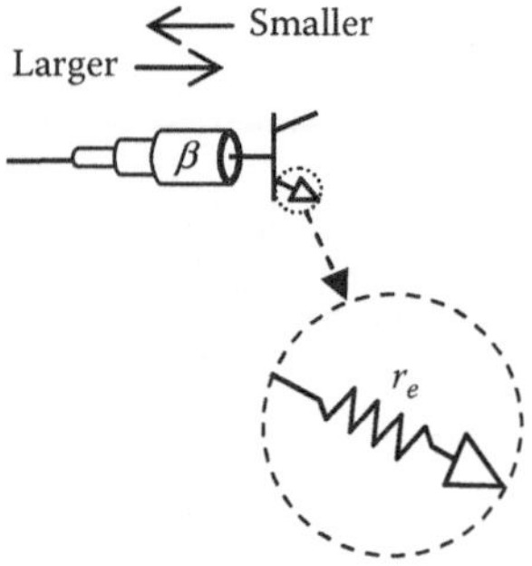

FIGURE 4.36
Illustration of two assumptions that can be helpful in finding the virtual resistances seen from the base and the emitter viewpoints.

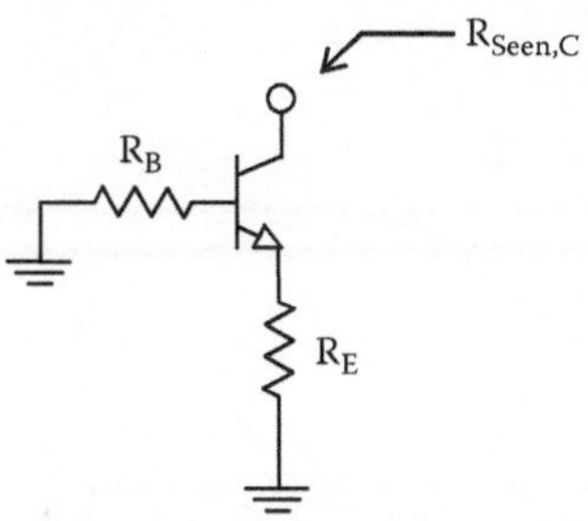

FIGURE 4.37
The most general case when looking into the collector of a bipolar transistor.

By replacing the transistor with its small-signal equivalent circuit, it can be shown that the virtual resistance seen into the collector of the transistor is:

$$R_{Seen,C} \cong r_o\left(1+\frac{\beta R_E}{R_E+R_B+r_\pi}\right). \tag{4.60}$$

Three special cases are worth noting because they are usually encountered in the analysis of bipolar analog circuits. If the emitter is grounded (i.e., the *emitter degeneration* resistance connecting the emitter to the ground (R_E) is zero), the resistance seen into the collector will be r_o, no matter how much resistance we have in the base (connecting the base to the ground):

$$R_E = 0 \Rightarrow R_{Seen,C} = r_o. \tag{4.61}$$

If R_E is not zero, there will be two cases: the resistance seen into the collector will be $(1 + g_m R_E)\, r_o$ for small R_E (much smaller than r_π), and $(1 + \beta)\, r_o$ for large R_E (much greater than r_π):

$$R_E \neq 0 \,\&\, R_B = 0 \Rightarrow \begin{cases} R_E \ll r_\pi \Rightarrow R_{Seen,C} = (1+g_m R_E)r_o \\ R_E \gg r_\pi \Rightarrow R_{Seen,C} = (1+\beta)r_o \cong \beta r_o \end{cases} \tag{4.62}$$

In other words, when the base is grounded and $R_E = 0$, the resistance seen into the collector ($R_{Seen,C}$) is r_o. Now, if R_E is swept from zero to infinity, at the beginning $R_{Seen,C}$ starts increasing proportionally to R_E, and then gradually saturates to approximately βr_o. Figure 4.38 visualizes the three cases expressed by equations (4.61) and (4.62).

Whereas the resistance seen into the collector of a bipolar transistor cannot go beyond βr_o, based on Equation (4.37), it was already explained that the resistance seen into the drain of an MOS transistor does not have a limit.

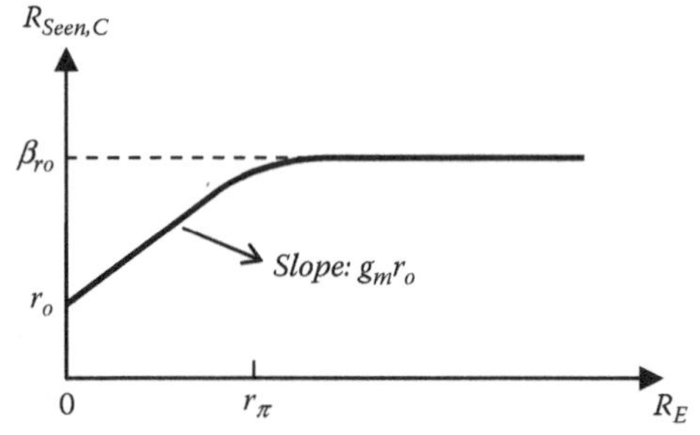

FIGURE 4.38
Visualization of the resistance seen into the collector as a function of R_E when $R_B = 0$.

Similarly to $R_{Seen,\,C}$, it starts from r_o when the source degeneration resistance, R_S, is zero, and keeps increasing with the slope of $g_m r_o$ as far as the source degeneration resistance, R_S, increases.

Figure 4.39 compares the previously mentioned resistances of a bipolar transistor and an MOS transistor assuming that both transistors have the same g_m and r_o.

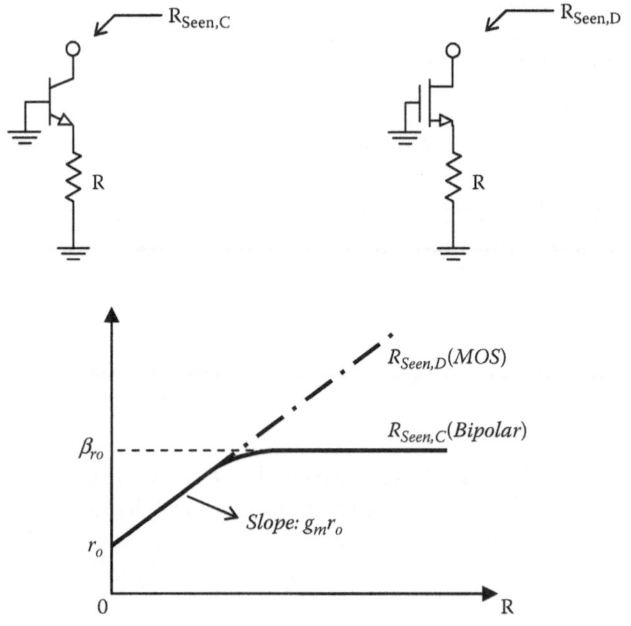

FIGURE 4.39
Comparison between the output resistances of bipolar and MOS transistors with a resistance R in the emitter/source.

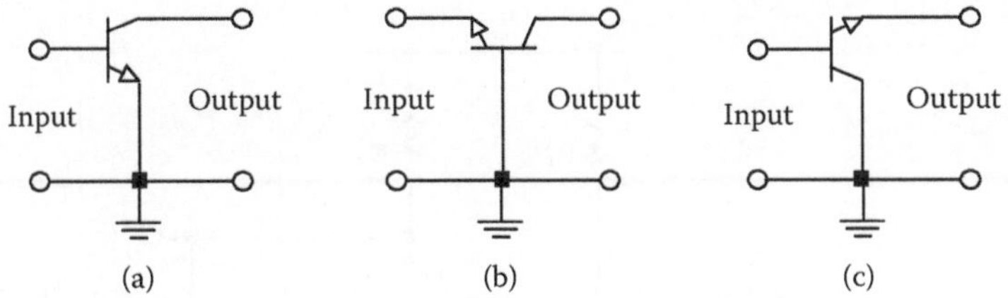

FIGURE 4.40
Single-stage configurations for a bipolar amplifier: (a) common-emitter, (b) common-base, and (c) common-collector or emitter follower.

4.8.2 Single-Stage Bipolar Configurations

A bipolar transistor biased in the active mode can be used as an amplifier in three configurations based on where the input signal is applied and where the output signal is taken from. As depicted in Figure 4.40, the common-emitter (CE) configuration is the case where the input signal is applied to the base and the output is taken from the collector. For the common-base (CB) configuration, the input and output signals are coupled to the emitter and the collector, respectively. In the third configuration, the common-collector (CC) or emitter follower (EF), the transistor receives the input signal at the base, and delivers the output signal at the emitter.

As was mentioned earlier for the MOS configurations, it is worth repeating that the issue of where the input and output signal is coupled to and from determines the amplifier's configuration. This is independent from the type of the amplifier, which is determined by the nature of the input and output signals (i.e., whether they are voltage or current). Again, voltage amplifiers are studied first, and then the other three types of amplifiers will be analyzed.

4.8.2.1 Common-Emitter Configuration

Figure 4.41 is a voltage-divider biased bipolar transistor used as a voltage amplifier in the CE configuration. As can be seen, it is capacitively coupled to the signal source from one end and to the load from the other end. Similar to the common-source configuration for an MOS voltage amplifier, the purpose of the bypass capacitor connected from the emitter to the AC ground is to increase the gain of the amplifier.

To analyze the amplifier in the AC mode, let us draw its AC schematic as depicted in Figure 4.42. In both Figure 4.41 and Figure 4.42, the dashed line indicates the case where a bypass capacitor is used. With no bypass capacitor, the input and output resistances are obtained by inspection as:

$$R_i = R_B \| \beta(r_e + R_E), \tag{4.63}$$

and

$$R_o = R_C \left\| r_o \left(1 + \frac{\beta R_E}{R_E + R_B + r_\pi} \right) \right. . \tag{4.64}$$

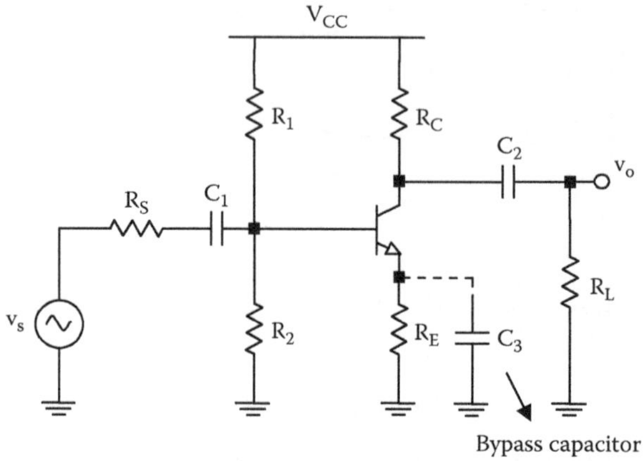

FIGURE 4.41
Voltage amplifier with CE configuration.

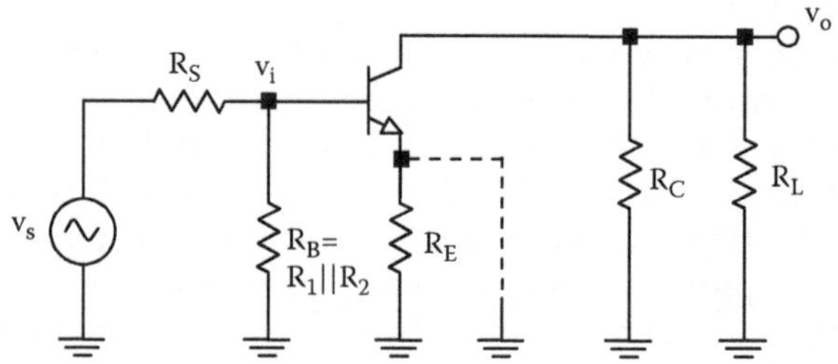

FIGURE 4.42
AC schematic of the CE amplifier of Figure 4.41.

The overall gain consists of two terms representing the fraction of the input signal that is delivered to the amplifier and the amplification performed by the amplifier:

$$A_{VS} = \frac{v_o}{v_s} = \frac{v_i}{v_s} \cdot \frac{v_o}{v_i}. \tag{4.65}$$

The former simply comes from a voltage division between the internal resistance of the signal source and the input resistance of the amplifier:

$$\frac{v_i}{v_s} = \frac{R_i}{R_i + R_S}. \tag{4.66}$$

To determine the latter, let us study the circuit from the emitter viewpoint. As illustrated in Figure 4.43, the emitter current (approximately equal to the

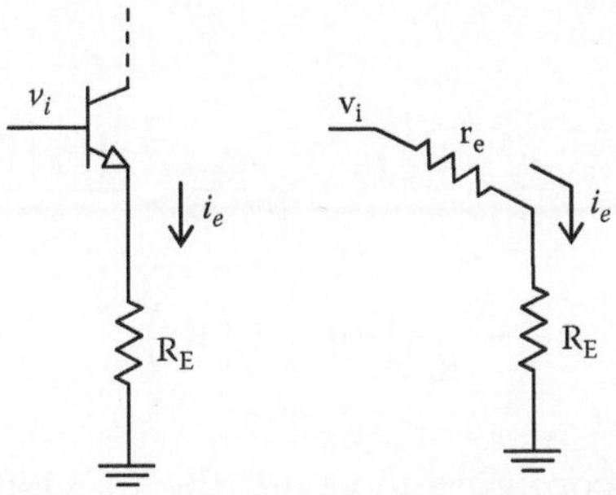

FIGURE 4.43
The virtual resistance connected from the base to the ground from the emitter viewpoint.

collector current) is the current through *the resistance seen from the base (input) through the base-emitter junction to the ground* across which the voltage v_i is applied:

$$i_e = \frac{v_i}{r_e + R_E} = \frac{i_c}{\alpha} \cong i_c. \tag{4.67}$$

On the other hand, the output voltage is the result of passing this current through *the resistance seen from the collector (output) to the ground*:

$$v_o = -\left(R_o \| R_L\right).i_c. \tag{4.68}$$

Using Equation (4.67) and Equation (4.68) the second term of the voltage gain of (4.65) is obtained as:

$$\frac{v_o}{v_i} = \frac{-\left(R_o \| R_L\right)}{r_e + R_E}. \tag{4.69}$$

Now, the overall gain of the amplifier can be written as:

$$A_{VS} = \frac{v_o}{v_s} = \frac{v_i}{v_s}.\frac{v_o}{v_i} = \frac{R_i}{R_i + R_S}.\frac{-\left(R_o \| R_L\right)}{r_e + R_E}. \tag{4.70}$$

For the case where R_E is bypassed, we will have:

$$R_i = R_B \| \beta r_e = R_B \| r_\pi, \tag{4.71}$$

$$R_o = r_o, \tag{4.72}$$

and

$$A_{VS} = \frac{v_o}{v_s} = \frac{v_i}{v_s} \cdot \frac{v_o}{v_i} = \frac{R_i}{R_i + R_S} \cdot \frac{-(R_o \| R_L)}{r_e}$$
$$= \frac{R_i}{R_i + R_S} \cdot \left(-g_m (R_o \| R_L)\right). \tag{4.73}$$

Comparing the CE configuration with the previously studied common-source configuration, it can be said that the larger transconductance of a bipolar transistor (compared with an MOS transistor with similar size and bias current) provides more pure voltage gain. However, the finite input resistance seen into the base of the bipolar transistor can cause a lower input resistance for the amplifier compared with its MOS counterpart, especially when R_E is bypassed. In some cases, this can lead to a large attenuation at the input, which lowers the overall gain.

4.8.2.2 Common-Base Configuration

In Figure 4.44, where the input signal is applied to the emitter and the output signal is taken from the collector, the common-base configuration, CB, is formed.

Similar to the CE configuration, the spirit of voltage amplification in the CB configuration can also be explained as follows. First, a base-emitter voltage is made out of the input signal. This voltage is then converted to a proportional collector current by the transistor. By having this current pass through the resistances at the collector side, the output voltage is produced. The difference with this configuration is that the input is applied to the

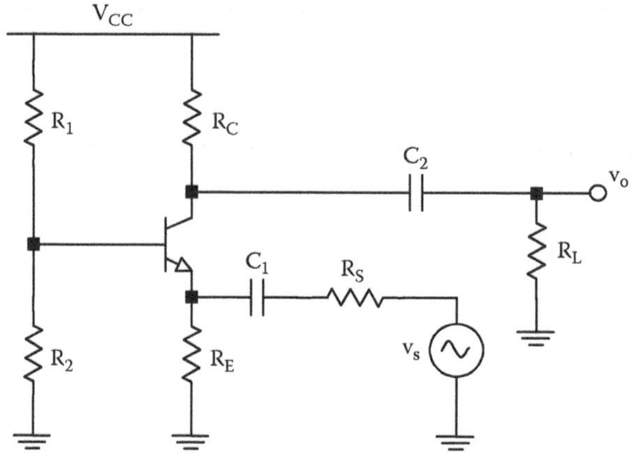

FIGURE 4.44
A common-base voltage amplifier.

emitter with the base connected to the ground, either directly or through a resistor. This way, any increase in the emitter voltage caused by the input signal reduces the base-emitter voltage. Thus, the base-emitter voltage will be 180° out of phase with the input signal. As the result of this additional sign inversion, the voltage gain of the CB configuration will be positive. The input resistance in this configuration is usually the parallel combination of the resistance seen into the emitter of the transistor and the resistance externally connected from the emitter to the ground. As we know, the former is usually small itself, and having it in parallel with another not-so-large resistance (the latter), makes the input resistance of a CB amplifier small. This is the drawback of this configuration when used as a voltage amplifier. Because the output is taken from the collector, the output resistance of the CB configuration is almost the same as that of the CE configuration.

Performing the AC analysis for the CB amplifier of Figure 4.44, the amplifier's AC schematic would be as is depicted in Figure 4.45.

The input and output resistances and the overall voltage gain can be obtained by inspection as:

$$R_i = R_E \left\| \left(r_e + \frac{R_B}{\beta} \right) \right., \tag{4.74}$$

$$R_o = R_C \left\| r_o \left(1 + \frac{\beta (R_E \| R_S)}{(R_E \| R_S) + R_B + r_\pi} \right) \right., \tag{4.75}$$

and

$$A_{VS} = \frac{v_o}{v_s} = \frac{v_i}{v_s} \cdot \frac{v_o}{v_i} = \frac{R_i}{R_i + R_S} \cdot \frac{R_o \| R_L}{r_e + \frac{R_B}{\beta}}, \tag{4.76}$$

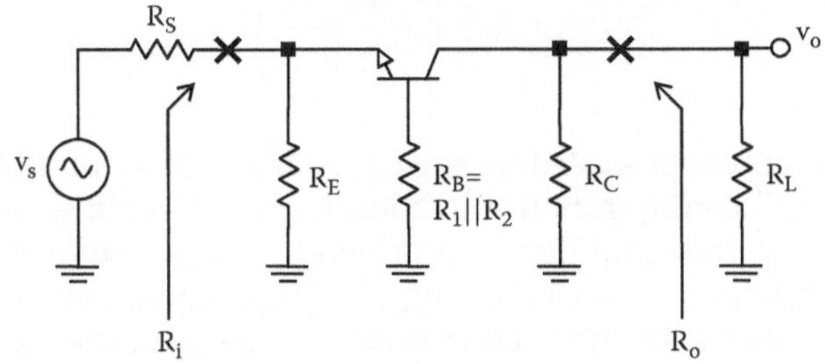

FIGURE 4.45
AC schematic of the CB amplifier of Figure 4.44.

To increase the voltage gain of a CB amplifier one may connect a capacitor from the base to the ground to bypass R_B. In this case we will have:

$$R_i = R_E \| r_e, \tag{4.77}$$

$$R_o = R_C \| r_o \left(1 + \frac{\beta(R_E \| R_S)}{(R_E \| R_S) + r_\pi}\right), \tag{4.78}$$

and

$$\begin{aligned} A_{VS} &= \frac{v_o}{v_s} = \frac{v_i}{v_s} \cdot \frac{v_o}{v_i} = \frac{R_i}{R_i + R_S} \cdot \frac{(R_o \| R_L)}{r_e} \\ &= \frac{R_i}{R_i + R_S} \cdot \left(g_m (R_o \| R_L)\right). \end{aligned} \tag{4.79}$$

4.8.2.3 Common-Collector or Emitter Follower Configuration

It was observed that the CE configuration, which is capable of providing a large voltage gain, will have a small input resistance if a bypass capacitor is used, and its output resistance is also not so small. The CB configuration suffers from the same problem at the output, as well, and its input resistance is also too low. Only the CC configuration can meet the requirements of a good voltage amplifier in terms of the input and output resistances. Figure 4.46 is a CC voltage amplifier, along with its AC schematic. The input and output resistances for this circuit are as follows:

$$R_i = R_B \| \beta(r_e + (R_E \| R_L)) \tag{4.80}$$

and

$$R_o = R_E \| \left(r_e + \frac{R_B \| R_S}{\beta}\right). \tag{4.81}$$

Similar to what was observed for the CD configuration for an MOS voltage amplifier, the CC configuration also cannot provide voltage amplification. Instead, by using this configuration, a good voltage buffer can be implemented, which is perfect for buffering a voltage signal source from a load with a smaller resistance than the internal resistance of the signal source.

Performing the AC analysis by inspection, the voltage delivered to the base of the transistor faces a voltage division between ($R_E \| R_L$) and r_e to reach the output node, and thus the overall voltage gain can be written as:

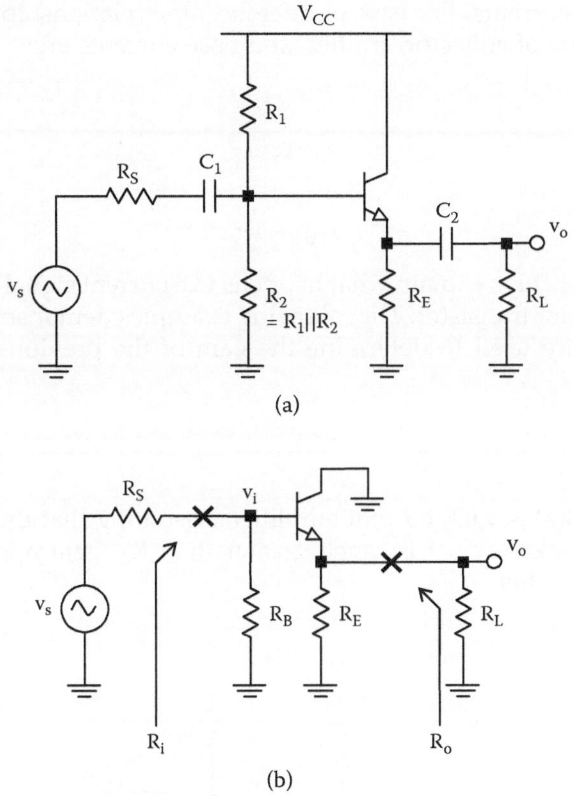

FIGURE 4.46
(a) A voltage amplifier with CC configuration, (b) AC schematic.

$$A_{VS} = \frac{v_o}{v_s} = \frac{v_i}{v_s} \cdot \frac{v_o}{v_i} = \frac{R_i}{R_i + R_S} \cdot \frac{R_L \| R_E}{\left(R_L \| R_E\right) + r_e}, \tag{4.82}$$

which is always smaller than unity. When designing CC amplifiers, the voltage gain is usually set close to 1.

4.8.3 Other Basic Types of Bipolar Amplifiers

The analysis of the other three basic types of amplifiers (current amplifiers, transconductance amplifiers, and transresistance amplifiers) has been studied in detail previously for MOS technology. Therefore, in this section, only a few issues are discussed that are exclusively encountered when analyzing bipolar amplifiers.

In these amplifiers, we deal with *currents* either at the input or at the output or both. Unlike the MOS transistor, all the three leads of a bipolar transistor have nonzero currents. For ease of reference, the relationships between the AC components of collector, emitter, and base currents are:

$$i_c = \beta . i_b, \quad (4.83(a))$$

$$i_c = \alpha . i_e, \quad (4.83(b))$$

and

$$i_e = (\beta + 1) i_b. \quad (4.83(c))$$

These relationships explain what happens to a current signal when passing through a bipolar transistor. The following examples demonstrate how these relationships are used to determine the gain of the previously mentioned amplifiers.

Example 4.13

Figure Ex. 4.13(a) is a CE current amplifier. Assuming that the output resistance of the transistor (r_o) is much greater than R_C, determine the current gain of the amplifier.

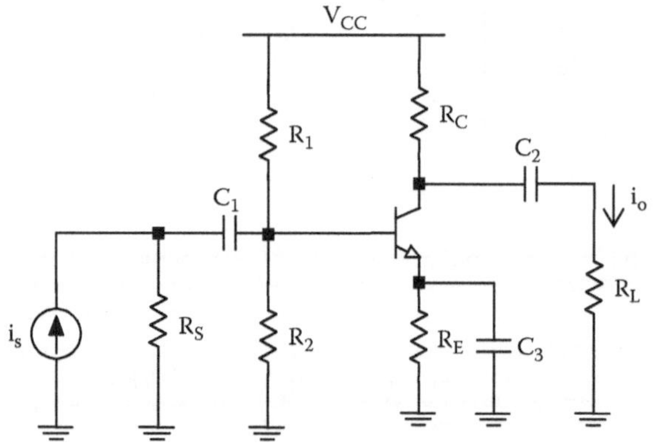

FIGURE EX. 4.13(a)

Solution:

Figure Ex. 4.13(b) is the AC schematic of the amplifier and illustrates how the source signal travels through the amplifier to reach the output. Because of a current division at the input, only a fraction of the input signal (i_s) reaches the transistor's input lead (i_b). The transistor then amplifies this current ($i_c = \beta i_b$), and after a current division at the output, only a fraction of the amplified signal is delivered to the load (i_o):

$$A_{IS} = \frac{i_o}{i_s} = \frac{i_b}{i_s} \cdot \frac{i_c}{i_b} \cdot \frac{i_o}{i_c} = \frac{R_B \| R_S}{(R_B \| R_S) + \beta r_e} . \beta . \frac{-R_C}{R_C + R_L}.$$

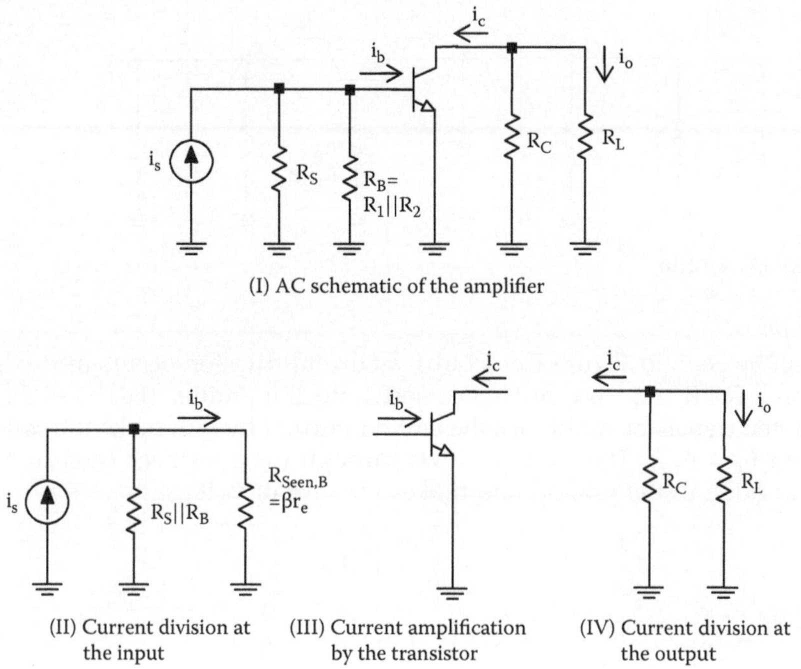

FIGURE EX. 4.13(b)

Example 4.14

Figure Ex. 4.14(a) is a transresistance amplifier with CB configuration. Find the transresistance of the amplifier, and make a general suggestion to improve the gain of transresistance amplifiers with this configuration.

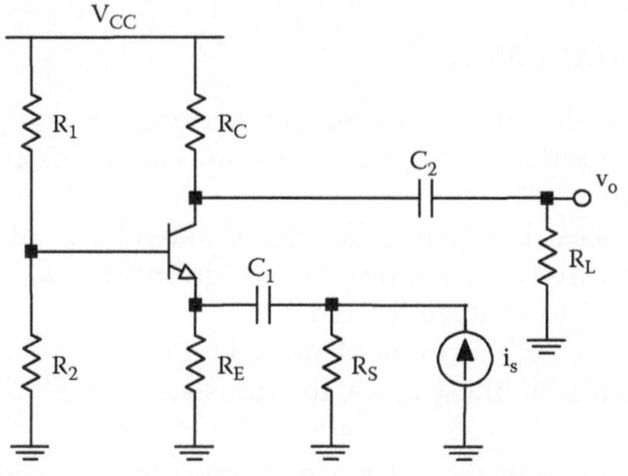

FIGURE EX. 4.14(a)

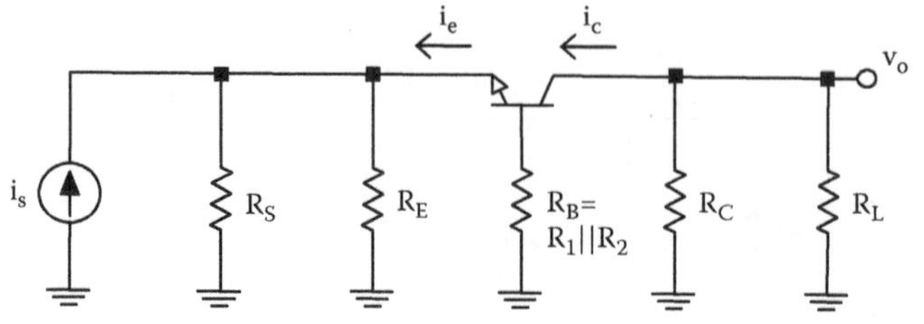

FIGURE EX. 4.14(b)

Solution:

As can be seen in Figure Ex. 4.14(b), a current division occurs at the input between $R_S||R_E$ and the resistance seen into the emitter (i.e., $r_e + (R_B/\beta)$). Then, the transistor multiplies the emitter current by α to make the collector current ($i_c = \alpha . i_e$). This current flows through the resistance seen from the output node to the ground and makes the output voltage:

$$R_{MS} = \frac{v_o}{i_s} = \frac{i_e}{i_s} \cdot \frac{i_c}{i_e} \cdot \frac{v_o}{i_c} = \frac{R_E \| R_S}{\left(R_E \| R_S\right) + \left(r_e + \frac{R_B}{\beta}\right)} . \alpha . \left(-R_C \| R_L\right)$$

A general suggestion to increase the gain of this amplifier is to bypass R_B by adding a bypass capacitor between the base and the ground. This causes the first term in the transresistance to be closer to one and reduces the attenuation at the input. In other words, the addition of the previously mentioned bypass capacitor reduces the input resistance of the amplifier, and enables the amplifier to absorb a larger fraction of the input signal.

4.9 An Important Note

This chapter introduced the inspection method to analyze small-signal amplifiers. It is important to note that, in some cases, this analysis technique cannot be applied.

One of these cases is where a feedback network is placed around the amplifier. In this case, which is not covered in this book, a specific method should be followed to analyze the circuit.

A second case is where a circuit element is connected between two terminals of the transistor.* In these cases, the transistor is replaced with its small-

* Sometimes, these elements are also used to apply feedback on the amplifier.

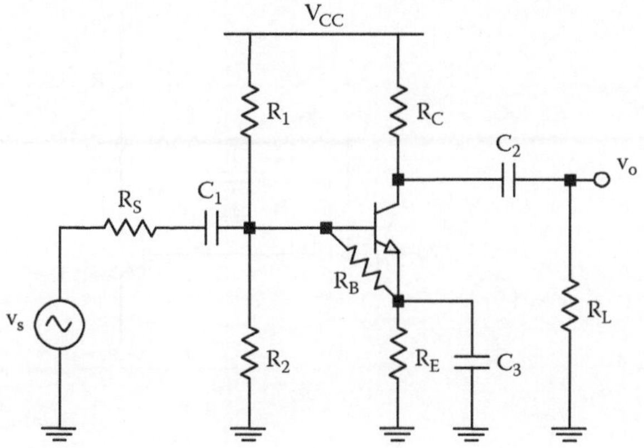

FIGURE 4.47
CE voltage amplifier with an additional resistor connected between the base and the emitter.

signal model, and the circuit is analyzed using the traditional approaches for electric circuit analysis. As an example, Figure 4.47 is a CE amplifier, in which the resistor R_B is connected between the base and the emitter. The purpose of this resistor is to reduce the effect of the transistor's β parameter.

Another example is the amplifier in Figure 4.48, where the resistor R_B is connected between the collector and the base. Sometimes, this resistor is used primarily for DC bias stability, but it can be shown that it also decreases the gain. To keep this resistor across the transistor in DC mode and avoid gain reduction effect in the AC mode, a capacitor is added to the circuit (see Figure 4.49(a)), which results in the desired DC and AC schematics (see Figure 4.49(b) and (c)).

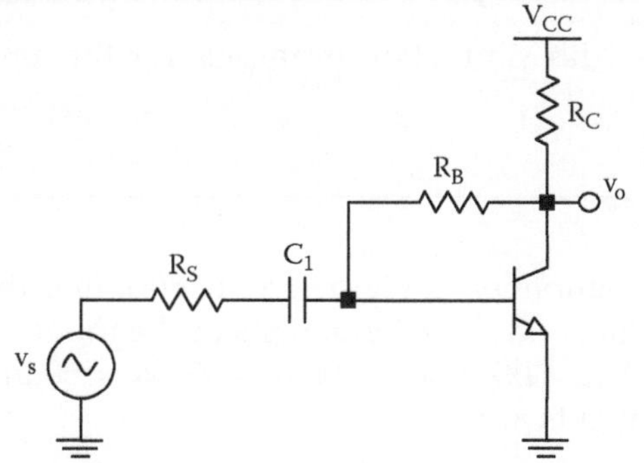

FIGURE 4.48
A CE amplifier with a resistor between the collector and the base for DC biasing.

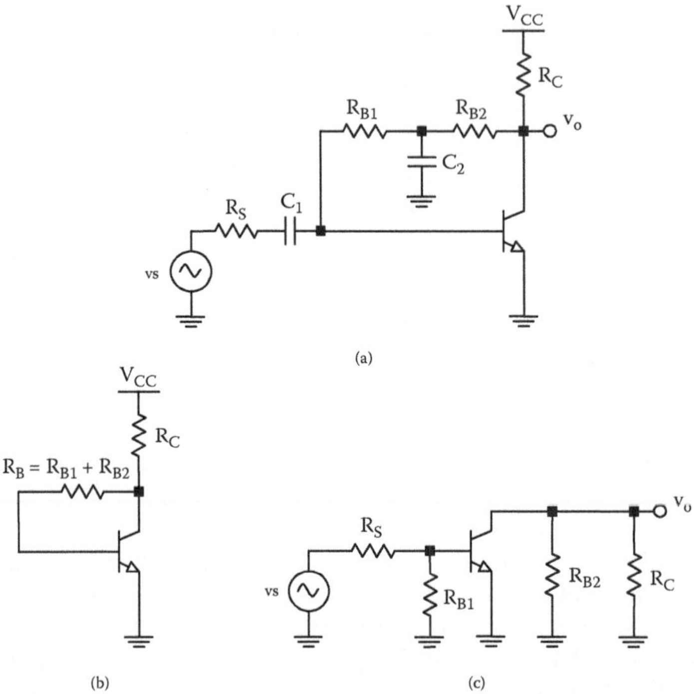

FIGURE 4.49
(a) Adding a capacitor to the amplifier of Figure 4.48, (b) DC schematic, and (c) AC schematic.

4.10 Simulation Examples

This section deals with simulation examples for the amplifiers studied in this chapter.

Example 4.15

For the CS voltage amplifier in Figure Ex. 4.15(a), find the amplifier's pure gain, the overall gain, and the attenuations at the input and output. For the transistor assume $V_{TH} = 0.7$ V, $k' = 200\ \mu A/V^2$, W = 50 μm, and L = 0.5 μm, and the circuit elements are:

$$V_{DD} = 5\text{ V},\ V_{SRC} = 1^{mV}\text{Sin}(2\pi \times 10\text{ kHz} \times t),\ R_{SRC} = 100\ \Omega,\ R_1 = 20\ k\Omega,$$
$$R_2 = 10\ k\Omega,\ R_D = 5\ k\Omega,\ R_S = 1.2\ k\Omega,\ R_L = 100\ k\Omega,\ C_1 = C_2 = C_3 = 1\ \mu F.$$

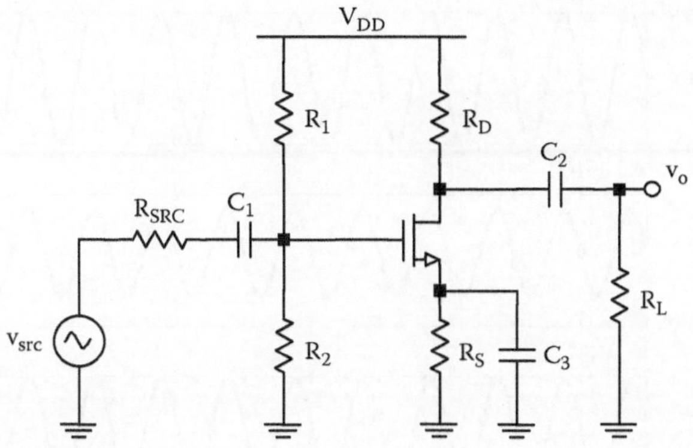

FIGURE EX. 4.15(a)

Solution:

```
M1 D G S S NMOSmodel W = 50u L = 0.5u
RSRC 1 SIG 0.1k
R1 G VDD 20k
R2 0 G 10k
RD D VDD 5k
RS 0 S 1.2k
RL   0 OUT 100k
C1   1 G 1u
C2 D OUT 1u
C3 0 S 1u
VSupply VDD 0 DC 5V
VSignal SIG 0 SIN (0 1m 10k)
.TRAN  0 1ms 0 0.5us
.PROBE
.OP
.MODEL NMOSmodel NMOS(Level = 1 VTO = 0.7 Kp = 0.2e-3)
.END
```

Figure Ex. 4.15(b) illustrates the simulated waveforms. The 2-$mV(p\text{-}p)$ signal applied at the input causes 1.97 $mV(p\text{-}p)$ variations at the gate. Thus, the attenuation at the input is obtained:

$$\frac{v_i}{v_{src}} = \frac{v_g}{v_{src}} = \frac{R_i}{R_i + R_{SRC}} = \frac{1.97^{mV}}{2^{mV}} = 0.985,$$

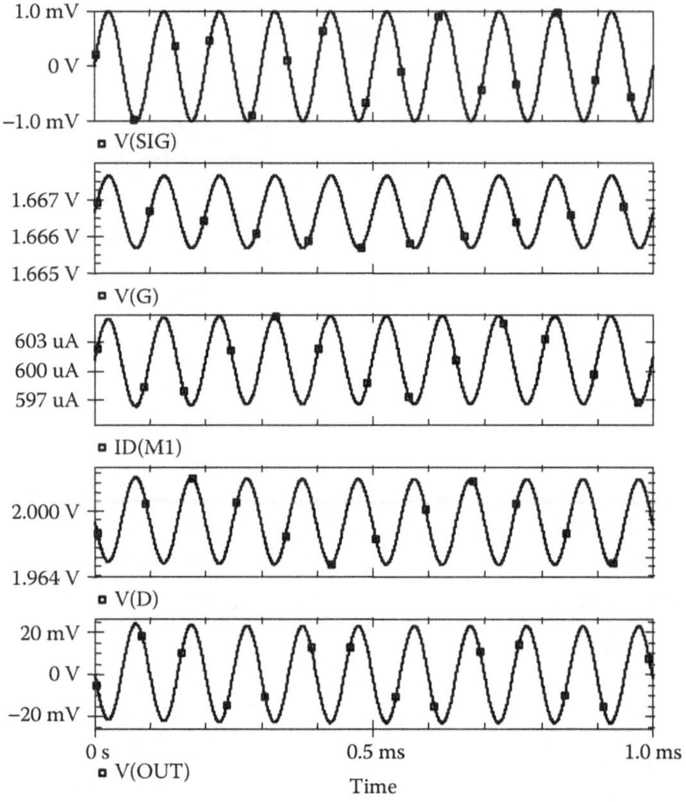

FIGURE EX. 4.15(b)

which can be used to find the input resistance of the amplifier:

$$\frac{R_i}{R_i + R_{SRC}} = \frac{R_i}{R_i + 100^{\Omega}} = 0.985$$

$$\Rightarrow R_i \cong 6.6^{k\Omega}.$$

Peak-to-peak amplitude at the output is measured as $v_{out}(p\text{-}p) = 45.7\ mV$, and thus the overall gain is obtained:

$$|A_{VS}| = \frac{v_o}{v_{src}} = \frac{45.7^{mV}}{2^{mV}} = 22.85.$$

To find the pure gain of the amplifier and the attenuation at the output, the amplifier must be simulated with a different load resistance. The calculations will be easy if the amplifier is simulated with no load resistance. In this case, the output amplitude is measured 47.9 mV. Because in this case there is no load resistance, the amplified signal is not attenuated at the output, and thus the ratio of the output voltage to the input voltage at the gate gives the pure gain of the amplifier:

$$|A_V| = \frac{v_o(No\ Load)}{v_g} = \frac{47.9^{mV}}{1.97^{mV}} = 24.3.$$

Observing that removing the load resistance causes the output voltage to increase from 45.7 mV to 47.9 mV, the attenuation at the output is easily found:

$$\frac{R_L}{R_L + R_o} = \frac{v_o(Loaded)}{v_o(Unloaded)} = \frac{45.7^{mV}}{47.9^{mV}} = 0.95.$$

Now, knowing the value of R_L the output resistance is achieved:

$$\frac{R_L}{R_L + R_o} = \frac{100^{k\Omega}}{100^{k\Omega} + R_o} = 0.95$$

$$\Rightarrow R_o \cong 5.3^{k\Omega}.$$

Example 4.16

Determine the input and output resistances of the source follower in Figure Ex. 4.16(a). For the transistor assume V_{TH} = 0.7 V, k' = 200 $\mu A/V^2$, W = 50 μm, and L = 0.5 μm, and the circuit elements are:

$$V_{DD} = 5\text{ V},\ V_{SRC} = 100^{mV}\text{Sin}(2\pi \times 10\text{ kHz} \times t),\ R_{SRC} = 100\ \Omega,\ R_1 = 20\ k\Omega,$$
$$R_2 = 10\ k\Omega,\ R_S = 100\ \Omega,\ R_L = 10\ k\Omega,\ C_1 = C_2 = 1\ \mu F.$$

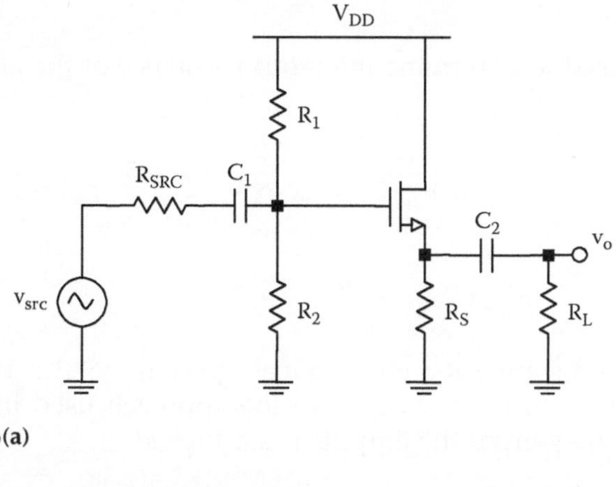

FIGURE EX. 4.16(a)

Solution:

```
M1 VDD G S S NMOSmodel W = 50u L = 0.5u
RSRC 1 SIG 0.1k
R1 G VDD 20k
R2 0 G 10k
RS 0 S 1.2k
RL 0 OUT 10k
C1 1 G   1u
C2 S OUT 1u
VSupply VDD 0 DC 5V
VSignal SIG 0 SIN (0 100m 10k)
.TRAN 0 500us 0 0.5us
.PROBE
.OP
.MODEL NMOSmodel NMOS(Level = 1 VTO = 0.7 Kp = 0.2e-3)
.END
```

Applying an input signal with a 200-mV peak-to-peak amplitude causes 165.5-mV peak-to-peak variations at the output as demonstrated in Figure Ex. 4.16(b). Therefore, the overall gain is:

$$A_{VS} = \frac{v_o}{v_{src}} = \frac{165.5^{mV}}{200^{mV}} = 0.83.$$

Measuring 197 $mV(p\text{-}p)$ variations at the gate, the attenuation at the input is obtained:

$$\frac{v_i}{v_{src}} = \frac{v_g}{v_{src}} = \frac{R_i}{R_i + R_{SRC}} = \frac{197^{mV}}{200^{mV}} = 0.985,$$

which can be used to determine the input resistance of the amplifier:

$$\frac{R_i}{R_i + R_{SRC}} = \frac{R_i}{R_i + 100^{\Omega}} = 0.985$$

$$\Rightarrow R_i \cong 6.6^{k\Omega}.$$

Running a second simulation with no load gives us 168.1 mV peak-to-peak amplitude at the output. According to the approach used in the previous example, the pure gain of the amplifier is achieved:

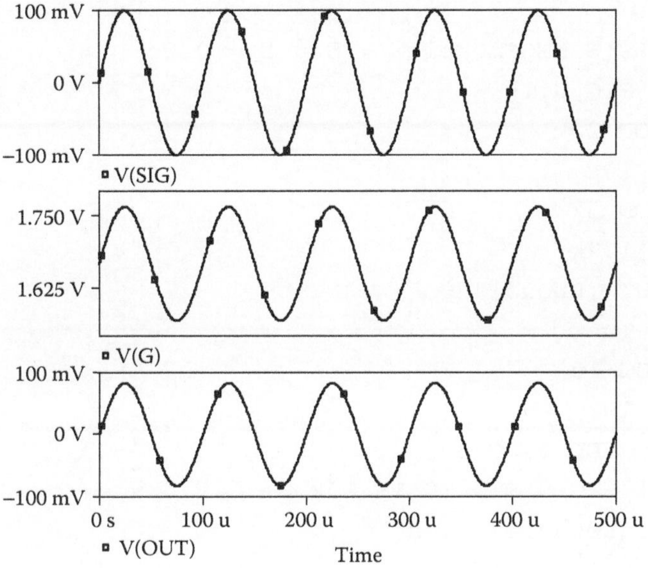

FIGURE EX. 4.16(b)

$$A_V = \frac{v_o(No\ Load)}{v_g} = \frac{168.1^{mV}}{197^{mV}} = 0.85,$$

and the attenuation at the output is found:

$$\frac{R_L}{R_L + R_o} = \frac{v_o(Loaded)}{v_o(Unloaded)} = \frac{165.5^{mV}}{168.1^{mV}} = 0.98.$$

Finally, knowing the value of R_L the output resistance is obtained:

$$\frac{R_L}{R_L + R_o} = \frac{10^{k\Omega}}{10^{k\Omega} + R_o} = 0.98$$

$$\Rightarrow R_o \cong 200^{\Omega}.$$

Example 4.17

For the amplifier studied in Example 4.15,

(a) Observe the loading effect by reducing the load resistance from $100\ k\Omega$ to $10\ k\Omega, 1\ k\Omega$, and $100\ \Omega$.

(b) Determine the input resistance.

Solution:

```
M1 D G S S NMOSmodel W = 50u L = 0.5u
RSRC 1 SIG 0.1k
R1 G VDD 20k
R2 0 G 10k
RD D VDD 5k
RS 0 S 1.2k
RL 0 OUT {RLoad}
C1 1 G 1u
C2 D OUT 1u
C3 0 S 1u
VSupply VDD 0 DC 5V
VSignal SIG 0 SIN (0 1m 10k)
.TRAN 0 500us 0 0.1us
.PROBE
.OP
.PARAM RLoad = 100k
.STEP PARAM RLoad List 100 1k 10k 100k
.MODEL NMOSmodel NMOS(Level = 1 VTO = 0.7 Kp = 0.2e-3)
.END
```

(a) The output voltage traces for different load resistance values are given in Figure Ex. 4.17.

(b) As an alternative method to determine the input resistance of the amplifier, one can divide the AC voltage at the input node of the amplifier (i.e., the gate) by the AC current that is fed into the amplifier by the signal source. In this example, the input resistance is achieved by dividing the gate voltage by the current flowing through C_1:

$$i_{c1}(p\text{-}p) = 0.295\ \mu A,\ v_g(p\text{-}p) = 1.97\ mV$$

$$\Rightarrow R_i = \frac{1.97^{mV}}{0.295^{\mu A}} \cong 6.7^{k\Omega}.$$

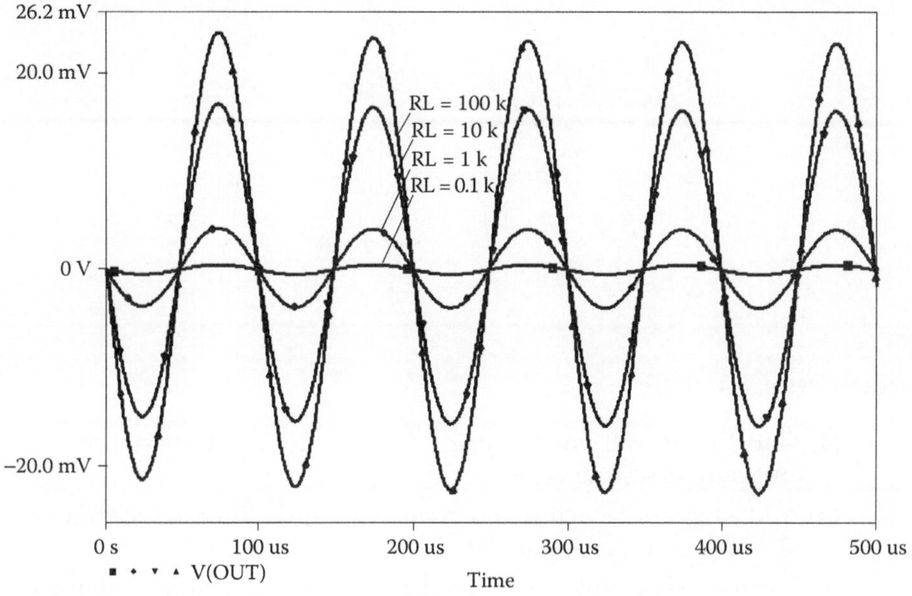

FIGURE EX. 4.17

4.11 Problems

4.1 A bipolar transistor with V_A = 100 V is biased in active mode at I_C = 1 mA. Find the values of g_m and r_o.

4.2 For an MOS transistor with V_{TH} = 0.7 V and k' = 100 $\mu A/V^2$, λ = 0.01 V^{-1}, W = 10 μm, and L = 0.25 μm, find g_m and r_o if it is biased in saturation mode with I_D = 1 mA.

4.3 *Intrinsic gain* of a transistor is defined as its $g_m r_o$ product. This is a measure of the maximum possible voltage gain that the transistor can exhibit if placed in an ideal amplification configuration. Determine the intrinsic gain for an MOS transistor, and compare it with that of a bipolar transistor.

4.4 A common-emitter bipolar amplifier is given in Figure Prob. 4.4, in which the input signal source, v_s, is directly coupled to the amplifier. Assuming that the internal resistance of the input signal source (R_S) is 100 Ω, and for the transistor we have $V_{BE}(ON)$ = 0.7 V, $V_{CE}(Sat.)$ = 0.1 V, V_A = 100 V, and β = 100,

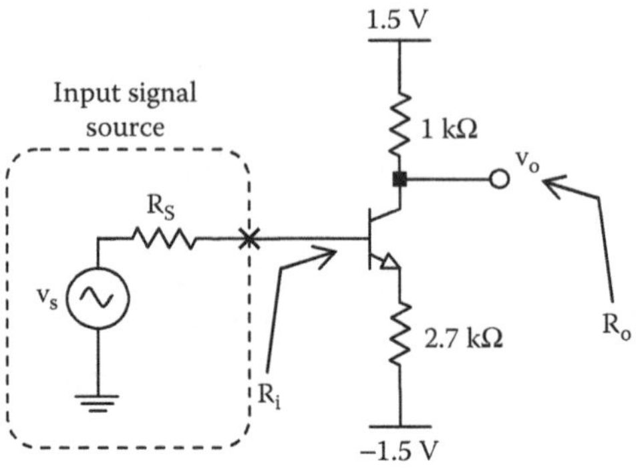

FIGURE PROB. 4.4

(a) Find the overall voltage gain, $A_{VS} = v_o/v_s$, input resistance, R_i, and output resistance, R_o.

(b) Add a bypass capacitor in parallel with the resistor in the emitter (2.7 *k*Ω), and repeat part (a).

(c) Capacitively couple a 1-kΩ load resistance to the circuit obtained in part (b), and recalculate the overall gain and input and output resistances.

4.5 A common-emitter amplifier is given Figure Prob. 4.5, in which both the input signal source and the load (R_L) are capacitively coupled to the amplifier. Assuming $V_{BE}(ON) = 0.7$ V, $V_{CE}(Sat.) = 0.1$ V, $V_A = 100$ V, and $\beta = 100$, for the transistor, and that C_{1-3} are large enough that can be assumed short circuits in AC analysis,

(a) What are the recommended ranges for R_S and R_L to have negligible signal attenuations when coupling the signal to/from the amplifier?

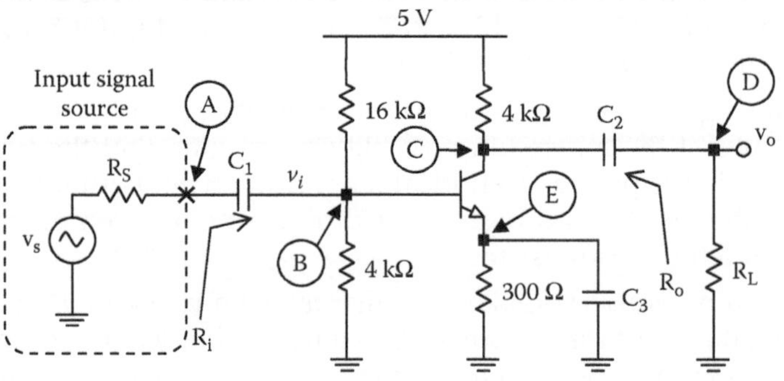

FIGURE PROB. 4.5

(b) Let $R_S = 100\ \Omega$, and $R_L = 10\ k\Omega$, and find the amplifier's gain, $A_V = v_o/v_i$, and the overall voltage gain, $A_{VS} = v_o/v_s$.

(c) Assuming $v_s(t) = 1^{mV}\ Sin(2\pi ft)$, find the whole voltages (including both DC and AC components) at nodes A–E pictured on the schematic.

4.6 For the circuit analyzed in Problem 4.5, what is the maximum input amplitude for which the amplifier is still in the active mode?

4.7 The same circuit components in Figure Prob. 4.5 are rearranged in Figure Prob. 4.7(a) and (b) to realize emitter follower and common-base configurations. Repeat Problem 4.5 for the circuits in Figure Prob. 4.7(a) and (b), and compare the results.

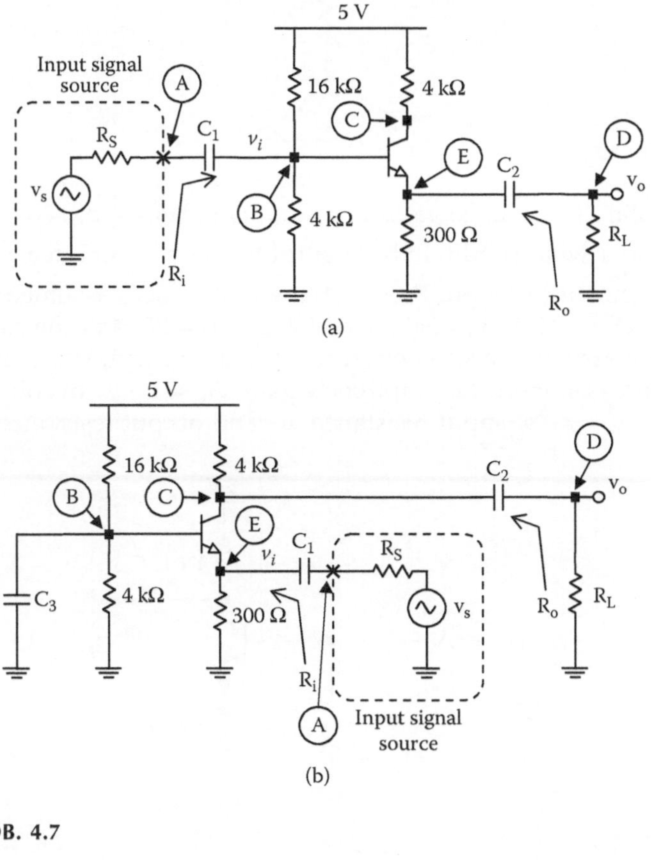

FIGURE PROB. 4.7

4.8 The amplifier in Figure Prob. 4.8 has two outputs. It performs as an emitter-follower to provide v_{o1}, and acts as a common-emitter amplifier to prepare v_{o2}. This circuit is sometimes used to convert the input signal to two signals at the output with the same amplitude and 180° out of phase.

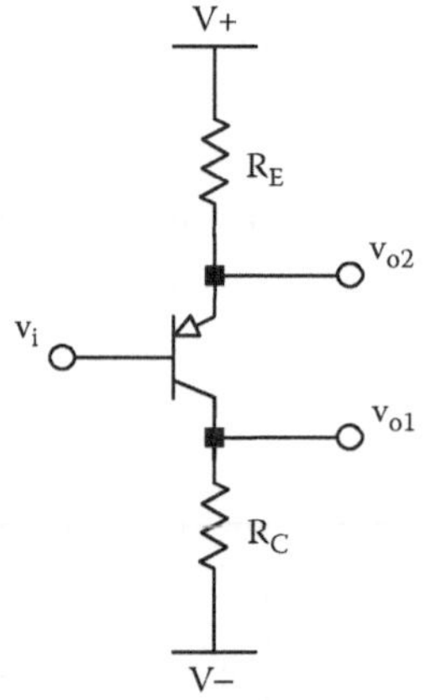

FIGURE PROB. 4.8

(a) Obtain the voltage gains $A_{V1} = v_{o1}/v_i$ and $A_{V2} = v_{o2}/v_i$.

(b) Under what condition both outputs will be equal in amplitude?

4.9 In the circuit in Figure Prob. 4.9, the transistor is assumed to have $V_{BE}(ON) = -0.7$ V, $V_{CE}(Sat.) = -0.1$V, and $\beta = 70$, and the capacitors C_{1-3} are assumed large enough to be replaced with short circuits in AC analysis. Find the amplifier's gain, $A_V = v_o/v_i$, overall voltage gain, $A_{VS} = v_o/v_s$, input resistance, R_i, and output resistance, R_o.

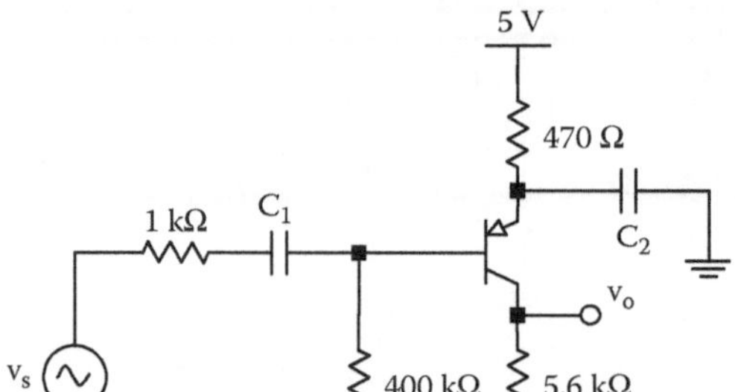

FIGURE PROB. 4.9

4.10 A common-gate amplifier is shown in Figure Prob. 4.10. Assuming $k' = 100\ \mu A/V^2$, $(W/L) = 40$, $V_{TH} = 0.7V$, and $\lambda = 0.01\ V^{-1}$ for the transistor,

(a) Find the amplifier's gain, $A_V = v_o/v_i$, overall voltage gain, $A_{VS} = v_o/v_{sig}$, input resistance, R_i, and output resistance, R_o.

(b) Find the intrinsic gain (introduced in Problem 4.3) for the transistor in this example.

(c) Discuss on the reasons for not achieving a voltage gain close to the intrinsic gain calculated in part (b), and try to increase the gain with minor modifications in the circuit still with the same transistor at the same bias point.

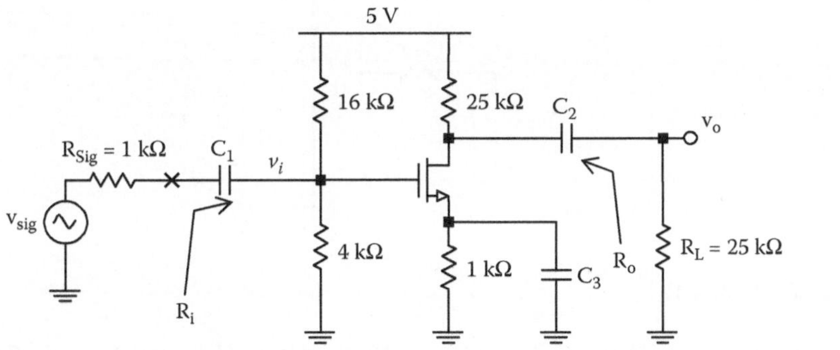

FIGURE PROB. 4.10

4.11 Find the AC equivalent circuits for bipolar and MOS diode-connected configurations, pictured in Figure Prob. 4.11, by replacing the transistor with its small-signal model.

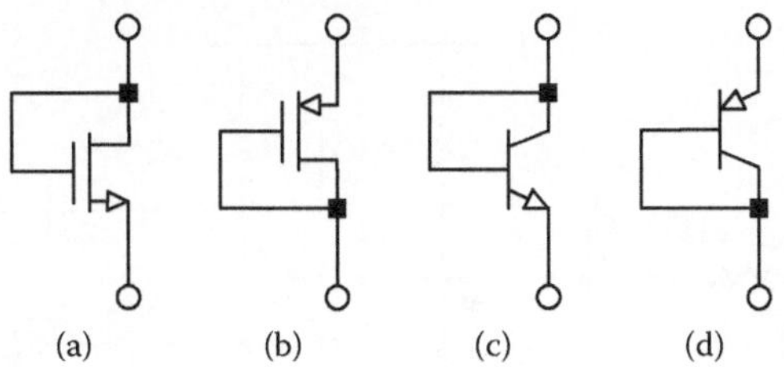

FIGURE PROB. 4.11

4.12 Figure Prob. 4.12 is a common-source amplifier. For the transistor, assume that $k' = 120\ \mu A/V^2$, $(W/L) = 50$, $\lambda = 0.01\ V^{-1}$, and $V_{TH} = 0.8$ V.

(a) Find the output, $v_o(t)$, if $v_{sig}(t) = 1^{mV}\text{Sin}(2\pi f_s t)$ is applied at the input.

(b) What happens to the output if V_{DD} is prepared by a regulated power supply with 1-mV(peak) sinusoidal ripple (i.e., $V_{DD} = 1.5^V + 1^{mV}\text{Sin}(2\pi f_r t)$)?

(c) What if V_{SS} includes the same AC contents (i.e., $V_{SS} = -1.5^V + 1^{mV}\text{Sin}(2\pi f_r t)$)?

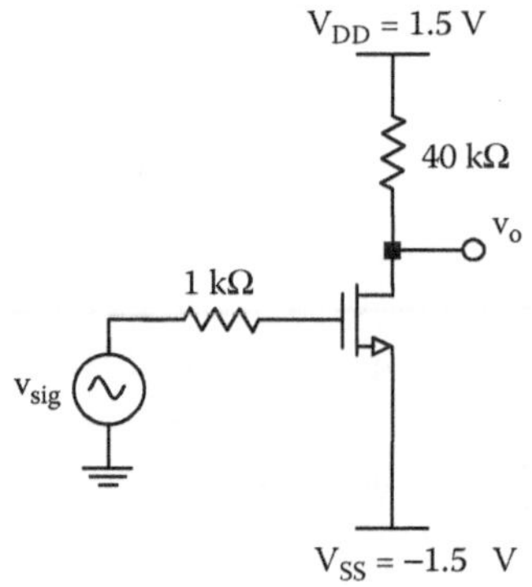

FIGURE PROB. 4.12

4.13 Given $k' = 70\ \mu A/V^2$, $(W/L) = 20$, $\lambda = 0.02\ V^{-1}$, and $V_{TH} = -0.8$ V, find the input and output resistances, transconductance

$$G_M = \frac{i_o}{v_i}\bigg|_{v_o=0},$$

and the overall transconductance $G_{MS} = i_o/v_{sig}$ for the circuit in Figure Prob. 4.13.

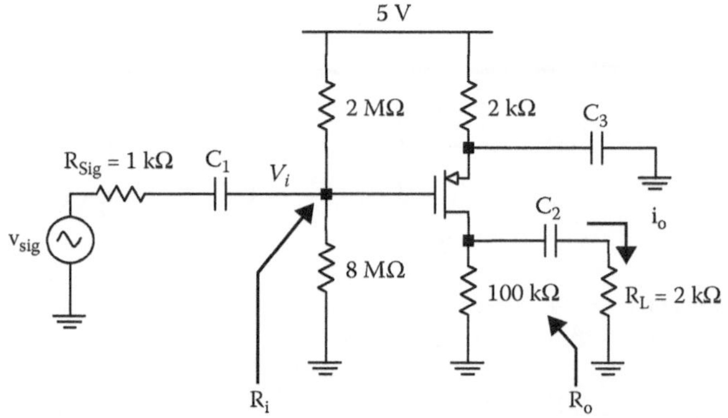

FIGURE PROB. 4.13

4.14 Figure Prob. 4.14 is a PNP transistor in common-base configuration. Given $V_{EB}(ON) = 0.7V$, $V_{EC}(Sat.) = 0.1$ V, $V_A = 80$ V, and $\beta = 50$,

(a) Determine the input and output resistances R_i and R_o.

(b) Calculate the voltage gain $A_V = v_o/v_i$, current gain $A_I = i_o/i_i$, transconductance $G_M = i_o/v_i$, and transresistance $R_M = v_o/i_i$.

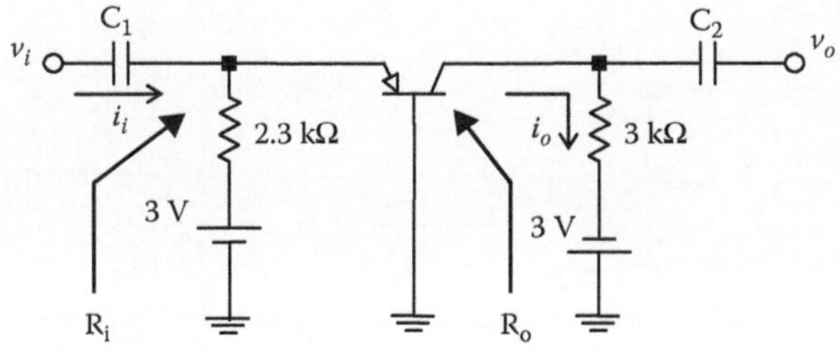

FIGURE PROB. 4.14

Chapter 5

MultiStage Amplifiers

5.1 Introduction

When designing an amplifier, as studied in the previous chapter, several requirements must be met simultaneously. In a real amplifier design, sometimes the desired specifications are defined such that no single-stage configuration can be introduced as a good candidate. For example, none of the single-stage configurations studied so far can fulfill the requirements for a high-performance voltage amplifier. These requirements are high input resistance, low output resistance, and high voltage gain, as studied previously. In such cases, designing a *multistage amplifier* would be the solution. In multistage amplifier design, input and output requirements are usually fulfilled by the input and output stages, respectively. If more gain is needed than provided by these two stages, then additional stages can be inserted.

This chapter deals with the analysis of multistage amplifiers and discusses some important considerations.

5.2 Biasing/Coupling

Perhaps the easiest and most straightforward way to design a multistage amplifier is to individually design each stage, and then capacitively couple them together. This way, each stage has its own bias network, and the AC signal will be transferred from one stage to another via the coupling capacitors. As an example, Figure 5.1(a) is a two-stage amplifier that consists of a common-source (CS) stage for high input resistance and high gain, followed by a source follower (SF) stage for low output resistance. In this circuit, the first coupling capacitor, C_1, couples the AC component of the input signal to the first stage. The first stage amplifies the input signal. Then, the coupling capacitor, C_2, rejects the DC component at the output of the first stage and delivers it to the second stage.

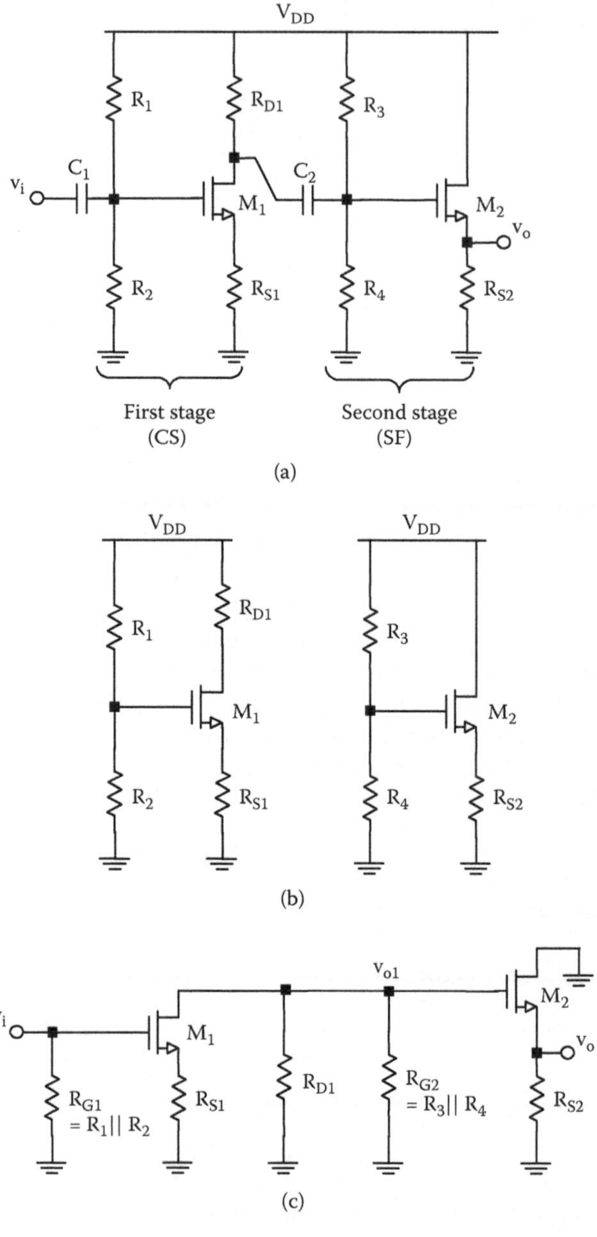

FIGURE 5.1
(a) A two-stage amplifier with capacitive coupling (b) DC schematic (c) AC schematic.

In DC analysis, the capacitors are treated as open circuits, disconnecting the stages from each other. This splits the large multistage circuit into single stages, as illustrated in Figure 5.1(b). In AC mode, the coupling capacitors are replaced with short circuits, connecting the cascaded stages together and carrying the signal from one stage to another (see Figure 5.1(c)).

Some of the major drawbacks for this method of multistage amplifier design are listed as follows:

- Per-stage bias networks increase circuit complexity and size.
- The DC currents that the bias networks draw from the power supply increase the power dissipation of the amplifier.
- For integrated circuits, the implementation of typical coupling capacitors, usually on the order of microfarads and even nanofarads, is not feasible.

In the capacitively coupled, multistage amplifiers represented by the circuit in Figure 5.1(a), the AC signal at the output of one stage rides on a DC component. The coupling capacitor rejects this DC component from the signal and superposes the pure AC signal on the DC voltage of the input node of the next stage. Why isn't the whole voltage, including both DC and AC components, at the output of the first stage directly coupled to the next stage? The answer is that the DC component of the first stage then disturbs the biasing of the second stage. As an example, in Figure 5.1(a), R_3 and R_4 produce the gate DC voltage required to bias M_2. Now, one may ask: Why not use the DC voltage at the drain of M_1 to bias M_2? This way, as presented in Figure 5.2, the coupling capacitor C_2, R_3, and R_4 will no longer be needed, no DC current will be drawn by R_3 and R_4 anymore, and the signal amplified by the first stage is directly coupled to the second stage. *Direct coupling* is

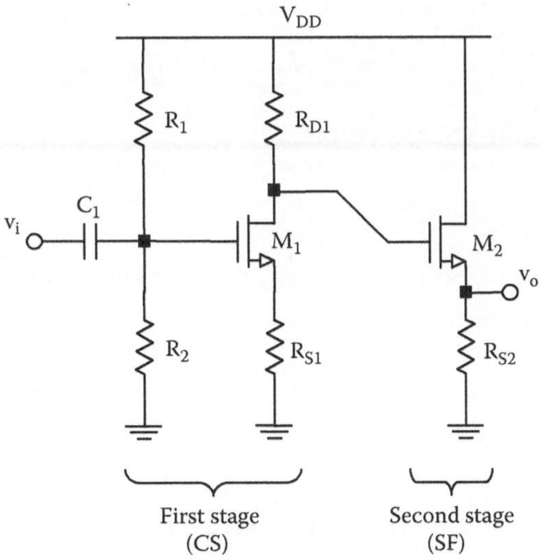

FIGURE 5.2
A two-stage amplifier with direct coupling.

indeed, the second approach of multistage amplifier design, which usually enables one to design the intended amplifier with higher quality in terms of the circuit size, power dissipation, frequency bandwidth, etc.

To analyze a directly coupled multistage amplifier in DC mode, one should usually start from the first stage. Then, using the currents and voltages that are obtained, the following stages can be analyzed; however, sometimes the next stages impose some constraints on, or even determine the bias point of the preceding stages. In the DC analysis and design of directly coupled multistage amplifiers, the interrelationship between the bias points of the consecutive stages makes the design procedure relatively more complicated.

Example 5.1

Assuming $V_{TH} = 0.5\ V$, and $\beta = 20\ mA/V^2$ for the transistors, and $V_{DD} = 5\ V$, $R_1 = 60\ k\Omega$, $R_2 = 90\ k\Omega$, $R_{D1} = 1\ k\Omega$, $R_{S1} = 1\ k\Omega$, and $R_{S2} = 1\ k\Omega$, find the operating points of the transistors in the amplifier of Figure 5.2.

Solution:

Starting with the first stage, which is biased by a voltage-divider bias network, we have:

$$V_{G1} = \frac{R_2}{R_1 + R_2}.V_{DD} = \frac{90k}{60k + 90k} \times 5V = 3V,$$

and the source voltage is:

$$V_{S1} = R_{S1}\, I_{D1} = (1^{k\Omega})\ I_{D1}.$$

So, replacing the gate-source voltage:

$$V_{GS1} = V_{G1} - V_{S1} = 3V - (1^{k\Omega})\ I_{D1}$$

in the quadratic equation for the drain current we have:

$$I_{D1} = \frac{\beta}{2}(V_{GS1} - V_{TH})^2 = \frac{20mA / V^2}{2}(3V - (1k)I_{D1} - 0.5)^2,$$

and

$$I_{D1}^2 - 5.1I_{D1} + 6.25 = 0$$

is the equation that should be solved for I_{D1}. Although 2 mA and 3 mA are two roots for this quadratic equation, only 2 mA is acceptable as it returns a reasonable gate-source voltage larger than the threshold voltage. Thus, the drain voltage will be:

$$V_{D1} = V_{DD} - R_{D1}I_{D1} = 5\ V - 2\ V = 3\ V.$$

which is also the gate voltage for M_2. This simplifies finding the drain current of M_2 as the transistors are the identical, their gate voltages are the same, and also $R_{S1} = R_{S2}$. Therefore, it can be said that the drain currents of the transistors are the same:

$$I_{D2} = I_{D1} = 2\ mA.$$

Example 5.2

Given $|V_{BE}(ON)| = 0.7\ V$, $|V_{CE}(Sat.)| = 0.1\ V$, $\beta_{NPN} = 100$, and $\beta_{PNP} = 70$ for the transistors, and $V_{CC} = 5\ V$, $R_1 = 32\ k\Omega$, $R_2 = 8\ k\Omega$, $R_{C1} = 4\ k\Omega$, $R_{E1} = 300\ \Omega$, $R_{E2} = 1\ k\Omega$, and $R_{C2} = 700\ \Omega$, find the operating points of the transistors in the two-stage amplifier depicted in Figure Ex. 5.2(a).

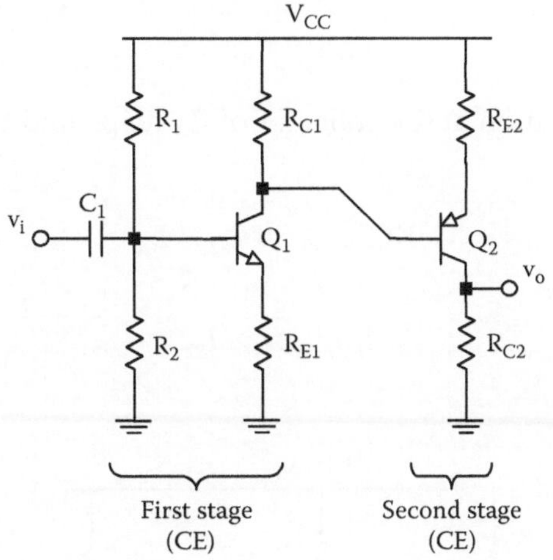

FIGURE EX. 5.2(a)

Solution:

The first stage has a voltage-divider bias network to bias Q_1, which sets its collector current as:

$$I_{C1} = \frac{V_{BB1} - V_{BE}(ON)}{R_{E1} + \dfrac{R_{B1}}{\beta_{NPN}}},$$

where

$$V_{BB1} = \frac{R_2}{R_1 + R_2} . V_{CC} = \frac{8k}{32k + 8k} \times 5V = 1V$$

and

$$R_{B1} = R_1 \| R_2 = 32k \| 8k = 6.4k\Omega.$$

Replacing the parameters with their values, the collector current of Q_1 is:

$$I_{C1} = \frac{1V - 0.7V}{0.3k + \frac{6.4k}{100}} = \frac{0.3V}{0.364k} = 0.82mA$$

As presented in Figure Ex. 5.2(b), the base voltage required for Q_2 to be biased in the active mode is indeed the voltage at the collector of Q_1, V_{C1}. It can also be said that the voltage across R_{C1}, V_{RC1}, is applied between the upper end of R_{E2} and the base of Q_2, which can be written as:

$$V_{RC1} = R_{C1} . I_{RC1}$$

Writing a KCL equation at the collector of Q_1 helps find the current through R_{C1}, I_{RC1}, as:

$$\text{KCL: } I_{RC1} + I_{B2} = I_{C1}$$

or

$$I_{RC1} = I_{C1} - I_{B2}.$$

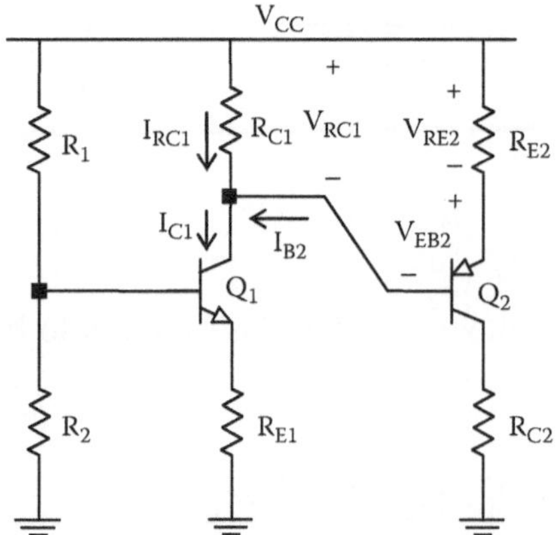

FIGURE EX. 5.2(b)

A simplifying assumption at this point is that the base current of Q_2 is much smaller than the collector current of Q_1, which gives:

$$I_{B2} << I_{C1} \Rightarrow I_{RC1} \cong I_{C1}.$$

The truth of this assumption will be examined later when I_{B2} is determined. As indicated, the voltage across R_{C1}, V_{RC1}, appears across the series combination of R_{E2} and the emitter-base of Q_2 and thus:

$$V_{RC1} = V_{RE2} + V_{EB2} = R_{E2}.I_{E2} + V_{EB}(ON).$$

This equation can be solved for I_{E2} as:

$$I_{E2} = \frac{V_{RC1} - V_{EB}(ON)}{R_{E2}} \cong \frac{R_{C1}I_{RC1} - V_{EB}(ON)}{R_{E2}} = \frac{3.28V - 0.7V}{1k} = 2.58mA,$$

which gives:

$$I_{C2} = \frac{\beta_{PNP}}{\beta_{PNP} + 1} I_{E2} = \frac{70}{70+1} \times 2.58mA = 2.54mA$$

and

$$I_{B2} = \frac{1}{\beta_{PNP} + 1} I_{E2} = \frac{1}{70+1} \times 2.58mA = 36\mu A.$$

With this small amount of I_{B2}, which is obviously much smaller than I_{C1}, the truth of the aforementioned simplifying assumption is verified.

The last step is to make sure that the transistors are in the active mode by finding

$$\begin{aligned} V_{CE1} &= V_{CC} - (I_{RC1}.R_{C1} + I_{E1}.R_{E1}) \\ &\cong V_{CC} - I_{C1}(R_{C1} + R_{E1}) \\ &= 5\ V - 0.82\ mA(4\ k + 0.3\ k) \\ &= 1.47\ V \end{aligned}$$

and

$$\begin{aligned} V_{EC2} &= V_{CC} - (I_{RE2}.R_{E2} + I_{C2}.R_{C2}) \\ &\cong V_{CC} - I_{C2}(R_{C2} + R_{E2}) \\ &= 5\ V - 2.54\ mA(0.7\ k + 1k) \\ &= 0.68\ V. \end{aligned}$$

Because both V_{CE1} and V_{EC2} are greater than $|V_{CE}(Sat.)| = 0.1\ V$, it is confirmed that both transistors are in the active mode.

5.3 AC Analysis

To analyze a multistage amplifier in the AC mode, its primitive single-stage amplifier stages should be identified first. This helps thinking of the large and somewhat complicated multistage circuit as a combination of small and simple circuits, as illustrated in Figure 5.3. The AC analysis of a multistage amplifier is comprised of two major parts: finding the required specifications of the primitive stages, and properly combining them to obtain the specifications of the whole amplifier. The primitive amplifier stages are analyzed in the AC mode based upon the same techniques described earlier for single-stage amplifiers.

The overall gain of an N-stage amplifier, illustrated in Figure 5.3, is the product of the gains of the primitive stages, provided that the loading effect of each stage on the preceding stage is taken into account:

$$A_{Overall} = \prod_{i=1}^{N} A_i^* \tag{5.1}$$

where A_i^* is the gain of the i-th stage when loaded by the next stage (or by the load for $i = N$). For voltage amplifiers and transresistance amplifiers where the output signal is a voltage, the loaded gain of the i-th stage becomes:

$$A_i^* = \frac{R_{L,i}}{R_{L,i} + R_{o,i}} .A_i, \tag{5.2}$$

where A_i, $R_{o,i}$, and $R_{L,i}$ are the unloaded gain, output resistance, and the load resistance for the i-th stage, respectively. $R_{L,i}$ can be the input resistance of the next stage ($R_{in,i+1}$) or the amplifier's output load (R_L) if $i = N$. For current

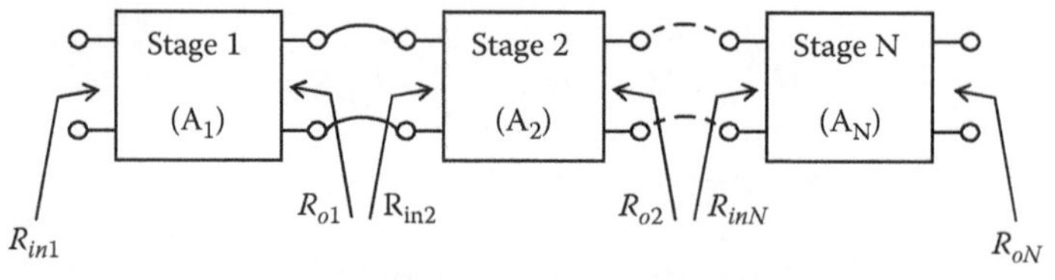

FIGURE 5.3
Block representation of a multistage amplifier comprised of primitive stages.

amplifiers and transconductance amplifiers where the output signal is a current, the loaded gain of the *i*-th stage becomes:

$$A_i^* = \frac{R_{o,i}}{R_{L,i} + R_{o,i}}.A_i. \tag{5.3}$$

One can always use Equation (5.2) and Equation (5.3) to calculate the overall gain using the unloaded gains and attenuations at the outputs of the primitive stages, but it is usually preferred to directly find the loaded gain, A_i^*, for each stage. This will be seen in the following examples.

The input resistance of a multistage amplifier is usually determined by the first stage. In some cases, however, when looking into the input port of the first stage to determine the amplifier's input resistance, the input resistance of the second stage is also seen. Similarly, the output resistance for a multi-stage amplifier is usually determined by the last stage unless it is somehow transparent (i.e., the preceding stage also contributes to the output resistance).

Example 5.3

Find the input and output resistances and the overall gain for the two-stage amplifier in Figure 5.1.

Solution:

The AC schematic of the amplifier is given in Figure Ex. 5.3 for reference convenience.

Notice that the output of the first stage is indeed the input signal for the second stage ($v_{i2} = v_{o1}$). The overall gain is written as:

$$A_V = \frac{v_o}{v_i} = A_{v1}.A_{v2} = \frac{v_{o1}}{v_i}.\frac{v_o}{v_{o1}}.$$

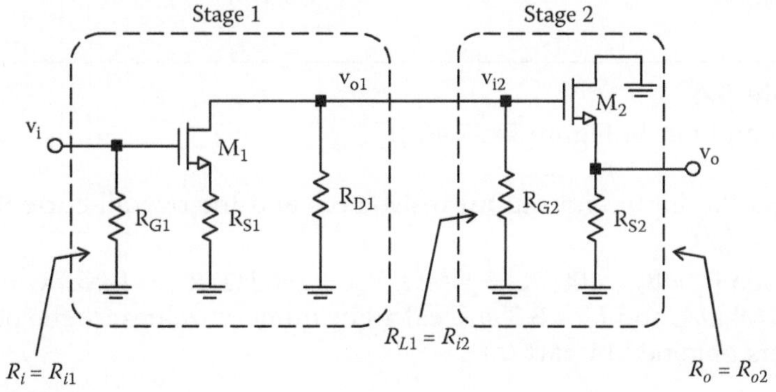

FIGURE EX. 5.3

The first stage is a common-source stage, loaded by the input resistance of the second stage. If the output resistance of the transistor, r_{o1}, is much larger than R_{D1}, loaded gain of the first stage is given as:

$$\frac{v_{o1}}{v_i} = \frac{-R_{D1} \| R_{i2}}{R_{S1} + \dfrac{1}{g_{m1}}} = \frac{-R_{D1} \| R_{G2}}{R_{S1} + \dfrac{1}{g_{m1}}}.$$

Gain of the second stage, which is a source follower, is:

$$\frac{v_o}{v_{o1}} = \frac{R_{S2}}{R_{S2} + \dfrac{1}{g_{m2}}},$$

and the overall gain becomes:

$$A_V = \frac{v_o}{v_i} = \frac{v_{o1}}{v_i} \cdot \frac{v_o}{v_{o1}} = \frac{-R_{D1} \| R_{G2}}{R_{S1} + \dfrac{1}{g_{m1}}} \cdot \frac{R_{S2}}{R_{S2} + \dfrac{1}{g_{m2}}}.$$

As depicted in Figure Ex. 5.3, the input resistance of the amplifier is the same as that of the first stage:

$$R_i = R_{i1} = R_{G1},$$

and its output resistance is determined by the second stage:

$$R_o = R_{o2} = R_{S2} \| (1/g_{m2}).$$

Example 5.4

For the amplifier in Figure Ex. 5.4(a),

(a) Find the input and output resistances and the overall current gain $A_{IS} = i_o/i_s$.

(b) Given $\beta_1 = \beta_2 = 100$, $R_S = 1\ M\Omega$, $R_{B1} = 150\ k\Omega$, $R_{E1} = 1\ k\Omega$, $R_L = 1\ k\Omega$, $I_{C1} = 4\ mA$, and $I_{C2} = 3.3\ mA$, calculate numeric values for the parameters obtained in part (a).

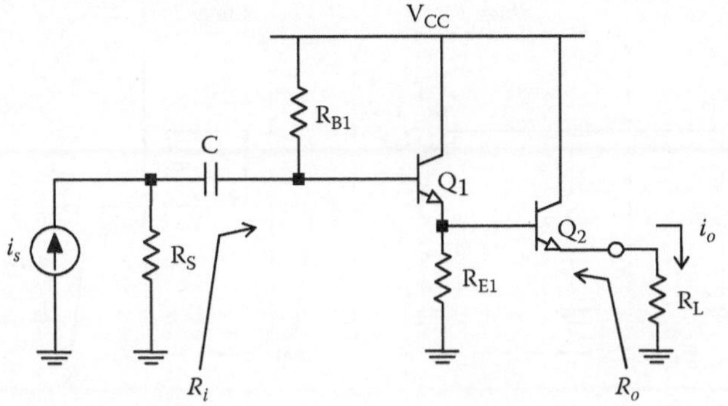

FIGURE EX. 5.4(a)

Solution:

Figure Ex. 5.4(b) illustrates the AC schematic of the amplifier. This is an example of the cases where the input and output resistances are not determined only by the input and output stages, respectively.

The input resistance, R_i, is obtained by inspection as:

$$R_i = R_{B1} \| \beta_1[r_{e1} + (R_{E1} \| R_{L1})]$$

where

$$R_{L1} = \beta_2(r_{e2} + R_L).$$

To find the output resistance, the input signal source, i_s, is replaced with an open circuit as depicted in Figure Ex. 5.4(c), and we have:

$$R_o = r_{e2} + \frac{1}{\beta_2}(R_{o1})$$

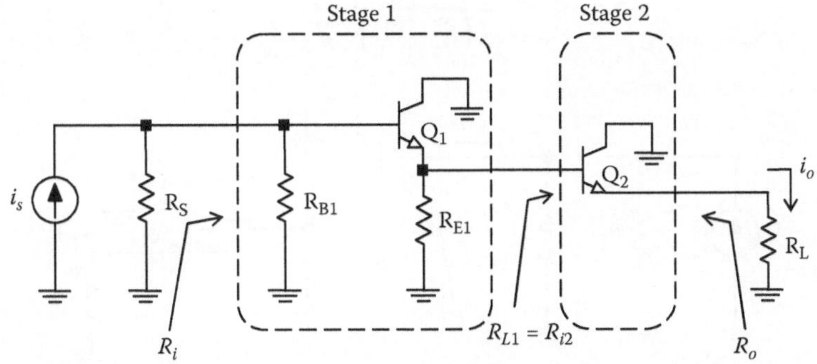

FIGURE EX. 5.4(b)

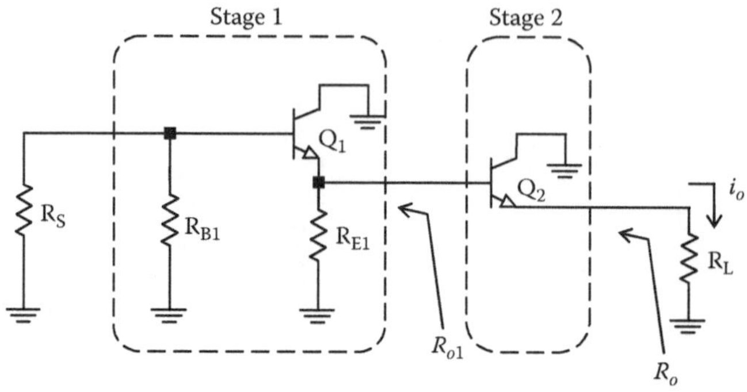

FIGURE EX. 5.4(c)

where

$$R_{o1} = R_{E1} \left\| \left[r_{e1} + \frac{1}{\beta_1}\left(R_{B1} \| R_S\right) \right] \right. .$$

Probably the most straightforward way to determine the overall current gain of this amplifier is to find out how the input signal is delivered to Q_1, amplified, delivered to Q_2, amplified more, and finally delivered to the load, R_L. Using the intermediate signals specified on the AC schematic in Figure Ex. 5.4(d), the overall current gain can be written as:

$$A_{IS} = \frac{i_o}{i_s} = \frac{i_{b1}}{i_s} \cdot \frac{i_{e1}}{i_{b1}} \cdot \frac{i_{b2}}{i_{e1}} \cdot \frac{i_o}{i_{b2}} .$$

The first term, i_{b1}/i_s, represents the current division at the input:

$$\frac{i_{b1}}{i_s} = \frac{R_{B1} \| R_S}{\left(R_{B1} \| R_S\right) + R_A}$$

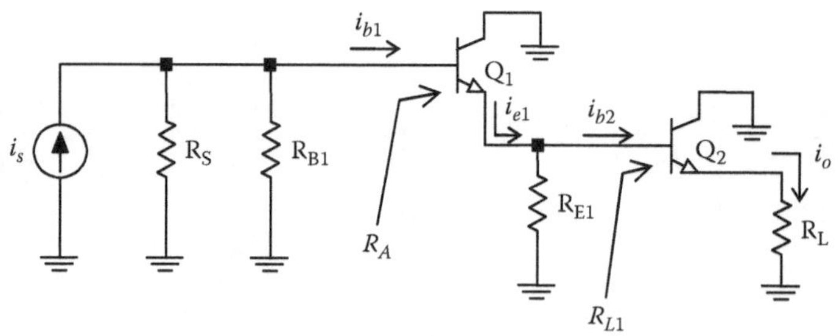

FIGURE EX. 5.4(d)

where $R_A = \beta_1[r_{e1} + (R_{E1} \| R_{L1})]$. This ratio is always less than 1. The second term is the current amplification with a gain of $(\beta_1 + 1)$ performed by Q_1:

$$\frac{i_{e1}}{i_{b1}} = \beta_1 + 1.$$

The emitter current of Q_1 should pass through a current division network comprised of R_{E1} and R_{L1} to be delivered to Q_2. The third term represents this attenuation as

$$\frac{i_{b2}}{i_{e1}} = \frac{R_{E1}}{R_{E1} + R_{L1}}.$$

The base current of Q_2 is then amplified with a factor of $\beta_2 + 1$:

$$\frac{i_{e2}}{i_{b2}} = \beta_2 + 1,$$

and directly delivered to the load:

$$i_o = i_{e2}.$$

The overall current gain of the amplifier can be finally written as:

$$A_{IS} = \frac{i_o}{i_s} = \frac{i_{b1}}{i_s}.\frac{i_{e1}}{i_{b1}}.\frac{i_{b2}}{i_{e1}}.\frac{i_o}{i_{b2}} = \frac{R_{B1} \| R_S}{\left(R_{B1} \| R_S\right) + R_A}.(\beta_1 + 1).\frac{R_{E1}}{R_{E1} + R_{L1}}.(\beta_2 + 1).$$

A good AC design with negligible signal attenuations (represented by the first and the third terms in the preceding relationship) can result in a very high current gain of β^2. With a typical β of around 100, this gain can easily be as high as 10,000.

(b) Using the given bias currents, r_{e1} and r_{e2} are achieved as:

$$r_{e1} = \frac{V_T}{I_{C1}} = \frac{25^{mV}}{4^{mA}} = 6.25\Omega$$

and

$$r_{e2} = \frac{V_T}{I_{C2}} = \frac{25^{mV}}{3.3^{mA}} = 7.6\Omega.$$

Then, substituting the given numeric values one obtains:

$$R_{L1} = \beta_2(r_{e2} + R_L) = 100(0.00625^{k\Omega} + 1^{k\Omega}) = 100.625^{k\Omega} \cong 100^{k\Omega}.$$

and therefore the input resistance of the amplifier is:

$$\begin{aligned} R_i &= R_{B1} \| \beta_1[r_{e1} + (R_{E1} \| R_{L1})] \\ &= 150^{k\Omega} \| 100[0.0076^{k\Omega} + (1^{k\Omega} \| 100^{k\Omega})] \\ &\cong 150^{k\Omega} \| 100^{k\Omega} \\ &= 60^{k\Omega}. \end{aligned}$$

To calculate the output resistance, R_{o1} should be determined first:

$$\begin{aligned} R_{o1} &= R_{E1} \left\| \left[r_{e1} + \frac{1}{\beta_1}\left(R_{B1} \| R_S\right)\right] \right. \\ &= 1^{k\Omega} \left\| \left[0.00625^{k\Omega} + \frac{1}{100}\left(150^{k\Omega} \| 1^{M\Omega}\right)\right] \right. \\ &\cong 1^{k\Omega} \| 1.3^{k\Omega} \\ &\cong 0.57^{k\Omega}. \end{aligned}$$

Then, the output resistance is obtained as:

$$\begin{aligned} R_o &= r_{e2} + \frac{1}{\beta_2}(R_{o1}) \\ &= 7.6^{\Omega} + \frac{1}{100}(0.57^{k\Omega}) \\ &= 13.3^{\Omega}. \end{aligned}$$

Knowing that:

$$\begin{aligned} R_A &= \beta_1[r_{e1} + (R_{E1} \| R_{L1})] \\ &= 100[0.00625^{k\Omega} + (1^{k\Omega} \| 100^{k\Omega})] \\ &\cong 100^{k\Omega}, \end{aligned}$$

the overall current gain of the amplifier is also calculated as:

$$A_{IS} = \frac{R_{B1} \| R_S}{(R_{B1} \| R_S) + R_A}(\beta_1 + 1)\frac{R_{E1}}{R_{E1} + R_{L1}}(\beta_2 + 1)$$

$$= \frac{150^{k\Omega} \| 1^{M\Omega}}{(150^{k\Omega} \| 1^{M\Omega}) + 100^{k\Omega}}(101)\frac{1^{k\Omega}}{1^{k\Omega} + 1^{k\Omega}}(101)$$

$$\cong 0.57 \times 101 \times 0.5 \times 101$$

$$\cong 2907.$$

5.4 Useful Compound Configurations

Among the multistage amplifiers, combinations of transistors are available, which provide specific features and are worth being assigned specific names and studied in more detail. The following are three of the well-known two-transistor configurations.

5.4.1 Darlington Pair

Figure 5.4(a) depicts the popular *Darlington pair*, a configuration that is capable of providing a very large current gain using two transistors. This compound can also be thought as a super transistor with the three terminals named as C for the collector, *B* for the base, and *E* for the emitter, as illustrated in Figure 5.4(b).

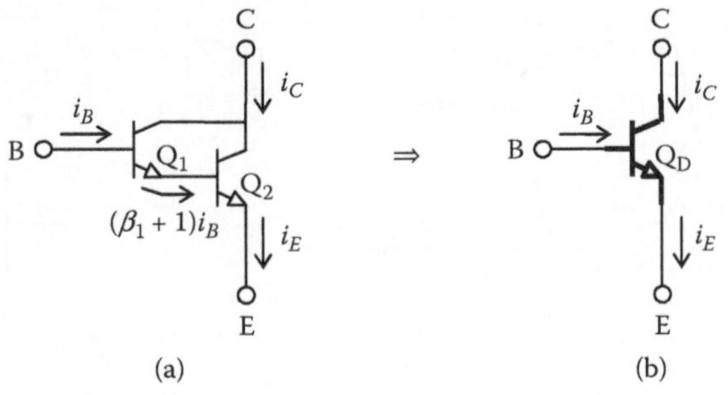

FIGURE 5.4
Darlington pair.

The key point in this connection is that the base current, i_B, is first amplified by Q_1 with a factor of $\beta_1 + 1$, and then enters the base terminal of Q_2 to be boosted up once again, this time with a factor of $\beta_2 + 1$:

$$\begin{aligned} i_E &= (\beta_2 + 1)i_{B2} \\ &= (\beta_2 + 1).(\beta_1 + 1)i_B \\ &= (1 + \beta_1 + \beta_2 + \beta_1.\beta_2)i_B \end{aligned} \tag{5.4(a)}$$

The collector current, i_C, is obviously the summation of the collector currents of Q_1 and Q_2:

$$i_C = i_{C1} + i_{C2}$$

which can be written in terms of the base current, i_B, as:

$$i_C = (\beta_1 + \beta_2 + \beta_1.\beta_2)i_B. \tag{5.4(b)}$$

As a result, the current gain β_D for the super transistor can be written as:

$$\beta_D = \beta_1 + \beta_2 + \beta_1.\beta_2 \cong \beta_1.\beta_2. \tag{5.5}$$

The Darlington pair is really useful in applications where large current gain is needed. Moreover, huge input resistance and very low output resistance is also possible using a Darlington pair. As illustrated in Figure 5.5, the input resistance can be very large if the input signal is applied to the base terminal of a Darlington pair:

$$R_{Seen,\,B} = \beta_1[r_{e1} + \beta_2(r_{e2} + R_E)] \tag{5.6}$$

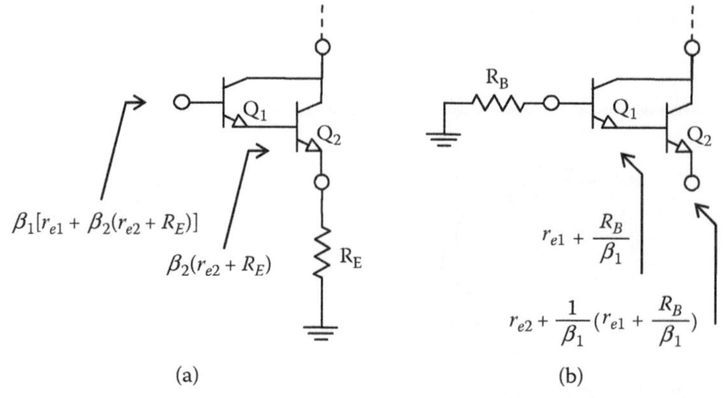

FIGURE 5.5
Achieving (a) very high input resistance and (b) very low output resistance by a Darlington pair.

and the output resistance will be very small if the output is taken from the emitter:

$$R_{Seen,E} = r_{e2} + \frac{1}{\beta_2}\left(r_{e1} + \frac{R_B}{\beta_1} \right). \tag{5.7}$$

If the Darlington pairs in Figure 5.5(a) and (b) are replaced by their equivalent super transistor, Q_D, Equation (5.6) and Equation (5.7) can be rewritten as:

$$R_{Seen,\,B} = \beta_D(r_{eD} + R_E) \tag{5.8}$$

and

$$R_{Seen,E} = r_{eD} + \frac{R_B}{\beta_D}, \tag{5.9}$$

in which r_{eD} is the AC resistance seen between the base and the emitter of the Darlington pair from the emitter viewpoint. This resistance can be written as:

$$r_{eD} = r_{e2} + \frac{r_{e1}}{\beta_2}. \tag{5.10}$$

If both Q_1 and Q_2 have the same β (i.e., $\beta = \beta_1 = \beta_2$), it can be easily shown that:

$$r_{eD} = 2r_{e2} = 2\frac{r_{e1}}{\beta}. \tag{5.11}$$

This means that the equivalent transconductance for a Darlington pair, g_{mD}, is:

$$g_{mD} = \frac{g_{m2}}{2} = \frac{\beta.g_{m1}}{2}. \tag{5.12}$$

The reduced transconductance of Q_D compared with Q_2 is known as a drawback for the Darlington connection when used as a voltage amplifier.

Example 5.5

Given $\beta_1 = \beta_{1A} = 100$, $R_B = 200\ k\Omega$, $R_E = 1\ k\Omega$, $R_L = 5\ k\Omega$, and $I_{C1} = 1\ mA$ for each of the amplifiers in Figure Ex. 5.5(a), find the overall current and voltage gains $A_I = i_o/i_i$ and $A_V = v_o/v_i$.

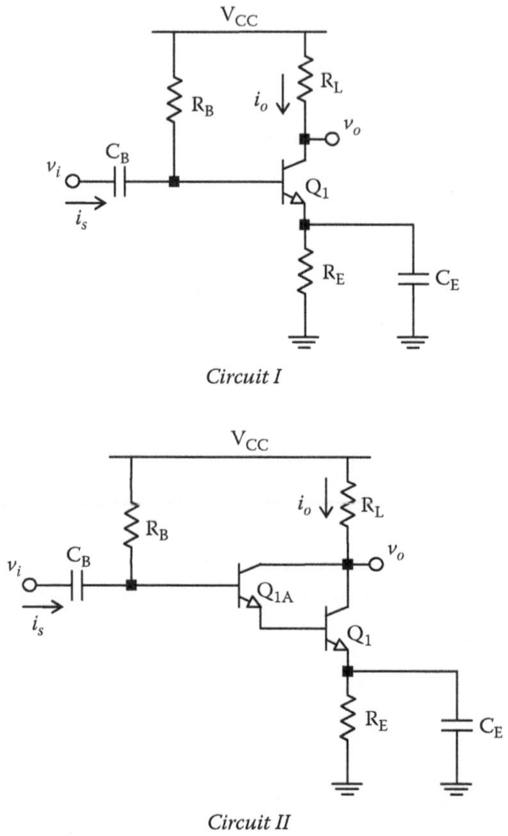

FIGURE EX. 5.5(a)

Solution:

AC schematics of the amplifiers are presented in Figure Ex. 5.5(b). In both circuits, for Q_1 we have:

$$r_{e1} = \frac{V_T}{I_{C1}} = \frac{25^{mV}}{1^{mA}} = 25\Omega,$$

and in Circuit II, because the collector current of Q_{1A} is smaller than I_{C1} by a factor of β_1, its r_e is expected to be larger by the same factor:

$$r_{e1A} = \frac{V_T}{I_{C1A}} = \frac{V_T}{I_{C1A}/\beta_1} = \frac{25^{mV}}{1^{mA}/100} = 2.5^{k\Omega}.$$

The current gain for Circuit I is then written as:

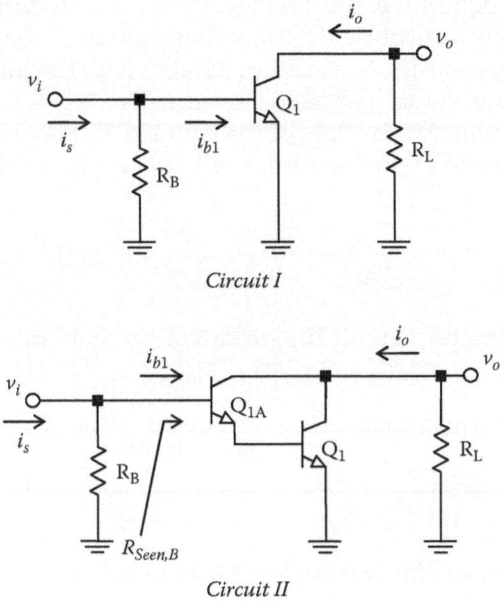

FIGURE EX. 5.5(b)

$$A_{I,I} = \frac{i_o}{i_i} = \frac{i_{b1}}{i_i}.\frac{i_o}{i_{b1}} = \frac{R_B}{R_B + \beta_1 r_{e1}}.\beta_1$$

$$= \frac{200^{k\Omega}}{200^{k\Omega} + (100 \times 0.025^{k\Omega})} \times 100 \cong 98.8.$$

To find the current gain of Circuit II, the resistance seen into the base of the Darlington pair should be determined. This resistance, illustrated in Figure Ex. 5.5(b), is obtained as:

$$R_{Seen,\,B} = \beta_{1A}(r_{e1A} + \beta_1 r_{e1}) = 2\beta_D r_{e1} = 2 \times 10^4 \times 25\ \Omega = 500\ k\Omega.$$

Remembering that the Darlington pair can be taken as a single transistor with a beta of $\beta_D \cong \beta^2 = 10{,}000$ and $r_{eD} = 2\,r_{e1} = 50\ \Omega$, it could also be said that:

$$R_{Seen,\,B} = r_{\pi D} = \beta_D r_{eD} = 2\beta_D r_{e1} = 2 \times 10^4 \times 25\ \Omega = 500\ k\Omega.$$

Now, the current gain is written as:

$$A_{I,II} = \frac{i_o}{i_i} = \frac{i_{b1}}{i_i}.\frac{i_o}{i_{b1}} = \frac{R_B}{R_B + R_{Seen,B}}.\beta_D$$

$$= \frac{200^{k\Omega}}{200^{k\Omega} + 500^{k\Omega}} \times 10^4 \cong 2857.$$

Before trying to determine the voltage gains, it is worth noting that the resistor R_B does not contribute to the voltage gains of the amplifiers at all. This is because this resistor is placed in parallel with the input signal source, and thus the input signal (which is assumed to be of voltage nature) is applied to the amplifier regardless of the value of R_B. Having the input signal applied to the base of Q_1 in Circuit I, the voltage gain is obtained as:

$$A_{V,I} = \frac{v_o}{v_i} = \frac{-R_L}{r_{e1}} = \frac{-5^{k\Omega}}{0.025^{k\Omega}} = -200\,.$$

The same reasoning for Circuit II gives a voltage gain of:

$$A_{V,II} = \frac{v_o}{v_i} = \frac{-R_L}{r_{eD}} = \frac{-R_L}{2r_{e1}} = \frac{-5^{k\Omega}}{0.05^{k\Omega}} = -100\,.$$

5.4.1.1 *Varieties of the Darlington Configuration*

For the Darlington pair introduced in Figure 5.4, the relationships provided for the currents were for both DC and AC components as well as whole currents. For the sake of simplicity, it is assumed that the transistors have the same β in both DC and AC modes, and independent of their bias currents. In this configuration, the bias current of Q_1 must be β times smaller than that of Q_2, but in certain cases, the designer prefers to bias Q_1 independent of Q_2. One simple way to increase the bias current of Q_1 is to add a resistor R_P in parallel with base-emitter of Q_2 as illustrated in Figure 5.6.

This way, the DC voltage across R_P is $V_{BE}(ON)$, and thus the bias current of Q_1 increases to:

$$I_{E1} = I_{B2} + \frac{V_{BE}(ON)}{R_P}. \tag{5.13}$$

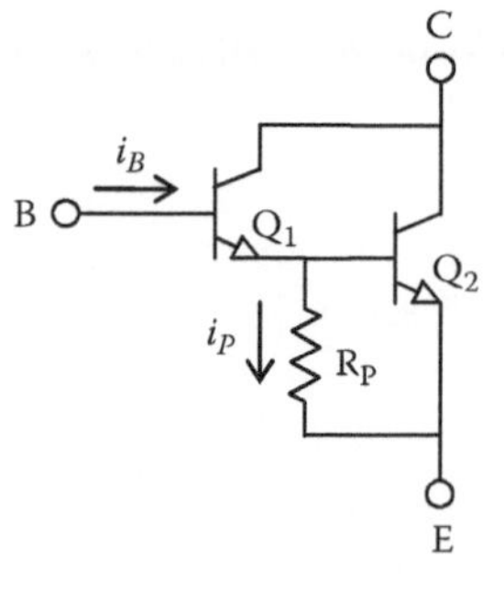

FIGURE 5.6
Darlington pair with an added resistor.

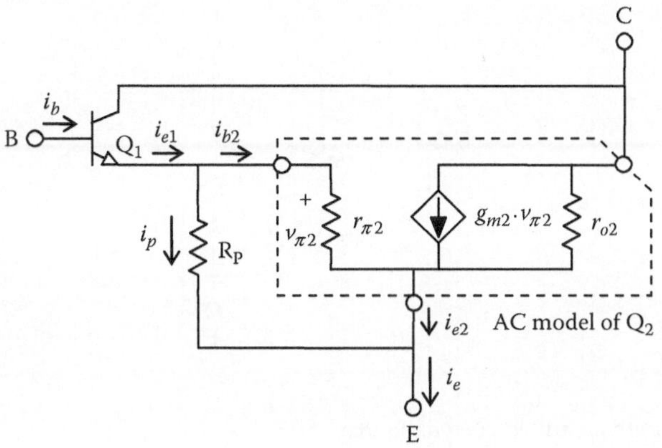

FIGURE 5.7
Illustration of the effect of R_P in AC mode.

The disadvantage of this approach is that the portion of I_{E1} that passes through R_P is not amplified by Q_2, and thus, the overall β of the circuit drops. Usually, the super-β of the Darlington configuration plays a critical role in AC mode. Therefore, let us concentrate on the AC function of R_P, illustrated in Figure 5.7, with Q_2 replaced by its small-signal AC model.

Because R_P is in parallel with $r_{\pi2}$, it can be said that a current division takes place between these two resistors, and only a fraction of the AC emitter current of Q_1 is delivered to Q_2 to be amplified:

$$i_{b2} = \frac{R_P}{R_P + r_{\pi2}}.i_{e1} = \frac{R_P}{R_P + r_{\pi2}}.(\beta_1 + 1).i_b\,,$$

and

$$i_{e2} = (\beta_2 + 1).i_{b2} = (\beta_1 + 1)(\beta_2 + 1)\frac{R_P}{R_P + r_{\pi2}}.i_b\,.$$

The emitter current, i_e, is then written as:

$$i_e = i_{e2} + i_p,$$

in which the unamplified current, i_p, can be usually neglected, and thus:

$$i_e \cong i_{e2} = (\beta_1 + 1)(\beta_2 + 1)\frac{R_P}{R_P + r_{\pi2}}.i_b\,. \tag{5.14}$$

Therefore, it can be concluded that adding the resistor R_P decreases the AC current gain of the Darlington configuration to:

$$\beta_D = (\beta_1 + 1)(\beta_2 + 1)\frac{R_P}{R_P + r_{\pi2}} \cong \beta^2.\frac{R_P}{R_P + r_{\pi2}}\,. \tag{5.15}$$

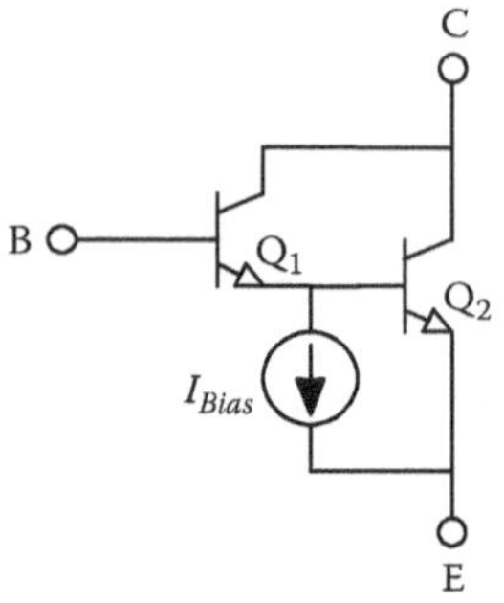

FIGURE 5.8
Darlington pair with an added current source.

Figure 5.8 presents another way to bias Q_1 independent of Q_2, and avoid the loss in the equivalent beta of the Darlington configuration. In this case, a current source is employed to provide Q_2 with the excess DC current it requires. In the AC mode, the current source does not exist anymore, and the circuit exhibits the same current gain as the original Darlington pair.

5.4.2 Cascode Amplifier

In bipolar technology, a *cascode amplifier* is a two-stage amplifier with the first stage in common-emitter configuration followed by a common-base stage. The cascode amplifier is realized in CMOS technology using a common-source stage followed by a common-gate stage. Simplified AC schematics of the cascode amplifier in both bipolar and CMOS technologies are depicted in Figure 5.9.

Input resistance of the cascode amplifier is determined by its input stage as

$$R_i = r_{\pi 1} \tag{5.16(a)}$$

for the bipolar cascode amplifier in Figure 5.9(a), and it is infinite

$$R_i = \infty \tag{5.16(b)}$$

for its MOS counterpart in Figure 5.9(b).

The output resistance is the resistance that R_L sees when looking into the collector/drain of Q_2/M_2. To find this resistance, Q_1/M_1 is taken as a large resistance, r_{o1}, connected between the emitter/source of Q_2/M_2 and the ground, as illustrated in Figure 5.10. The output resistance for the bipolar cascode amplifier becomes

$$R_o = \beta_2 r_{o2}, \tag{5.17(a)}$$

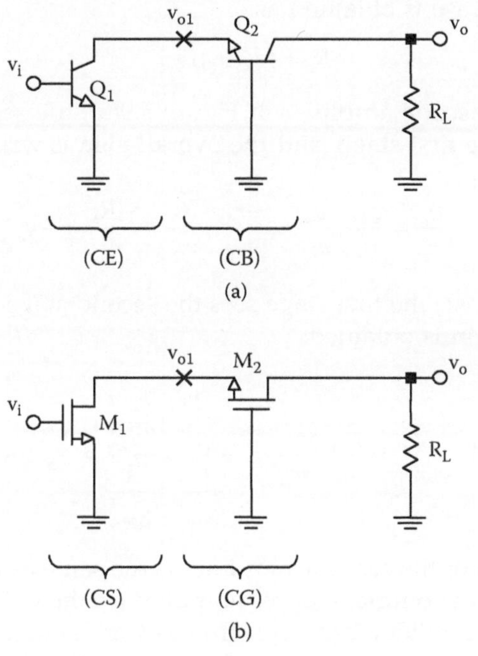

FIGURE 5.9
Simplified AC schematic of the cascode amplifier using (a) bipolar (b) MOS transistors.

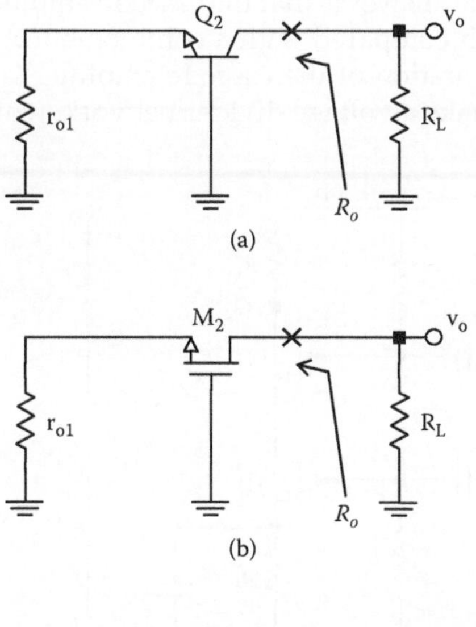

FIGURE 5.10
Simplified AC schematic of the cascode amplifier using (a) bipolar (b) MOS transistors.

and for the MOS case is obtained as

$$R_o = (g_{m2}.r_{o2}).r_{o1}. \quad (5.17(b))$$

In the bipolar cascode amplifier in Figure 5.9(a), the second stage acts as a load of r_{e2} for the first stage, and the overall gain is written as:

$$A_V = \frac{v_o}{v_i} = \frac{v_{o1}}{v_i}.\frac{v_o}{v_{o1}} = \frac{-r_{e2}}{r_{e1}}.\frac{R_L}{r_{e2}} = \frac{-R_L}{r_{e1}} = -g_{m1}R_L. \quad (5.18(a))$$

For the CMOS case, the first stage sees the second stage as a load of $1/g_{m2}$, and the overall gain is obtained:

$$A_V = \frac{v_o}{v_i} = \frac{v_{o1}}{v_i}.\frac{v_o}{v_{o1}} = \frac{-\frac{1}{g_{m2}}}{\frac{1}{g_{m1}}}.\frac{R_L}{\frac{1}{g_{m2}}} = -g_{m1}R_L. \quad (5.18(b))$$

The overall gain of the cascode amplifier is the same as that of a common-emitter amplifier (or common-source amplifier for the CMOS version). Now that the cascode amplifier has no significant advantage over a common-emitter amplifier (or common-source amplifier for the CMOS version), what is the added value of the second transistor (Q_2/M_2)? To answer this question in detail, we need to learn how to analyze amplifiers in frequency domain. Analysis of amplifiers at high frequencies in general, and a detailed discussion about the frequency response of the cascode amplifier will be presented in Chapter 8. A simple answer is that the cascode amplifier has a much wider frequency bandwidth compared with a common-emitter/source amplifier.

A CMOS implementation of the cascode amplifier is presented in Figure 5.11. R_1, R_2, and R_3 make a voltage divider network, which provides the gate

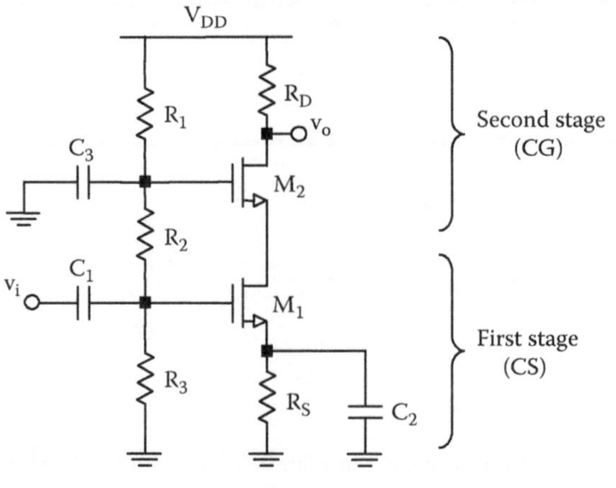

FIGURE 5.11
A CMOS implementation of the cascode amplifier.

voltages required to bias M_1 and M_2. C_1 is a coupling capacitor, and C_2 and C_3 are both bypass capacitors for the first and the second stages, respectively.

5.4.2.1 Folded-Cascode Amplifier

A drawback of the regular cascode amplifier in Figure 5.12(a) is that by stacking the cascode transistor M_2 on top of M_1, the DC component of the output becomes closer to the supply voltage, V_{DD}. This is because the transistor M_2 needs at least a voltage of $V_{DS}(Sat.)$ between its drain-source to remain in saturation. This can limit the output voltage swing especially in low-voltage applications where V_{DD} is not so high. The *folded-cascode configuration*, illustrated in Figure 5.12(b), is a solution to this problem, in which the upper part of the amplifier, comprised of the cascode transistor M_2, and the resistor R_D, is folded down. When folding the upper part, a current source is added, and the cascode transistor is changed from NMOS to PMOS to fulfill the biasing requirements of the transistor. The purpose of the current source in DC mode is to provide M_1 with the drain current it needs and to bias M_2 at:

$$I_{D2} = I_{DD} - I_{D1}.$$

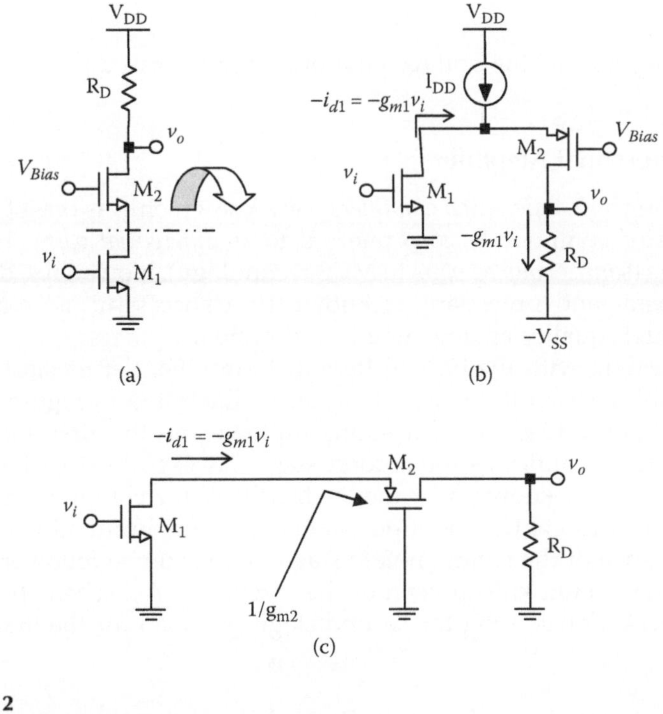

FIGURE 5.12
Folded cascode amplifier (a) Basic idea of folding the cascode transistor (b) The folded-cascode amplifier (c) AC schematic.

It should be added that M_1 is biased by its own bias network, which is not included in Figure 5.12(a) and (b). In AC mode, because of the application of the input signal, v_i, the drain current of M_1 contains an AC component of:

$$i_{d1} = g_{m1}v_i.$$

Assuming that M_1 provides an AC current of $i_{d1} = g_{m1}v_i$ at node A as illustrated in Figure 5.12(b), all the AC current goes into the source of M_2. This is simply because the current source must remain constant at I_{DD}. The cascode transistor, M_2, which is a common-gate stage, exhibits a current gain of 1, and delivers the same current to R_D. This can be clearly seen in the AC schematic of the amplifier in Figure 5.12(b), which is exactly the same as that of the regular cascode amplifier. The only difference is that the type of M_2 is changed from NMOS to PMOS, which does not change any of the AC results. Thus, the AC voltage at the output is obtained as:

$$v_o = i_{d2}R_D = (-g_{m1}v_i)R_D = (-g_{m1}R_D)v_i,$$

and the overall voltage gain of the amplifier can be written as:

$$A_V = \frac{v_o}{v_i} = -g_{m1}R_D,$$

which is the same as the voltage gain of a regular cascode.

5.4.3 Differential Amplifier

Not very long ago, *differential amplifiers* were known only as one of the useful two-transistor compounds, also referred to as *differential pairs* and *emitter-coupled pairs* (*source-coupled pairs* in MOS technology). Nowadays, differential signal transfer and processing is known as a successful approach in the design of high-quality analog circuits and systems.

Before dealing with the basic differential amplifier, let us start with two similar amplifier circuits. The first circuit is illustrated in Figure 5.13. This is an amplifier consisting of a source follower as the first stage, and a common-gate amplifier as the second stage. As explained earlier, the low input resistance is known as a major drawback for common-gate voltage amplifiers. In this circuit, the input signal source is buffered from the low input resistance of the common-gate stage by the source follower stage.

To find the overall voltage gain of this circuit, its AC schematic is drawn in Figure 5.14. Considering the second stage as a load for the first stage:

$$R_{L1} = R_{i2} = \frac{1}{g_{m2}},$$

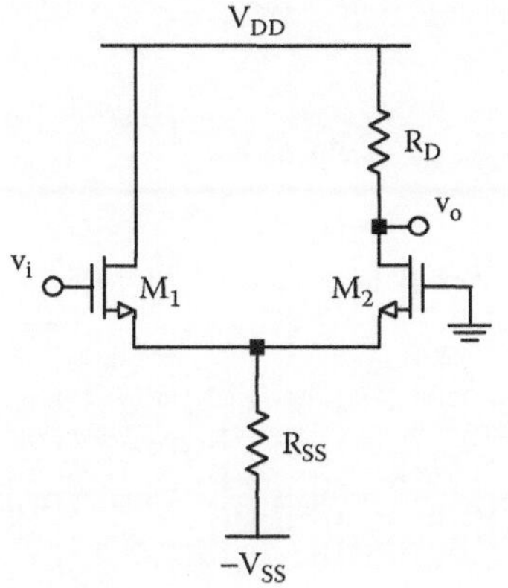

FIGURE 5.13
A SF-CG two-stage amplifier.

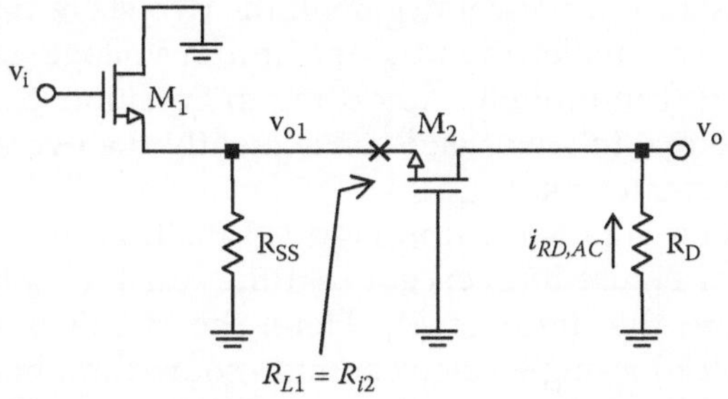

FIGURE 5.14
AC schematic of the amplifier of Figure 5.13.

the overall gain is written as

$$A_V = \frac{v_o}{v_i} = \frac{v_{o1}}{v_i} \cdot \frac{v_o}{v_{o1}} = \frac{R_{SS} \left\| \frac{1}{g_{m2}} \right.}{\left(R_{SS} \left\| \frac{1}{g_{m2}} \right. \right) + \frac{1}{g_{m1}}} \cdot \frac{R_D}{\frac{1}{g_{m2}}}, \tag{5.19}$$

which will be reduced to

$$A_V = \frac{v_o}{v_i} = \frac{g_{m1}}{g_{m1} + g_{m2}} \cdot g_{m2} R_D \tag{5.20}$$

if $R_{SS} >> 1/g_{m2}$. This condition is usually met as R_{SS} is chosen large enough to provide M_1 and M_2 with reasonable bias current levels. This means that the source follower successfully performs as a voltage buffer and provides

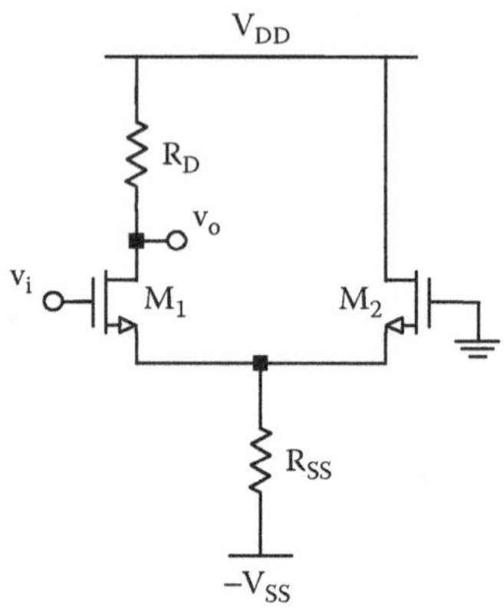

FIGURE 5.15
A CS amplifier with an additional transistor.

a huge input resistance. The overall gain of the two-stage circuit is now very high, and not much smaller than the maximum voltage gain that a single common-gate stage can provide. According to Equation (5.20), to maximize the gain of the source follower and consequently the overall gain, one can choose g_{m1} much greater than g_{m2}.

The second circuit is illustrated in Figure 5.15. This circuit is very similar to the amplifier of Figure 5.13, except that the resistor R_D has moved from the drain of M_2 to the drain of M_1. Here, the circuit is not a two-stage amplifier. It is indeed a single-stage common source amplifier with an additional transistor, M_2. As presented in the AC schematic of Figure 5.16, the main role of M_2 is to reduce the resistance connected to the source of M_1 from R_{SS} to $R_{SS}\|1/g_{m2}$. To bias the transistors M_1 and M_2 with reasonable current levels, R_{SS} is usually a large resistance. Thus, the addition of M_2 significantly increases the voltage gain of the circuit from:

$$A_V = \frac{v_o}{v_i} = \frac{-R_D}{R_{SS} + 1/g_{m1}} \tag{5.21}$$

to

$$A_V = \frac{v_o}{v_i} = \frac{-R_D}{\left(R_{SS}\|1/g_{m2}\right) + 1/g_{m1}}. \tag{5.22}$$

Taking the fact that usually $R_{SS} >> 1/g_{m2}$ into account, Equation (5.22) reduces to:

$$A_V = \frac{v_o}{v_i} = \frac{-R_D}{1/g_{m2} + 1/g_{m1}} = \frac{g_{m2}}{g_{m1} + g_{m2}}.(-g_{m1}R_D). \tag{5.23}$$

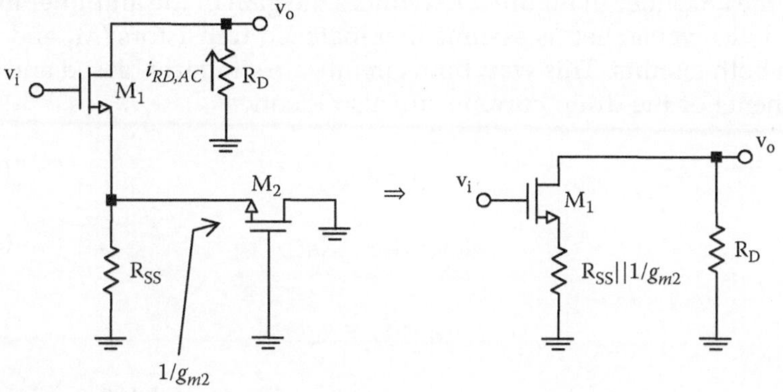

FIGURE 5.16
AC schematic of the amplifier of Figure 5.15.

This is a great result, which states that the circuit, without a bypass capacitor, is now capable of exhibiting a large gain, not much smaller than the maximum voltage gain of a common-source amplifier with bypass capacitor. According to Equation (5.23), to maximize the gain of the common-source stage and, consequently, the overall gain, one can choose g_{m2} much greater than g_{m1}.

It is interesting to note that the voltage gains of the amplifiers of Figure 5.13 and Figure 5.15 have the same absolute value.

As presented in the AC schematics of Figure 5.14 and Figure 5.16, the AC component of the output voltage is obtained as:

$$v_o = -R_D i_{R_D,AC} \tag{5.24}$$

for both circuits. The AC current flowing through R_D in the amplifier of Figure 5.13 is obtained from Equation (5.20) as:

$$v_o = \frac{g_{m1}g_{m2}}{g_{m1}+g_{m2}}.R_D.v_i \Rightarrow i_{R_D,AC} = i_{d2} = \frac{-g_{m1}g_{m2}}{g_{m1}+g_{m2}}.v_i. \tag{5.25(a)}$$

Similarly, Equation (5.23) gives the AC current flowing through R_D in the amplifier of Figure 5.15 as:

$$v_o = \frac{-g_{m1}g_{m2}}{g_{m1}+g_{m2}}.R_D.v_i \Rightarrow i_{R_D,AC} = i_{d1} = \frac{g_{m1}g_{m2}}{g_{m1}+g_{m2}}.v_i. \tag{5.25(b)}$$

This means that the AC currents flowing through R_D in both circuits are of the same amplitude, and going in opposite directions. It should be mentioned, however, that the condition previously presented to maximize the gain of the amplifier in Figure 5.13 reduces the gain of the amplifier in Figure 5.15 and vice versa. Let us assume that matched transistors (M_1 and M_2) are used in both circuits. This way, both circuits are similarly biased and the AC components of the drain currents are also identical:

$$I_{D1} = I_{D2}, \tag{5.26(a)}$$

$$g_{m1} = g_{m2} = g_m, \tag{5.26(b)}$$

and

$$i_{d1} = -i_{d2} = i_{ac} = g_m v_i/2. \tag{5.26(c)}$$

The AC currents i_{d1} and i_{d2}, now equal in amplitude and 180 degrees out of phase, both are present in both circuits. The only issue is that in each circuit, only one of them passes through a resistor (R_D) to generate the output voltage.

This is illustrated in Figure 5.17. This illustration leads us to the following question: *Now that one i_{ac} is in each side of each of the circuits, why not convert both of them to voltage?* This suggests the addition of a resistor R_D to either the drain of M_1 in Figure 5.13 or the drain of M_2 in Figure 5.15, which generates two outputs, as presented in Figure 5.18. In this circuit, the voltage gains from the input to the outputs are obtained using Equation (5.20) and Equation (5.23) as:

$$A_V = \frac{v_{o2}}{v_i} = \frac{-v_{o1}}{v_i} = \frac{g_{m1} g_{m2}}{g_{m1} + g_{m2}} .R_D. \tag{5.27}$$

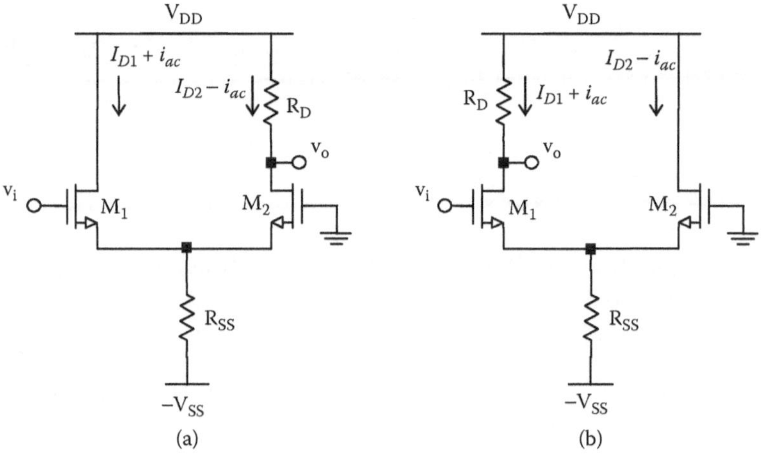

FIGURE 5.17
AC currents in the amplifiers of (a) Figure 5.13 and (b) Figure 5.15.

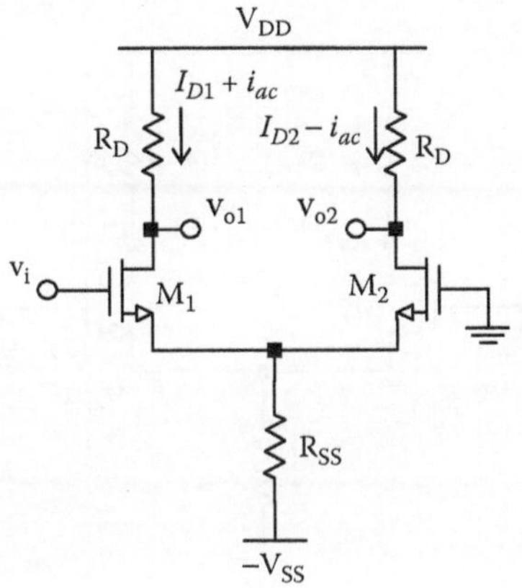

FIGURE 5.18
Generating two outputs by adding a second R_D to the circuits of Figure 5.17.

If the transistors, M_1 and M_2, are assumed identical, then the voltage gains for the outputs v_{o1} and v_{o2}, which are referenced to the ground, will be:

$$A_V = \frac{v_{o2}}{v_i} = \frac{-v_{o1}}{v_i} = \frac{1}{2} g_m R_D. \tag{5.28}$$

This means that for a certain input signal, v_i, the AC components of the outputs v_{o1} and v_{o2} are equal in amplitude and 180° out of phase.

Example 5.6

Given $\beta_1 = \beta_2 = 10\ mA/V^2$, $V_{TH} = 0.6\ V$, $R_{SS} = 100\ k\Omega$, $R_D = 150\ k\Omega$, and $V_{DD} = V_{SS} = 5\ V$ in the amplifier in Figure 5.18,

(a) Find the gains $A_{V1} = v_{o1}/v_i$ and $A_{V2} = v_{o2}/v_i$.
(b) Sketch the whole output voltages $v_{O1}(t)$ and $v_{O2}(t)$ if an input voltage of $v_I(t) = 1^{mV} Sin(\omega t)$ is applied.

Solution:

(a) DC schematic of the amplifier is drawn in Figure Ex. 5.6(a). Because of the circuit topology,

$$V_{GS1} = V_{GS2},$$

and noting that the transistors are identical and the circuit is symmetric, one can write:

$$I_{D1} = I_{D2}.$$

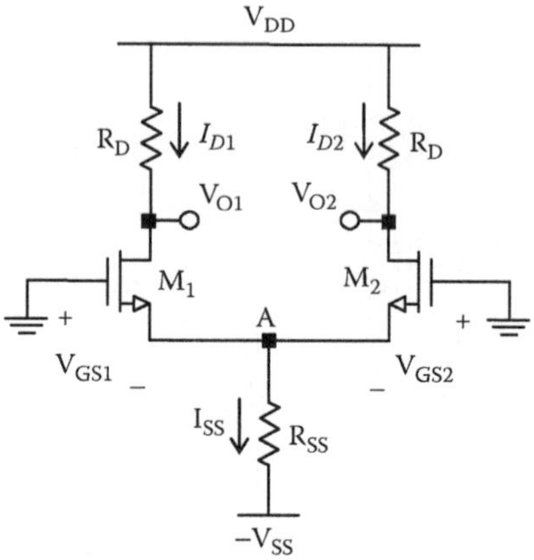

FIGURE EX. 5.6(a)

A KCL at node A gives:

$$I_{SS} = I_{D1} + I_{D2},$$

which with equal I_{D1} and I_{D2} means:

$$I_{D1} = I_{D2} = I_{SS}/2.$$

The current through R_{SS} can be found by dividing the voltage across it by its resistance:

$$I_{SS} = \frac{(0 - V_{GS1}) - (-V_{SS})}{R_{SS}} = \frac{V_{SS} - V_{GS1}}{R_{SS}} = \frac{5^V - V_{GS1}}{100^{k\Omega}}.$$

This gives the following relationship between I_{D1} and V_{GS1}:

$$I_{D1} = \frac{I_{SS}}{2} = \frac{5^V - V_{GS1}}{200^{k\Omega}}$$

or

$$V_{GS1} = 5^V - 200^{k\Omega} I_{D1}.$$

Replacing V_{GS1} from the preceding equation into the square-law $I_D - V_{GS}$ characteristic equation one obtains:

$$I_{D1} = \frac{\beta}{2}(V_{GS1} - V_{TH})^2 = \frac{10^{mA/V^2}}{2}(5^V - 200^{k\Omega} I_{D1} - 0.6^V)^2$$

or

$$I_{D1} = 5(4.4^V - 200I_{D1})^2.$$

$I_{D1} = 22.3\ \mu A$ and $21.7\ \mu A$ are the two roots for this quadratic equation. The former is not an acceptable value for the drain current because it corresponds to a V_{GS1} that is smaller than the threshold voltage. Therefore, the latter gives the bias current for both transistors:

$$I_{D1} = I_{D2} = I_{SS}/2 = 21.7\ \mu A,$$

and their transconductances will be:

$$g_{m1} = g_{m2} = \sqrt{2\beta I_{D1}} = \sqrt{2\times 10^{mA/V^2}\times 0.0217^{mA}} = 0.66^{mA/V}.$$

Finally, the voltage gains A_{V1} and A_{V2} are found as:

$$A_V = \frac{v_{o2}}{v_i} = \frac{-v_{o1}}{v_i} = \frac{1}{2}g_{m1}R_D = \frac{1}{2}\times 0.66\times 150^{k\Omega} \cong 50$$

(b) DC component of the output voltages is calculated as:

$$\begin{aligned} V_{O1} = V_{O2} &= V_{DD} - R_D I_{D1} \\ &= 5\ V - (150^{k\Omega}\times 21.7^{\mu A}) \\ &\cong 1.75^V. \end{aligned}$$

Having an input signal of $v_I(t) = 1^{mV}Sin(\omega t)$, and knowing that the voltage gains are ±50, the AC components of the output voltages are obtained as:

$$v_{o1}(t) = A_{V1}v_I(t) = -50^{mV}Sin(\omega t)$$

and

$$v_{o2}(t) = A_{V2}v_I(t) = 50^{mV}Sin(\omega t).$$

The whole output voltages can then be written as:

$$v_{O1}(t) = V_{O1} + v_{o1}(t) = 1.75^V - 50^{mV}Sin(\omega t)$$

and

$$v_{O2}(t) = V_{O2} + v_{o2}(t) = 1.75^V + 50^{mV}Sin(\omega t),$$

and sketched as presented in Figure Ex. 5.6(b).

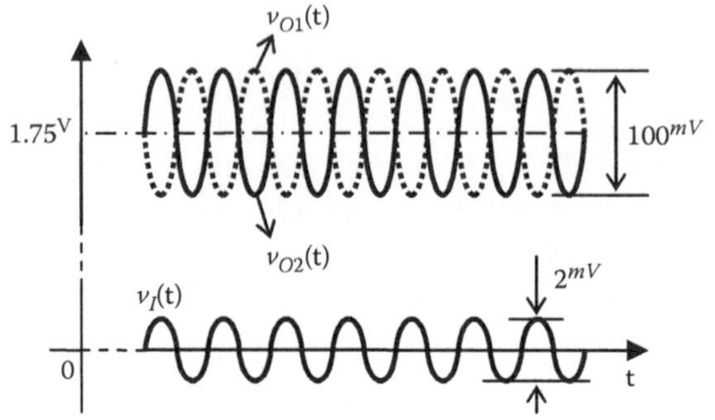

FIGURE EX. 5.6(b)

5.4.3.1 *The Basic Differential Amplifier*

Now, it is time to introduce the famous differential amplifier (see Figure 5.19), which is very similar to the circuits that have been studied so far. The only difference between this circuit and the amplifier of Figure 5.18 is the second input that is applied to the gate of M_2, v_{i2}. This circuit is symmetric, and the mirrored elements (the resistors R_D and the transistors M_1 and M_2) are assumed identical. As a result, DC conditions and small-signal model parameters (g_m and r_o) will be the same for both transistors. Because the circuit is symmetric, the voltage gain from v_{i2} to v_{o1} will be the same as the gain from v_{i1} to v_{o2}, and similarly the gain from v_{i2} to v_{o2} is the same as the gain from v_{i1} to v_{o1}:

$$v_{o1} = -v_{o2} = \frac{1}{2} g_m R_D . v_{i2} \quad \textit{if} \quad v_{i1} = 0, \tag{5.29(a)}$$

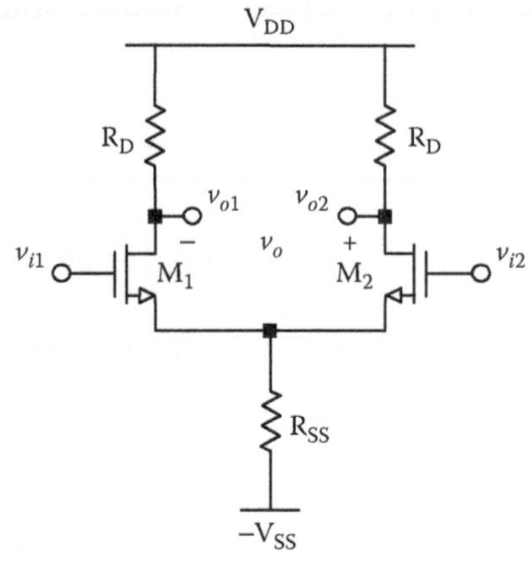

FIGURE 5.19
Basic differential amplifier.

as well as:

$$v_{o2} = -v_{o1} = \frac{1}{2} g_m R_D . v_{i1} \quad \text{if} \quad v_{i2} = 0. \tag{5.29(b)}$$

For the case where both inputs are applied to the amplifier, the effects of both inputs are superposed and the outputs become:

$$v_{o1} = -v_{o2} = \frac{1}{2} g_m R_D (v_{i2} - v_{i1}), \tag{5.30}$$

which indicates that the circuit amplifies the difference between the inputs.

Interestingly, if one takes the difference between the two outputs, which are of the same amplitude and 180° out of phase, the output amplitude will be doubled:

$$v_o = v_{o2} - v_{o1} = 2v_{o2} = -2v_{o1} = g_m R_D (v_{i1} - v_{i2}). \tag{5.31}$$

In differential circuits, when the output signal is taken from one of the output nodes referenced to the ground, it is called a *single-ended* output. If the difference between the two single-ended outputs is used as the output of the circuit, it is referred to as a *double-ended* or *differential* output. For instance, in Figure 5.19 v_{o1} and v_{o2} are single-ended outputs, whereas v_o is a differential output.

The traditional description for the function of a differential amplifier, illustrated in Figure 5.20(a), is that the circuit first subtracts one input from the other, and then amplifies the difference. This is why this circuit was sometimes referred to as the *difference* amplifier.

Nowadays, in a wide variety of analog circuits and systems it is preferred to differentially transfer, amplify, and process the signals. This is because of the advantages that this approach has over the traditional single-ended signal approach, most importantly, better noise immunity and larger voltage swing. All the amplifiers studied so far are designed to amplify single-ended signals. The differential amplifier can be thought of as an amplifier that is

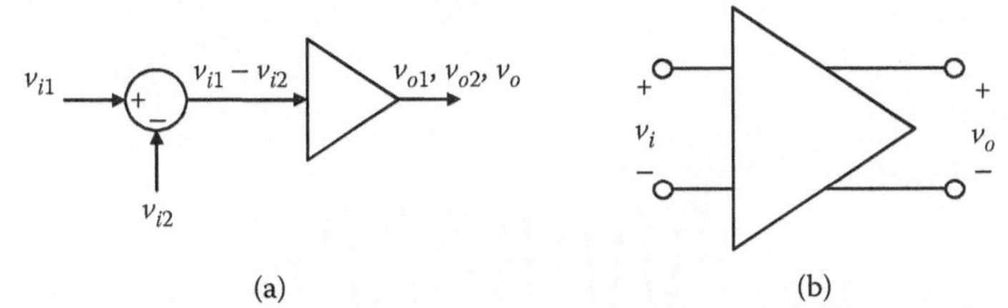

FIGURE 5.20
Function of a differential amplifier (a) subtracting one input from the other and then amplifying the difference, (b) amplifying a differential signal.

capable of amplifying a differential signal and providing its output in differential form. This is illustrated in Figure 5.20(b).

Each differential signal is imagined to be composed of two components: the *common-mode* and the *differential-mode* components, as depicted in Figure 5.21. For the differential voltage, v_A, between the nodes $A1$ and $A2$, the common-mode component is defined as the average of the voltages at $A1$ and $A2$:

$$v_{CM} = \frac{v_{A1} + v_{A2}}{2}, \tag{5.32(a)}$$

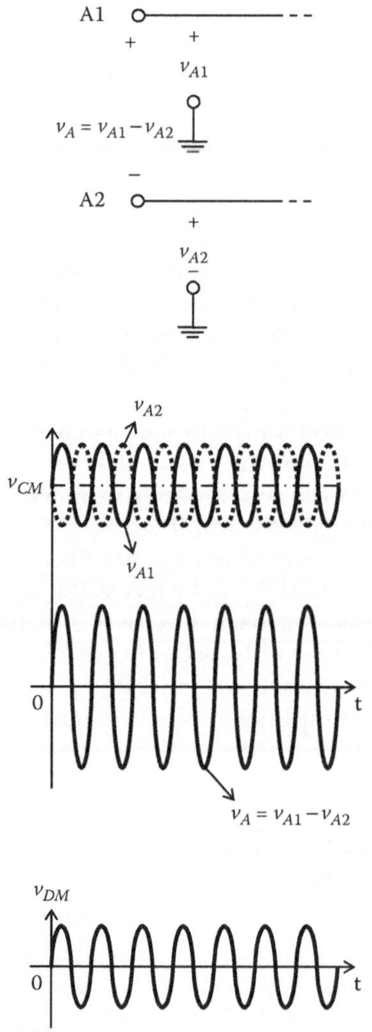

FIGURE 5.21
Common-mode and differential-mode components of a differential signal.

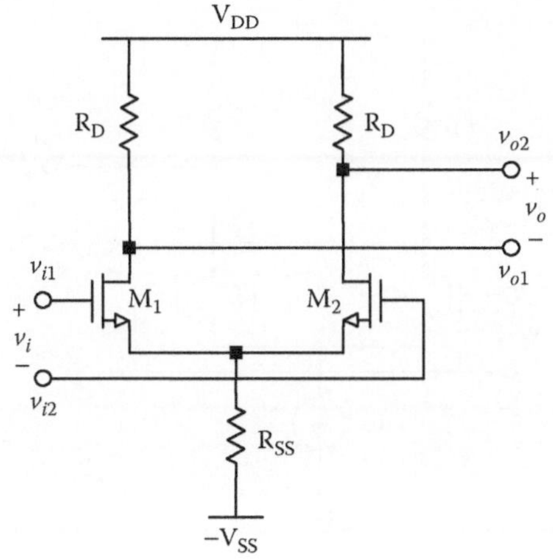

FIGURE 5.22
Differential amplifier amplifying a differential signal.

and the differential-mode component is half the difference between the single-ended voltages at $A1$ and $A2$:

$$v_{DM} = \frac{v_{A1} - v_{A2}}{2}. \tag{5.32(b)}$$

Although the common-mode component is depicted as a DC value in Figure 5.21, in general it can be a time-variant signal containing both AC and DC components. It should be noted that usually the signal is carried by the differential-mode component, while the DC conditions and all the noises that affect both input lines equally (e.g., the 60-Hz noise available in the environment) appear in the common-mode component.

With the previous definitions for common-mode and differential-mode components, the single-ended signals v_{A1} and v_{A2} can now be expressed as:

$$v_{A1} = v_{CM} + v_{DM} \tag{5.33(a)}$$

and

$$v_{A2} = v_{CM} - v_{DM}. \tag{5.33(b)}$$

This interpretation will be useful in the next sections where we think of the differential amplifier as a circuit that amplifies a differential input signal, v_i, and provides a differential output, v_o, as illustrated in Figure 5.22.

5.4.3.2 Differential Amplifier with Ideal Current Source

The biasing of the circuit in Figure 5.22 strongly depends on the common-mode component of the input, which sometimes contains unwanted and

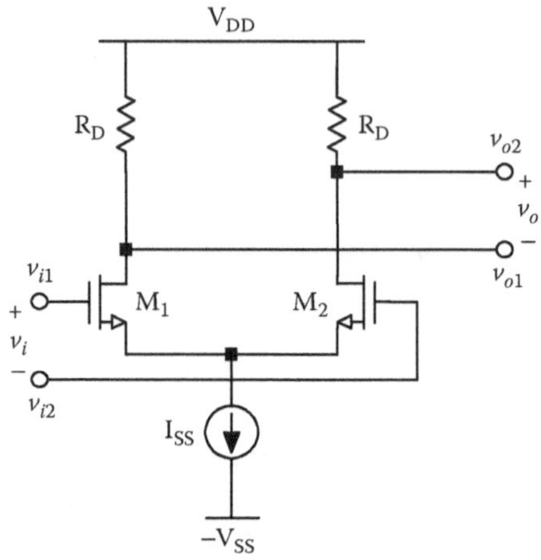

FIGURE 5.23
Differential amplifier with an ideal current source.

unpredictable variations. An effective solution is to replace R_{SS} with a current source, I_{SS}, taking advantage of the fact that a current source, by definition, generates its current independent of the voltage that appears across it. This current source in a differential amplifier is called the *tail current source*. The resulting circuit (see Figure 5.23) is still symmetric. Thus, both transistors are biased at

$$I_{D1} = I_{D2} = I_{SS}/2, \quad (5.34)$$

and the DC component of the output voltages is:

$$V_{O1} = V_{O2} = V_{DD} - R_D I_{D1} = V_{DD} - R_D I_{SS}/2. \quad (5.35)$$

In the differential amplifier in Figure 5.23, the common-mode and differential-mode components of the input signals are written as:

$$v_{CM} = \frac{v_{i1} + v_{i2}}{2} \quad (5.36(a))$$

and

$$v_{DM} = \frac{v_{i1} - v_{i2}}{2} = \frac{v_i}{2}. \quad (5.36(b))$$

The signals at the gates of M_1 and M_2 can also be expressed in terms of their common-mode and differential-mode components as:

$$v_{i1} = v_{CM} + v_{DM} \quad (5.37(a))$$

and

$$v_{i2} = v_{CM} - v_{DM}, \tag{5.37(b)}$$

and the differential input, v_i, is

$$v_i = v_{i1} - v_{i2} = 2\, v_{DM}. \tag{5.37(c)}$$

For small-signal operation, where the circuit is assumed to be linear and time-invariant, the superposition theorem can help simplify the analysis by studying the amplifier in two modes, *common-mode* (CM) and *differential-mode* (DM), and then concluding the overall response. In the CM analysis, two identical signal sources, v_{CM}, are applied to the inputs, and the behavior of the amplifier is studied, whilst in the DM analysis, two signal sources, v_{DM}, of the same amplitude but with different polarities are applied as the input signals. In the end, the results obtained from CM and DM analyses are superposed. This is illustrated in Figure 5.24.

In the CM, it is expected to have common-mode components for the drain currents of M_1 and M_2 as a result of the application of v_{CM} at the inputs. Because the circuit is symmetric and the applied inputs are the same, the common-mode components of the drain currents, i_{CM}, are the same:

$$i_{D1} = I_{D1} + i_{CM} = I_{SS}/2 + i_{CM} \tag{5.38(a)}$$

and

$$i_{D2} = I_{D2} + i_{CM} = I_{SS}/2 + i_{CM}. \tag{5.38(b)}$$

An interesting fact about this circuit in the common-mode (Figure 5.24(a)) is that the tail current source does not allow the drain currents of M_1 and M_2 both increase or decrease at the same time, and thus the drain currents must stay at $I_{SS}/2$:

$$\text{KCL at node A: } I_{SS} = i_{D1} + i_{D2} = I_{SS} + 2i_{CM}$$

$$\Rightarrow i_{CM} = 0.$$

This means that the output voltages will remain unchanged. Consequently, the *common-mode voltage gain* of the amplifier, defined as the ratio of the common-mode output to the common-mode input:

$$A_{V,CM} = \frac{v_{o,CM}}{v_{i,CM}} = \frac{(v_{o1} + v_{o2})/2}{(v_{i1} + v_{i2})/2} \tag{5.39}$$

will be zero.

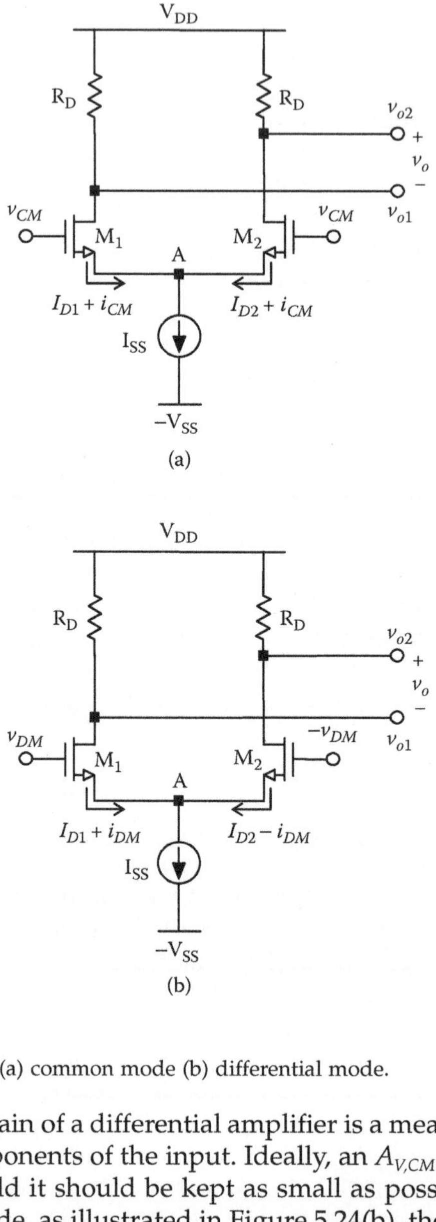

FIGURE 5.24
Differential amplifier in (a) common mode (b) differential mode.

Common-mode gain of a differential amplifier is a measure of how it treats the unwanted components of the input. Ideally, an $A_{V,CM}$ of zero is desirable, and in the real world it should be kept as small as possible.

In differential mode, as illustrated in Figure 5.24(b), the components of the drain currents of M_1 and M_2, i_{DM1} and i_{DM2}, are equal in amplitude and flow in opposite directions:

$$i_{D1} = I_{D1} + i_{DM} = I_{SS}/2 + i_{DM} \tag{5.40(a)}$$

and

$$i_{D2} = I_{D2} - i_{DM} = I_{SS}/2 - i_{DM} \tag{5.40(b)}$$

The tail current source in this mode allows the drain currents to equally deviate from their bias value by any amount as their differential mode components are always cancelled out in the KCL equation written for node A:

$$\text{KCL: } I_{SS} = i_{D1} + i_{D2} = I_{SS}/2 + i_{DM} + I_{SS}/2 - i_{DM}$$

$$\Rightarrow I_{SS} = I_{SS}.$$

Because the circuit is symmetric, and v_{DM} is applied to one side and $-v_{DM}$ to the other side, it can be said that whatever DM component is generated on one side will be present on the other side with the opposite polarity/direction. Interestingly, the voltage at any node on the axis of symmetry of the circuit (for example node A in Figure 5.24(b)) remains unaffected (i.e., contains no DM component). Therefore, such nodes are considered as *virtual grounds* in the differential mode, and the circuit is split into two halves, as illustrated in Figure 5.25.

In the basic differential amplifier, the differential-mode half-circuits are common-source amplifiers, for which we have:

$$i_{DM} = g_m v_{DM} = \frac{g_m v_i}{2} \tag{5.41}$$

and their voltage gains are obtained as:

$$v_{o1} = (-g_m R_D) v_{DM} = \frac{-g_m R_D}{2} v_i \tag{5.42(a)}$$

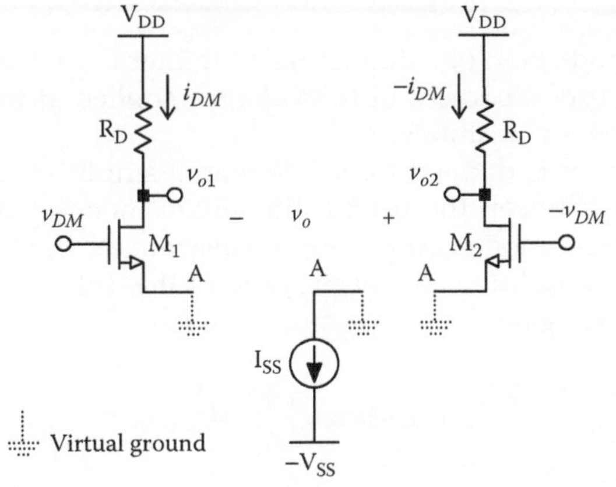

FIGURE 5.25
Half-circuits of the differential amplifier in differential mode.

and

$$v_{o2} = (-g_m R_D)(-v_{DM}) = \frac{g_m R_D}{2} v_i. \tag{5.42(b)}$$

The differential output of the amplifier is then written as:

$$v_o = v_{o2} - v_{o1} = 2(g_m R_D) v_{DM} = (g_m R_D) v_i. \tag{5.42(c)}$$

The *differential-mode voltage gain* of the amplifier, which is defined as the ratio of the differential output to the differential input:

$$A_{V,DM} = \frac{v_o}{v_i} = \frac{v_{o2} - v_{o1}}{v_{i1} - v_{i2}} \tag{5.43}$$

will then be:

$$A_{V,DM} = \frac{v_o}{v_i} = \frac{(g_m R_D) v_i}{v_i} = g_m R_D . \tag{5.44}$$

For a differential amplifier, *single-ended DM voltage gains* are also defined as:

$$A_{V,DM-SE} = \frac{v_{o2}}{v_i} = \frac{-v_{o1}}{v_i} , \tag{5.45}$$

which are obviously half of the amplifier's DM gain:

$$A_{V,DM-SE} = \frac{1}{2} A_{V,DM} = \frac{g_m R_D}{2} . \tag{5.46}$$

Differential-mode gain of a differential amplifier is a measure of how well it amplifies the useful content of the voltages applied at the input. Thus, a large $A_{V,DM}$ is usually desirable.

Another measure is defined for a differential amplifier which states how it discriminates between the useful differential mode component and the nonuseful common-mode component. *Common-mode rejection ratio* (*CMRR*) is defined as the absolute value of the ratio of the differential-mode gain to the common-mode gain:

$$CMRR = \left| \frac{A_{V,DM}}{A_{V,CM}} \right| . \tag{5.47}$$

CMRR is infinite for an ideal differential amplifier that does not amplify the common-mode component of the input at all, and it is desirable for it to be as large as possible in the real differential amplifiers. Typical value for CMRR is around 10,000 or more.

5.4.3.3 Differential Amplifier with Real Current Source

The tail current source is supposed to be eventually realized in the real world using electronic devices. Thus, it is better to be modeled by a Norton equivalent circuit, composed of an ideal current source, I_{SS}, in parallel with an internal resistance. The internal resistance usually has a negligible effect when biasing the amplifier and is therefore ignored. In AC analysis, however, where the ideal current source, I_{SS}, is replaced with an open circuit, the internal resistance of the current source is taken into account. To use the same equations so far derived for the differential amplifier, let us name the internal resistance of the current source R_{SS}. The differential amplifier with a real tail current source is depicted in Figure 5.26. It should be emphasized again that hereafter R_{SS} represents the internal resistance of the current source, typically on the order of 1–100 MΩ, and does not play the key biasing role anymore.

The differential-mode relationships already derived for the differential amplifier with an ideal current source will not be subject to any change if a real current source is used as the tail current source. As illustrated in Figure 5.27(a), this is simply because the sources of M_1 and M_2 are again both virtually grounded in the differential mode.

It is a different story in the common mode. As depicted in Figure 5.27(b), a second path is in parallel with I_{SS}, exclusively for the common-mode components of the drain currents to flow:

$$i_{D1} = I_{D1} + i_{CM} = I_{SS}/2 + i_{CM}, \tag{5.48(a)}$$

$$i_{D2} = I_{D2} + i_{CM} = I_{SS}/2 + i_{CM}, \tag{5.48(b)}$$

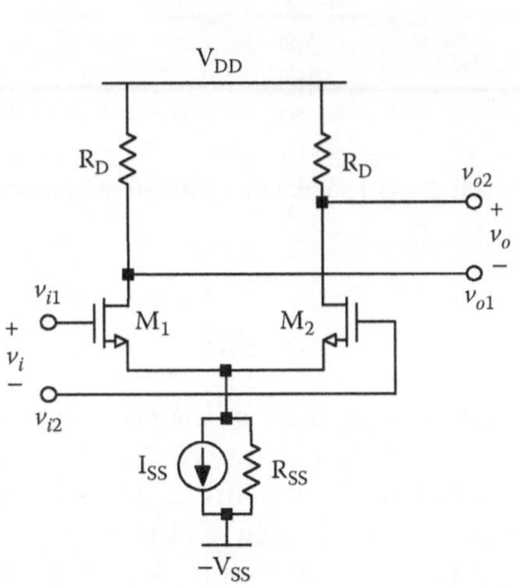

FIGURE 5.26
Differential amplifier with a real current source.

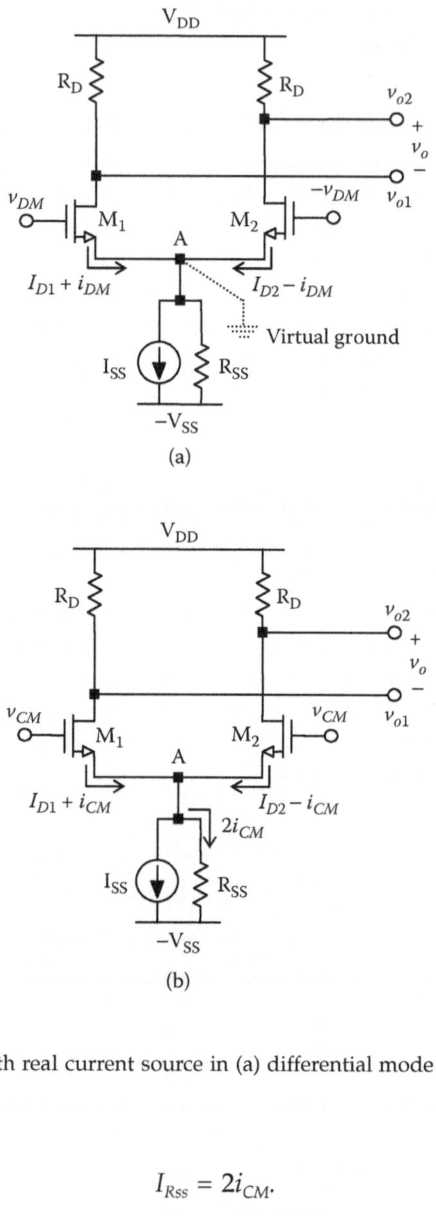

FIGURE 5.27
Differential amplifier with real current source in (a) differential mode (b) common mode.

and

$$I_{Rss} = 2i_{CM}. \tag{5.48(c)}$$

To analyze the differential amplifier in the common-mode more easily, let us draw it as a symmetric circuit comprised of two identical half-circuits connected to each other at node A using a dotted connection (see Figure 5.28). In this case, no current prefers to go through the dotted line because the half-circuits are identical both in topology and from the electrical standpoint. On the other hand, the sources of M_1 and M_2 will remain equipotential, even if the dotted line is cut. Thus, the circuit can be split into two

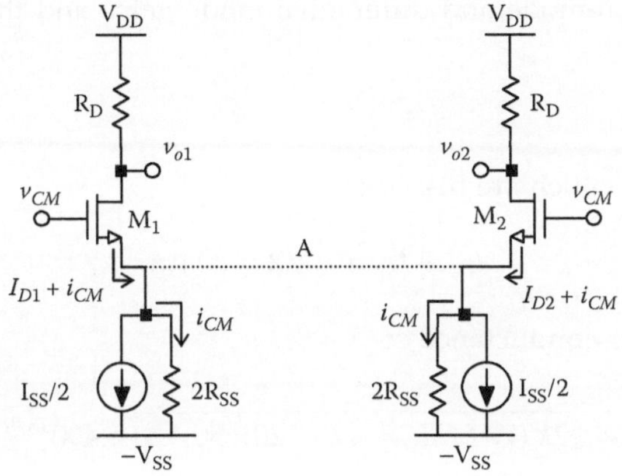

FIGURE 5.28
Half-circuits of the differential amplifier with real current source in common mode.

half-circuits in common-mode by cutting the dotted line without any change in its electrical conditions. For either one of the half-circuits, the gain, which is indeed the common-mode gain of the differential amplifier, is written as:

$$A_{V,CM} = \frac{v_{o1}}{v_{CM}} = \frac{v_{o2}}{v_{CM}} = \frac{-R_D}{2R_{SS} + \dfrac{1}{g_m}} = \frac{-g_m R_D}{1 + 2g_m R_{SS}}. \quad (5.49)$$

Remembering from Equation (5.44) that the differential gain of the amplifier is:

$$A_{V,DM} = \frac{v_o}{v_i} = g_m R_D,$$

the CMRR is obtained:

$$CMRR = \left|\frac{A_{V,DM}}{A_{V,CM}}\right| = \left|\frac{g_m R_D}{\dfrac{-g_m R_D}{1 + 2g_m R_{SS}}}\right| = 1 + 2g_m R_{SS} \cong 2g_m R_{SS}, \quad (5.50)$$

which is usually a large number.

Example 5.7

Assuming $k'_1 = k'_2 = 120\ \mu A/V^2$, $(W/L)_1 = (W/L)_2 = 50$, $I_{SS} = 40\ \mu A$, $R_{SS} = 1\ M\Omega$, $R_D = 150\ k\Omega$, and $V_{DD} = V_{SS} = 2.5\ V$ in the amplifier of Figure 5.26, find the common-mode and differential-mode gains and the CMRR of the circuit.

Solution:
The transistors, which are biased at

$$I_{D1} = I_{D2} = I_{SS}/2 = 20\ \mu A,$$

each have a transconductance of

$$g_m = \sqrt{2\beta I_D} = \sqrt{2k'(W/L)I_D} = \sqrt{2\times 120\times 50\times 20} \cong 490^{\mu A/V} = 0.49^{mA/V}.$$

Differential-mode gain is obtained as:

$$A_{V,DM} = \frac{v_o}{v_i} = g_m R_D = 0.49^{mA/V} \times 150^{k\Omega} = 73.5,$$

and the common-mode gain is calculated as:

$$A_{V,CM} = \frac{-g_m R_D}{1+2g_m R_{SS}} = \frac{-0.49^{mA/V}\times 150^{k\Omega}}{1+(2\times 0.49^{mA/V}\times 1^{M\Omega})} \cong -0.075.$$

As a result, CMRR of the circuit is found to be:

$$CMRR = \left|\frac{A_{V,DM}}{A_{V,CM}}\right| = \left|\frac{73.5}{-0.075}\right| \cong 980.$$

Example 5.8

Figure Ex. 5.8(a) is the differential amplifier with its differential output loaded by R_L. Derive the differential-mode and common-mode gains.

Solution:
To be able to analyze the amplifier using the half-circuit method, it should be fully symmetric. In the differential mode, if the load resistance is divided into two equal pieces, $R_L/2$, each of them with one side of the circuit, then the circuit will become symmetric (see Figure Ex. 5.8(b)). In this circuit, the differential-mode input, v_{DM}, is:

$$v_{DM} = v_i/2.$$

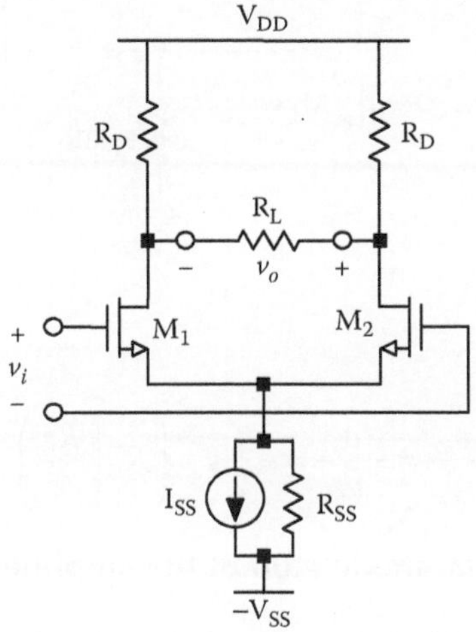

FIGURE EX. 5.8(a)

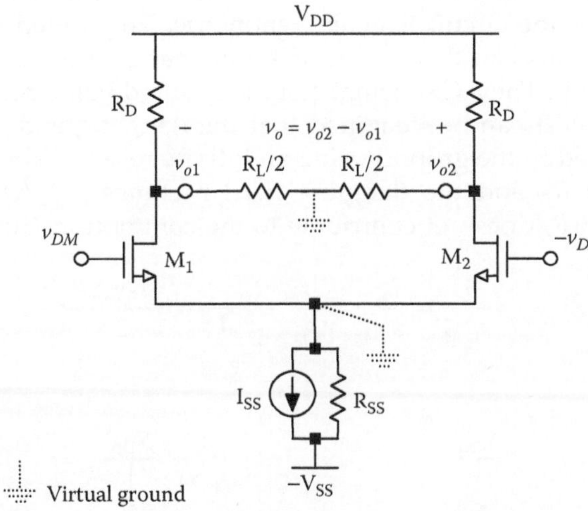

FIGURE EX. 5.8(b)

Now, all the nodes on the symmetry axis can be virtually grounded, and the circuit is split into two identical half-circuits. The only difference between the two half-circuits is the polarity of the input voltages, which causes 180° phase difference between the outputs v_{o1} and v_{o2}. Figure Ex. 5.8(c) depicts one of the differential-mode half-circuits, for which one can write:

$$\frac{v_{o1}}{v_{DM}} = -g_{m1}\left(R_D \middle\| \frac{R_L}{2}\right)$$

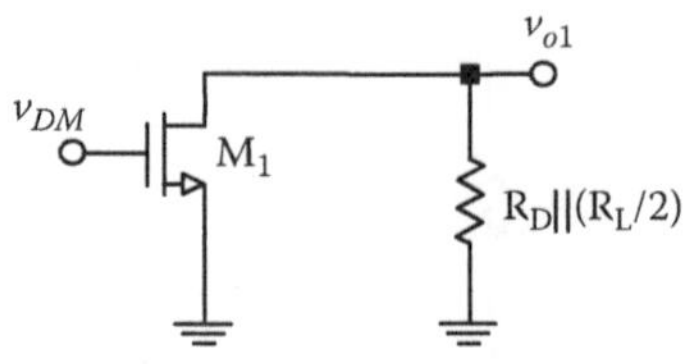

FIGURE EX. 5.8(c)

thus,

$$\frac{v_{o1}}{v_i} = \frac{-v_{o2}}{v_i} = -\frac{1}{2} g_{m1}\left(R_D \middle\| \frac{R_L}{2}\right).$$

Therefore, the differential-mode gain of the amplifier is:

$$A_{V,DM} = \frac{v_o}{v_i} = \frac{v_{o2} - v_{o1}}{v_i} = g_m\left(R_D \middle\| \frac{R_L}{2}\right).$$

In common mode, the circuit can be redrawn as illustrated in Figure Ex. 5.8(d). This way, the circuit is again symmetric. The dotted lines between the two half-circuits can then be cut without changing the electrical conditions of the circuit. The AC schematic of the resulted half-circuit is drawn in Figure Ex. 5.8(e). It can be imagined that the dangling end of the resistor $R_L/2$ is connected to the ground using an infinite resistor. Thus, the parallel combination of R_D and the dangling $R_L/2$ becomes R_D. As a result, the differential load R_L does not contribute to the common-mode gain:

$$A_{V,CM} = \frac{v_{o1}}{v_{CM}} = \frac{v_{o2}}{v_{CM}} = \frac{-R_D}{2R_{SS} + \dfrac{1}{g_{m1}}} = \frac{-g_{m1}R_D}{1 + 2g_{m1}R_{SS}}.$$

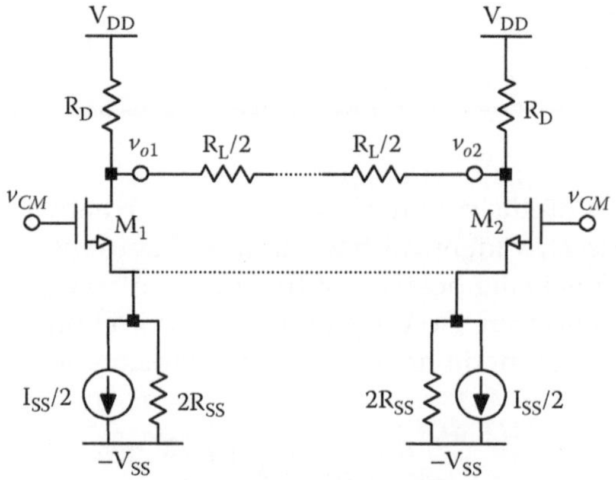

FIGURE EX. 5.8(d)

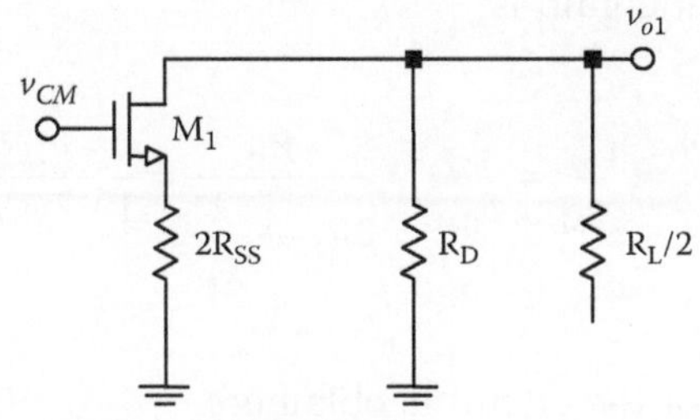

FIGURE EX. 5.8(e)

5.4.3.4 Bipolar Differential Amplifier

Bipolar version of the differential amplifier (see Figure 5.29) has exactly the same basics of operation as its CMOS counterpart. It can be shown that the differential-mode gains of the amplifier for differential and single-ended outputs are:

$$A_{V,DM} = \frac{v_o}{v_i} = \frac{v_{o2} - v_{o1}}{v_{i1} - v_{i2}} = g_m R_C \tag{5.51}$$

and

$$A_{V,DM\text{-}SE} = \frac{v_{o2}}{v_i} = \frac{-v_{o1}}{v_i} = \frac{1}{2} A_{V,DM} = \frac{g_m R_C}{2}, \tag{5.52}$$

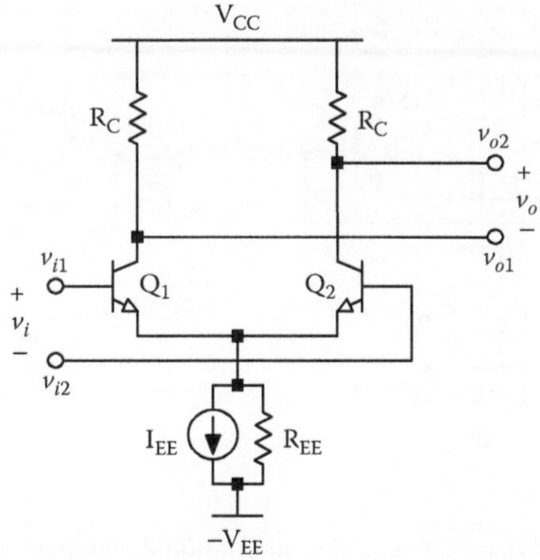

FIGURE 5.29
Bipolar differential amplifier with a real current source.

and the common-mode gain is:

$$A_{V,CM} = \frac{v_{o1}}{v_{CM}} = \frac{v_{o2}}{v_{CM}} = \frac{-R_C}{2R_{EE} + \dfrac{1}{g_m}} = \frac{-g_m R_C}{1 + 2g_m R_{EE}}. \tag{5.53}$$

Therefore, CMRR of the circuit is obtained:

$$CMRR = \left|\frac{A_{V,DM}}{A_{V,CM}}\right| = \left|\frac{g_m R_C}{\dfrac{-g_m R_C}{1 + 2g_m R_{EE}}}\right| = 1 + 2g_m R_{EE} \cong 2g_m R_{EE}. \tag{5.54}$$

Unlike the CMOS differential amplifier, which has infinite input resistance, two input resistances are defined for the bipolar differential amplifier: the input resistance in differential mode, $R_{i,DM}$, and the input resistance in common-mode, $R_{i,CM}$, illustrated in Figure 5.30(a) and (b), respectively. In the differential mode, node A is virtually grounded, and thus there will be a resistance of r_π seen from the base of each transistor to the ground, as presented in Figure 5.31(a):

$$R_{i,DM} = 2r_\pi. \tag{5.55}$$

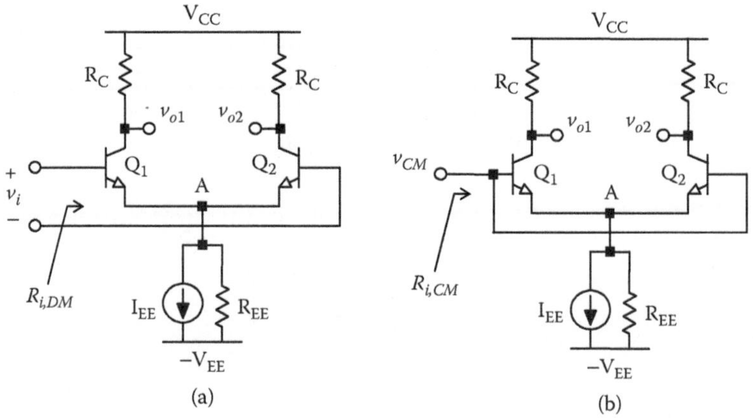

FIGURE 5.30
Input resistances of the differential amplifier in (a) differential mode (b) common mode.

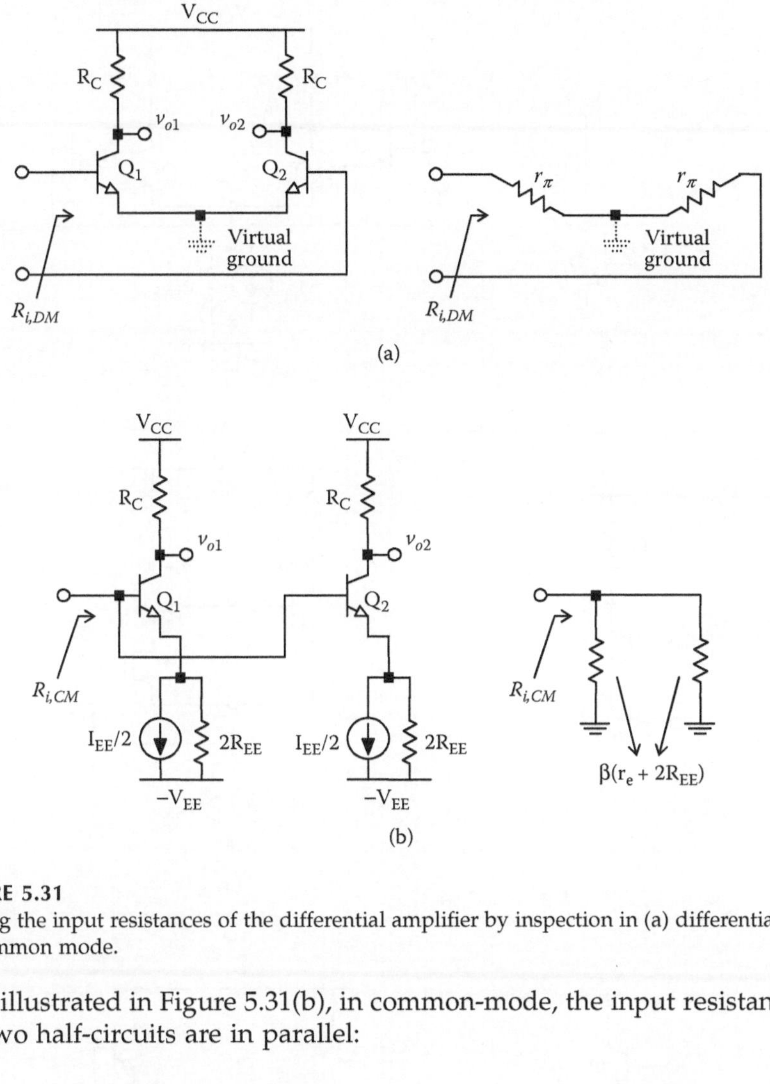

FIGURE 5.31
Finding the input resistances of the differential amplifier by inspection in (a) differential mode (b) common mode.

As illustrated in Figure 5.31(b), in common-mode, the input resistances of the two half-circuits are in parallel:

$$R_{i,CM} = [\beta(r_e + 2R_{EE}) \| \beta(r_e + 2R_{EE})] = \beta(r_e + 2R_{EE})/2 \cong \beta R_{EE}. \qquad (5.56)$$

5.4.3.5 Differential Amplifiers in General

The basic differential amplifier studied thus far was formed by properly connecting two identical common-emitter or common-source stages together and adding a current source for biasing. This idea can be generalized by using any other single- or multistage amplifier circuit as the primitive half-circuit and properly connecting and biasing them. Figure 5.32 gives examples of forming differential amplifiers using various primitive half-circuits.

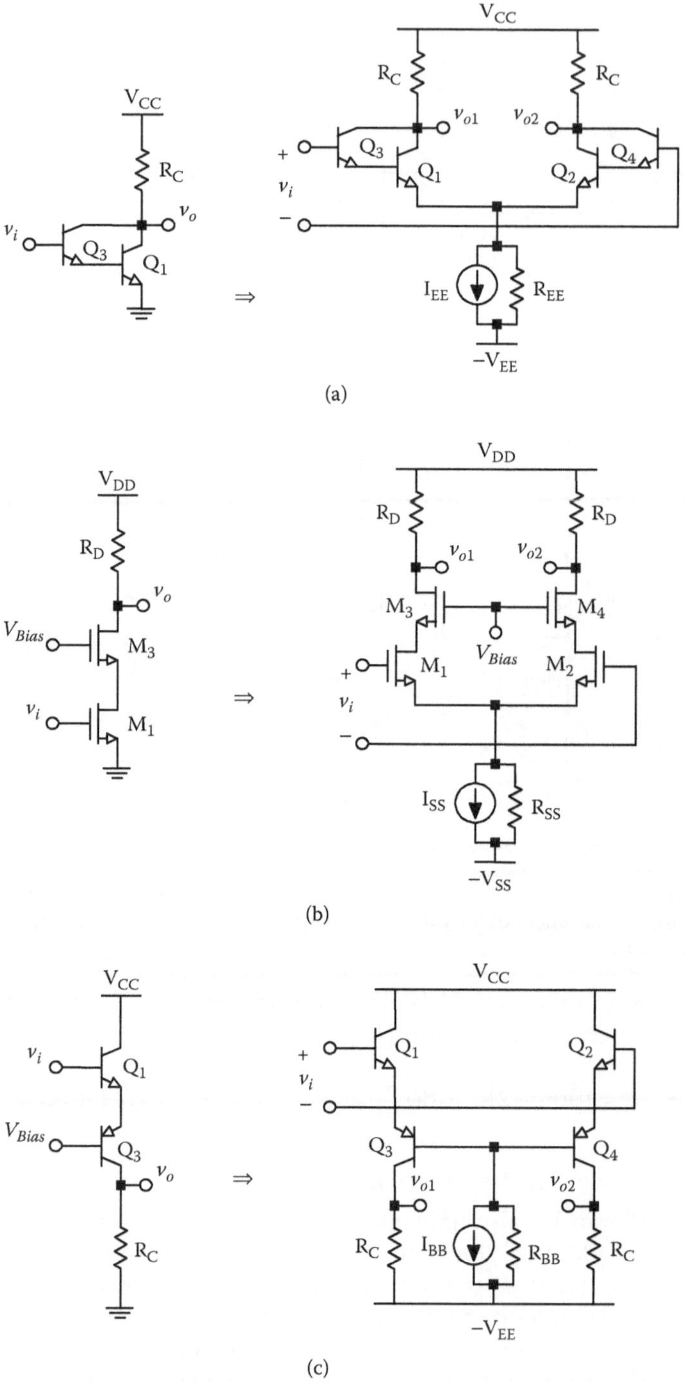

FIGURE 5.32
Differential amplifiers using (a) CE with Darlington pair (b) Cascode amplifier (c) EF-CB amplifier.

5.5 Simulation Examples

Example 5.9

For the two-stage amplifier in Figure Ex. 5.9(a), assume: $R_1 = 30\ k\Omega$, $R_2 = 10\ k\Omega$, $R_{C1} = 4\ k\Omega$, $R_{E1} = 330\ \Omega$, $R_{E2} = 800\ \Omega$, $R_{C2} = 2\ k\Omega$, $R_{SRC} = 1\ k\Omega$, $C_B = C_{E1} = C_{E2} = 1\ \mu F$, $V_{CC} = 3\ V$, Q_1: 2N3904, and Q_2: 2N3906. Apply a 10-kHz sinusoidal input voltage with peak amplitude of 0.1 mV, and find the voltage gain of each stage, and the overall gain of the amplifier. How are the gains resulted from simulations compared with the values obtained from analysis by inspection?

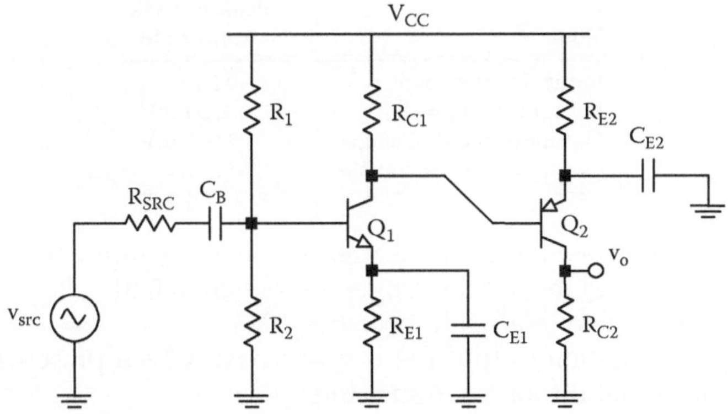

FIGURE EX. 5.9(a)

Solution:

```
VDC   VCC 0 3V
VSRC src 0 sin(0 0.1m 10k)
RSRC src A1 1k
CB  A1  B1 1U
R1  VCC B1 30K
R2  B1  0  10K
RC1 VCC C1 4K
RE1 E1  0  330
CE1 E1  0  1U
Q1  C1  B1 E1 Q2N3904
RE2 VCC E2 800
RC2 C2 0 2K
```

```
CE2 E2 0 1U
Q2  C2 C1 E2 Q2N3906
.LIB NOM.LIB
.TRAN 1U 500U 0 0.1U
.PROBE
.OP
.END
```

The outputs of the first and the second stages are presented in Figure Ex. 5.9(b). The measurements that are required to determine the voltage gains of the amplifier stages are listed in Table 5.1.

TABLE 5.1

Node	Peak-to-Peak Amplitude
Input signal source	0.2 mV
Input of the first stage	0.163 mV
Output of the first stage	4.996 mV
Output of the second stage	201.2 mV

Based on these measurements the attenuation at the input and the voltage gains for the first and the second stages are obtained 0.815, 30.65, and 40.27, respectively, and the overall gain becomes 1006.

Part of the simulation output file is given next, which presents the operating point information for the transistors:

```
**** OPERATING POINT INFORMATION  TEMPERATURE = 27.000 DEG C
*******************************************************

**** BIPOLAR JUNCTION TRANSISTORS

NAME          Q1           Q2
MODEL         Q2N3904      Q2N3906
IB            2.58E-06    -3.12E-06
IC            2.93E-04    -5.80E-04
VBE           6.33E-01    -6.91E-01
VBC          -1.11E+00     6.81E-01
VCE           1.74E+00    -1.37E+00
BETADC        1.14E+02     1.86E+02
GM            1.13E-02     2.23E-02
RPI           1.18E+04     8.29E+03
RX            1.00E+01     1.00E+01
```

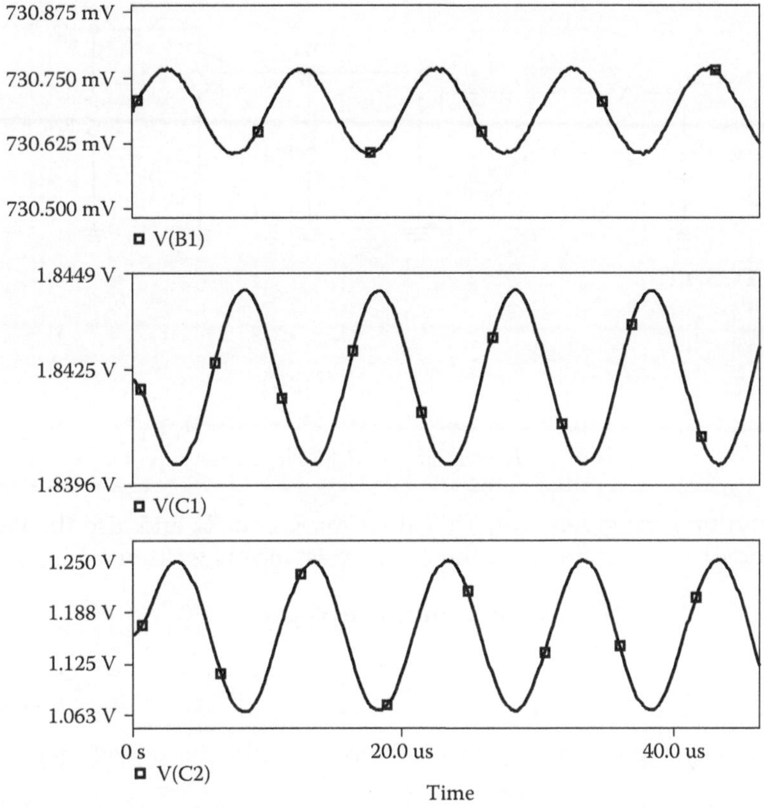

FIGURE EX. 5.9(b)

RO	2.57E+05	3.34E+04
CBE	9.73E-12	1.76E-11
CBC	2.75E-12	6.70E-12
CJS	0.00E+00	0.00E+00
BETAAC	1.33E+02	1.84E+02
CBX/CBX2	0.00E+00	0.00E+00
FT/FT2	1.44E+08	1.46E+08

From the AC schematic of the amplifier in Figure Ex. 5.9(c), the overall gain is written in terms of the attenuation at the input and the gains of the first and the second stages as:

$$A_V = \frac{v_o}{v_s} = \frac{v_i}{v_s} \cdot \frac{v_{o1}}{v_i} \cdot \frac{v_o}{v_{o1}} = \frac{R_i}{R_i + R_{SRC}} \cdot \frac{-\left(R_{C1} \| r_{\pi 2}\right)}{r_{e1}} \cdot \frac{-R_{C2}}{r_{e2}}$$

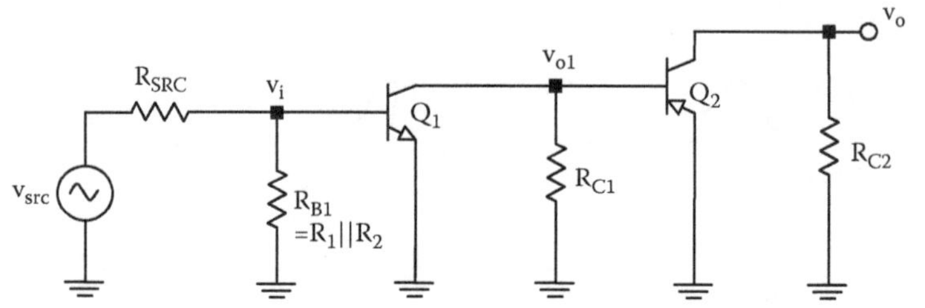

FIGURE EX. 5.9(c)

in which

$$R_i = R_B \| r_{\pi 1} = R_1 \| R_2 \| r_{\pi 1}.$$

Substituting the given numeric values for R_1 and R_2, and also the value of $r_{\pi 1}$ reported in the output file the input resistance is written as:

$$R_i = 30^{k\Omega} \| 10^{k\Omega} \| 11.8^{k\Omega}$$

$$\cong 4.6^{k\Omega}.$$

Thus, knowing g_{m1} and g_{m2} from the output file, the overall gain can be written as:

$$A_V = \frac{4.6^{k\Omega}}{4.6^{k\Omega} + 1^{k\Omega}} \cdot \frac{-\left(4^{k\Omega} \| 8.29^{k\Omega}\right)}{\dfrac{1}{11.3^{mA/V}}} \cdot \frac{-2^{k\Omega}}{\dfrac{1}{22.3^{mA/V}}}$$

$$= 0.821(-30.5)(-44.6) = 1115.$$

Table 5.2 compares the simulated gains with the values obtained by inspection, which are close enough to be approximated by the 10% rule explained before.

TABLE 5.2

Term	Simulation with SPICE	Analysis by Inspection
Attenuation at the input	0.815	0.821
First stage gain	–30.65	–30.5
Second stage gain	–40.27	–44.6
Overall gain	1006	1115

Example 5.10

Given $k' = 100\ \mu A/V^2$, $(W/L)_1 = (W/L)_2 = 50\ \mu m/0.5\ \mu m$, $V_{TH} = 0.6\ V$, $R_{SS} = 100\ k\Omega$, $R_D = 150\ k\Omega$, and $V_{DD} = V_{SS} = 5\ V$ in the differential amplifier in Figure Ex. 5.10(a),

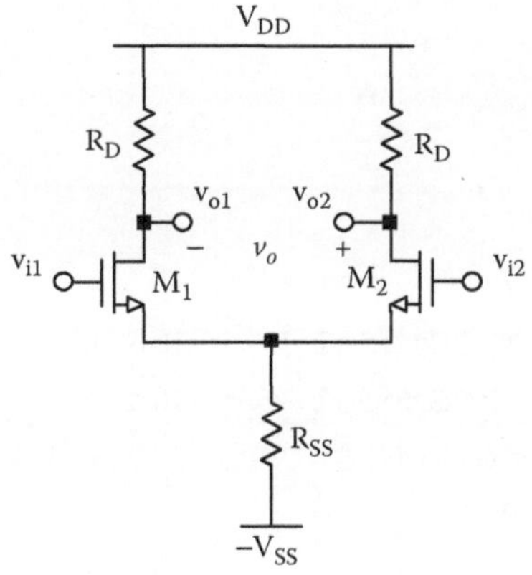

FIGURE EX. 5.10(a)

(a) Simulate the circuit when the inputs $v_{i1}(t)$ and $v_{i2}(t)$ have a common-mode component of $v_{CM}(t) = 1^{mV}Sin(2\pi\times 60^{Hz}t)$ and their differential-mode component is $v_{DM}(t) = 1^{mV}Sin(2\pi\times 1^{kHz}t)$. Plot the inputs $v_{i1}(t)$ and $v_{i2}(t)$, single-ended outputs $v_{o1}(t)$ and $v_{o2}(t)$, differential output $v_o(t) = v_{o2}(t) - v_{o1}(t)$, and the common-mode component of the output $v_{o,CM}(t) = (v_{o1}(t) + v_{o2}(t))/2$.

(b) Determine the common-mode gain $A_{VCM} = v_{o,CM}/v_{i,CM}$, differential-mode gain $A_{VDM} = v_o/(v_{i1} - v_{i2})$, and the CMRR of the amplifier.

Solution:

(a) This circuit with the given numeric values for the devices and components is indeed the circuit of Example 5.6; the inputs of this example are made as depicted in Figure Ex. 5.10(b).

```
VDC1 VDD 0 5V
VDC2 VSS 0 -5V
VCM C 0 Sin(0 1m 60)
VDM1 G1 C Sin(0 0.5m 1k)
VDM2 C G2 Sin(0 0.5m 1k)
RD1 VDD D1 150k
RD2 VDD D2 150k
```

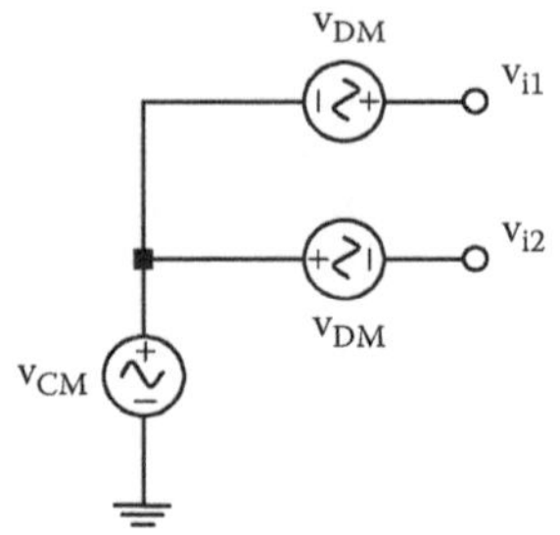

FIGURE EX. 5.10(b)

```
RSS S VSS 100k
M1 D1 G1 S S NMOSmodel W = 50u L = 0.5u
M2 D2 G2 S S NMOSmodel W = 50u L = 0.5u
.MODEL NMOSmodel NMOS (VTO = 0.6 kp = 100u)
.TRAN 1U 20m 0 1U
.PROBE
.OP
.END
```

The inputs and outputs resulted from simulation are presented in Figure Ex. 5.10(c).

(b) Table 5.3 lists the peak-to-peak amplitudes for both single-ended and differential outputs in the differential mode, and the common-mode component of the output along with the associated gains:

TABLE 5.3

Component	Input Amplitude (P-P)	Output Amplitude (P-P)	Voltage Gain
Differential-Mode, Single-ended	2 mV	99 mV	49.5
Differential-Mode, Differential	2 mV	198 mV	99
Common-mode	2 mV	1.49 mV	0.745

Common-mode rejection ratio of the amplifier is then calculated as:

$$CMRR = \left|\frac{A_{V,DM}}{A_{V,CM}}\right| = \left|\frac{198}{0.745}\right| = 133.$$

As the results indicate, simulation results are in close agreement with the results obtained from the analysis by inspection in Example 5.6. The interesting observation in this example is how the differential amplifier rejects the common-mode component of the inputs and amplifies the differential signal.

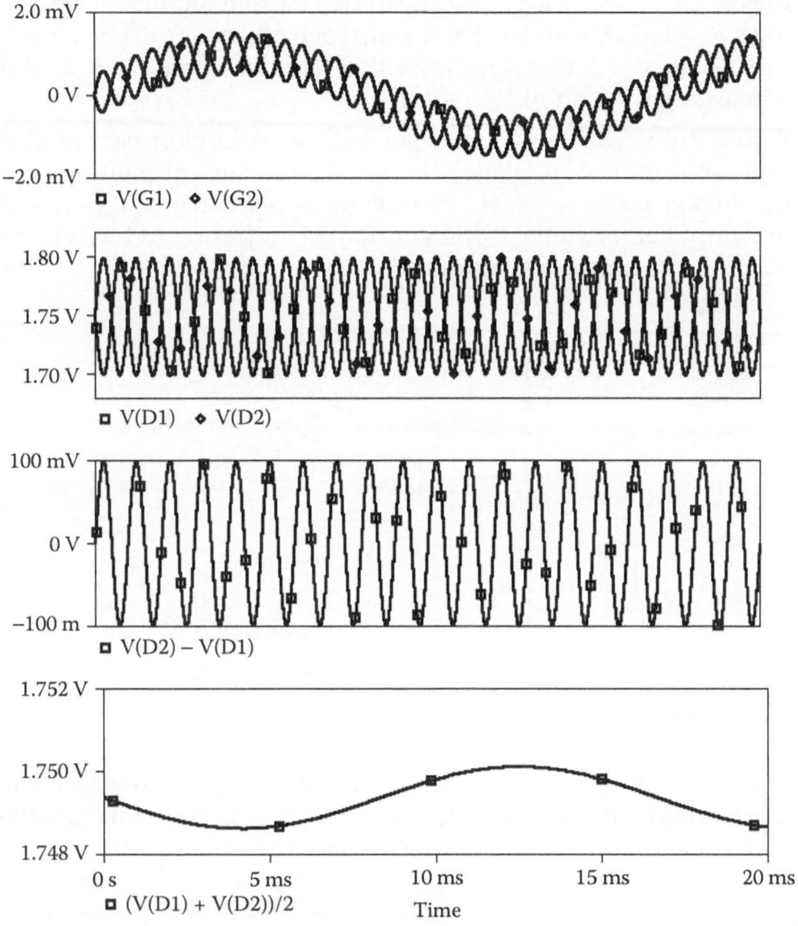

FIGURE EX. 5.10(c)

5.6 Problems

5.1 To achieve a high voltage gain, two identical amplifiers are cascaded. Given $R_i = 10\ k\Omega$, $R_o = 2\ k\Omega$, and $A_V = 100$ for each amplifier,

(a) Find the overall voltage gain for the two-stage amplifier.

(b) What will be the overall voltage gain if an input signal source of voltage type (v_s) with internal resistance of $R_S = 1\ k\Omega$ provides the input signal, and the output voltage is delivered to a resistive load of $R_L = 5\ k\Omega$?

5.2 Repeat problem 5.1 for a cascade of two current amplifiers with $R_i = 5\ k\Omega$, $R_o = 20\ k\Omega$, and $A_I = 100$ for each one of the amplifiers, an input signal source of current type (i_s) with internal resistance of $R_S = 40\ k\Omega$, and a resistive load of $R_L = 4\ k\Omega$.

5.3 Figure Prob. 5.3 is a CE amplifier with a Darlington pair. Find the input and output resistances R_i and R_o, the current gain $A_I = i_o/i_i$, the voltage gain $A_V = v_o/v_i$, and the transconductance $G_M = i_o/v_i$ for the amplifier. Assume $V_{BE}(ON) = 0.7\ V$, $V_{CE}(Sat.) = 0.1\ V$, $V_A \to \infty$, and $\beta = 100$ for the transistors.

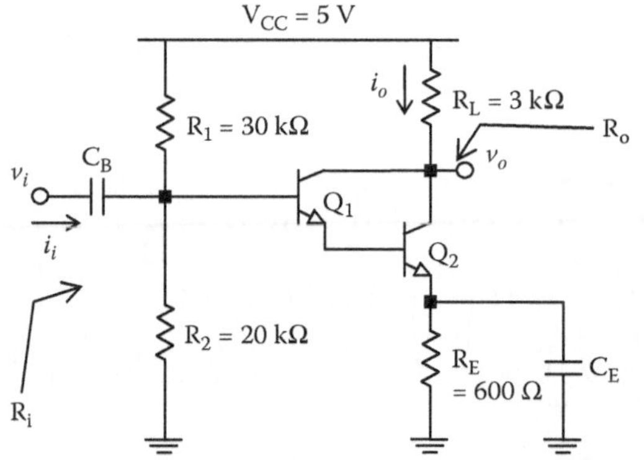

FIGURE PROB. 5.3

5.4 Repeat Problem 5.3 for the amplifier with quasi-Darlington configuration and current source in Figure Prob. 5.4. How this amplifier is compared with the circuit of problem 5.3?

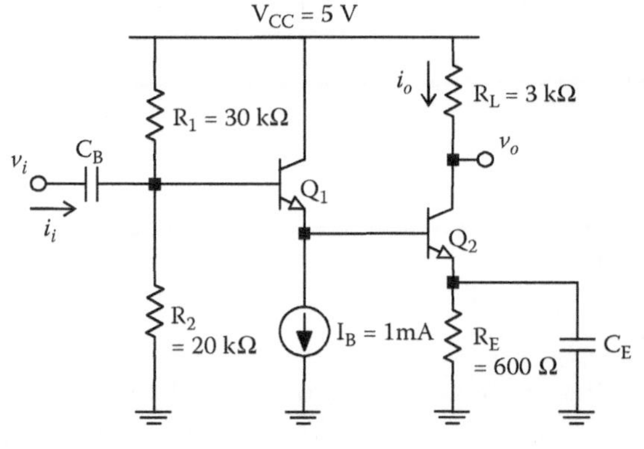

FIGURE PROB. 5.4

5.5 A super-β compound is introduced in Figure Prob. 5.5. Find the equivalent β, g_m, $V_{BE}(ON)$, and $V_{CE}(Sat.)$ for this compound, and compare them with those of a Darlington pair and a single transistor.

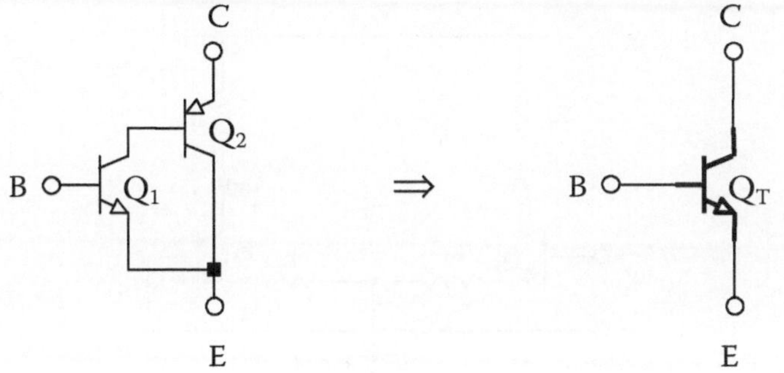

FIGURE PROB. 5.5

5.6 Figure Prob. 5.6 is a cascode amplifier. Assume that the transistors have $k' = 100\ \mu A/V^2$, (W/L) = 50, $V_{TH} = 0.7\ V$, and $\lambda = 0.01\ V^{-1}$, and are biased at 100 μA.

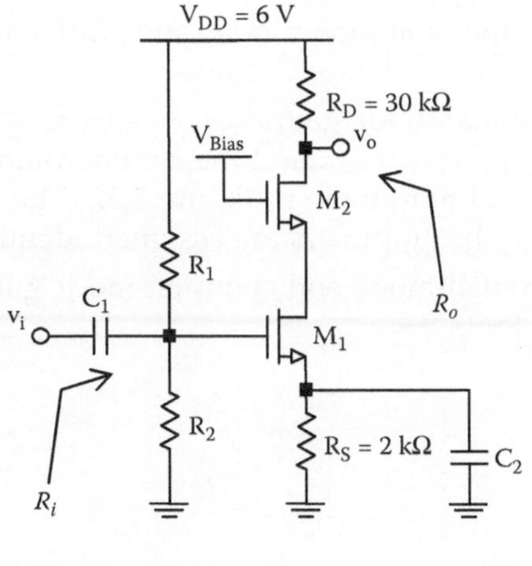

FIGURE PROB. 5.6

(a) Find the voltage gain of the amplifier $A_V = v_o/v_i$, input resistance, R_i, and output resistance, R_o.

(b) What is the acceptable range for V_{Bias} so that both transistors remain in saturation?

5.7 Assuming $\beta = 100$, $V_A = 120\text{V}$, $R_C = 50\text{k}\Omega$, $R_E = 5\ k\Omega$, $R_{EE} = 20\ M\Omega$, $I_{EE} = 200\ \mu A$, and $V_{CC} = V_{EE} = 5\ V$, for the differential amplifier in Figure Prob. 5.7,

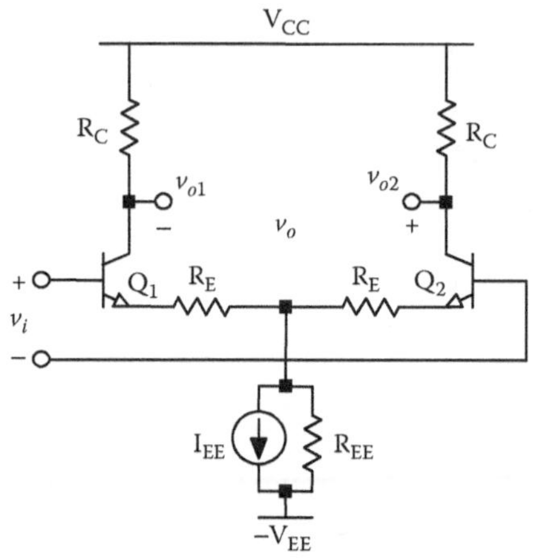

FIGURE PROB. 5.7

(a) Find the differential-mode gain $A_{VDM} = v_o/v_i$, common-mode gain $A_{VCM} = v_{oCM}/v_{iCM}$, and the common-mode rejection ratio.

(b) Determine the common-mode and differential-mode input resistances.

5.8 Find the differential-mode gain $A_{VDM} = v_o\ (= v_{o2} - v_{o1})/v_i$, common-mode gain $A_{VCM} = v_{oCM}/v_{iCM}$, and the common-mode rejection ratio for the differential amplifiers in Figure 5.32. The mirrored devices on both sides of the amplifiers are assumed identical.

5.9 Find the differential-mode and common-mode gains for the amplifiers in Figure Prob. 5.9.

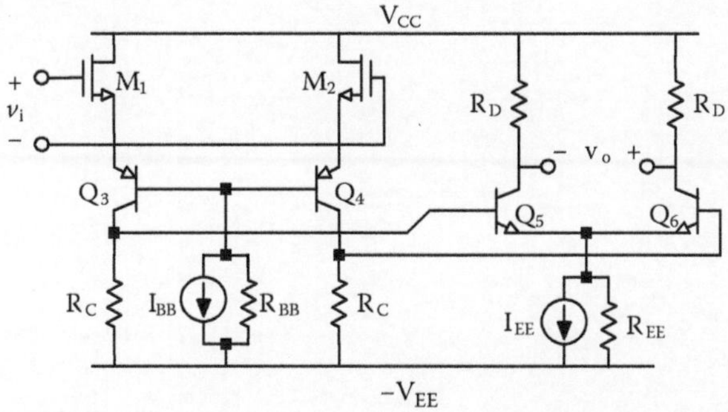

(a)

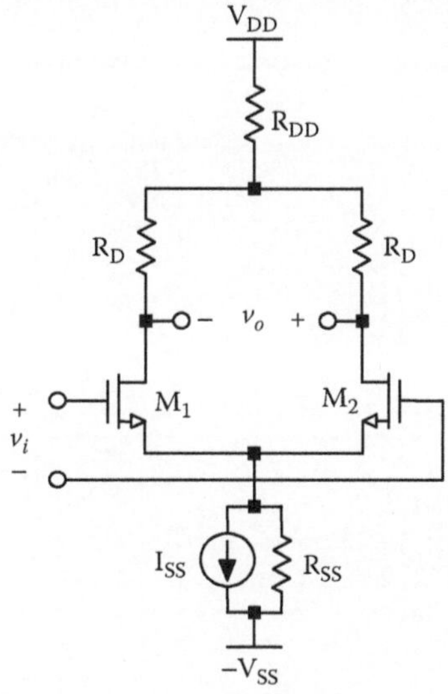

(b)

FIGURE PROB. 5.9

Chapter 6

Current Sources/Mirrors

6.1 Introduction

Current sources are one of the important building blocks that play a key role in almost all fields of electronic circuits, from analog to digital, from discrete to integrated, and from voltage-mode to current-mode. The main purpose of a current source is to keep the current flowing through a circuit branch as intended regardless of the voltage across it. If this current is defined independent from other voltages/currents in the circuit, the current source will be called an *independent current source.* In addition, *dependent current sources* are used to generate a current that is a (usually linear) function of another current or voltage in the circuit. The associated circuit symbols are given in Figure 6.1.

In electronic circuits, an independent current source providing a constant current is called a *current source*, and a dependent current source where the output current is proportional to another current is known as a *current mirror.*

In this chapter, basics of operation, analysis, and design of both current sources and current mirrors are explained. As will be seen, almost all current mirror configurations can also be used to make a current source. This is why sometimes the explanations for a current mirror and a similar current source of the same configuration are combined.

6.2 Simple Current Source/Mirror

To implement a current source or mirror in the real world we need to employ passive and active circuit elements in such a way that the intended functionality is achieved. The simplest implementation of a current source is depicted in Figure 6.2, in which a *negative-channel metal-oxide semiconductor* (NMOS) transistor is biased at a fixed current.

The first restriction in implementing a current source in the real world is that both ends of a real current source cannot be arbitrarily placed. A real

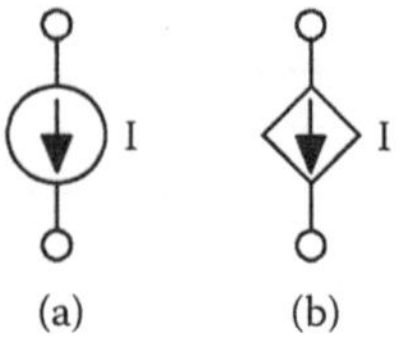

FIGURE 6.1
Circuit symbol for (a) an independent current source and (b) a dependent current source.

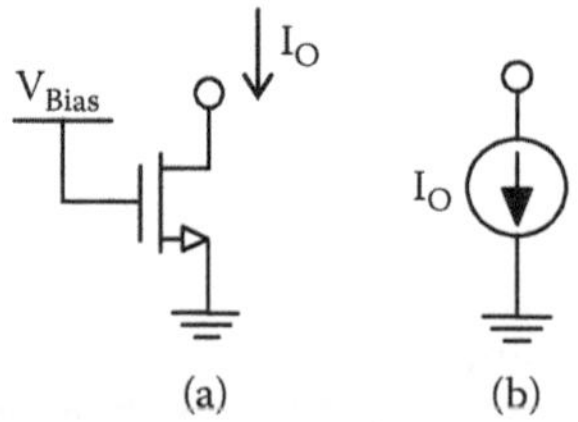

FIGURE 6.2
Basic concept of realizing a current source: (a) implementation, (b) ideal model.

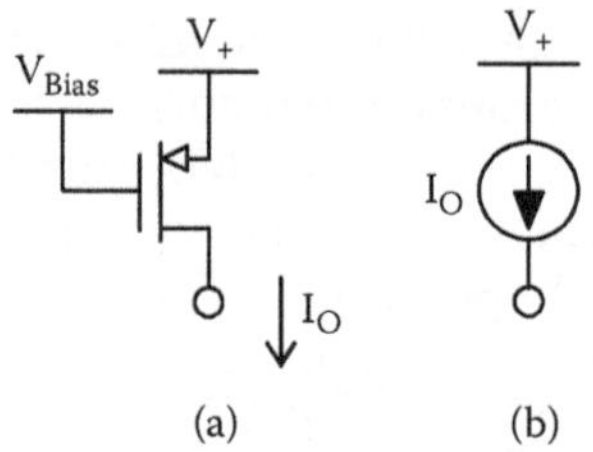

FIGURE 6.3
PMOS counterpart of the single-transistor current source of Figure 6.2: (a) implementation, (b) ideal model.

current source can either sink a certain current from the output node and pass it to the ground (or the negative power supply), or draw a predefined current from the positive power supply and source it to the output node. The latter can be realized by using a *positive-channel metal-oxide semiconductor* (PMOS) transistor (see Figure 6.3).

To operate as a good current source, the transistors in Figure 6.2 and Figure 6.3 should be biased in saturation mode. This implies that a voltage greater than the transistor's threshold voltage should be applied across the gate-source, and the drain-source voltage should be kept above the overdrive voltage. This is indeed the second deviation from the definition of an ideal current source that restricts the voltage across the current source. Assuming that the transistor is biased under the previously mentioned voltage conditions, the output current would be

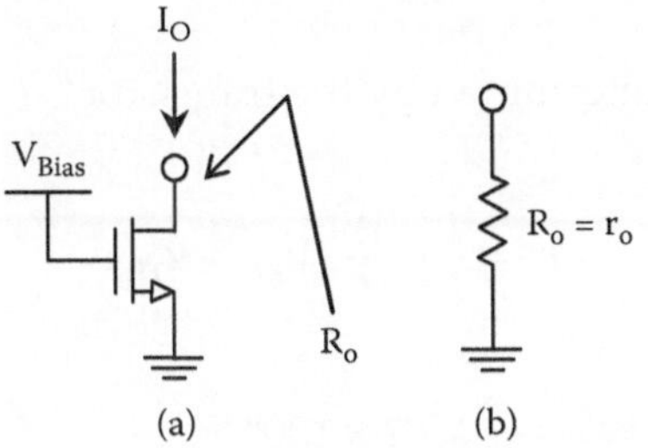

FIGURE 6.4
(a) Output resistance of a current source, (b) AC equivalent for a real current source.

$$I_O = \frac{1}{2} k'_N \frac{W}{L} (V_{Bias} - V_{THN})^2 \tag{6.1}$$

for the NMOS current source of Figure 6.2, and

$$I_O = \frac{1}{2} k'_P \frac{W}{L} \left(V_+ - V_{Bias} - |V_{THP}| \right)^2 \tag{6.2}$$

for its PMOS counterpart in Figure 6.3.

The third factor of nonideality for a real current source is its output resistance. Suppose that one of the current sources in Figure 6.2 and Figure 6.3 is used in a circuit where small-signal variations occur in currents and voltages. As depicted in Figure 6.4, the transistor can be modeled in AC mode by the resistance seen from its drain to the ground, r_o. The output resistance represents small-signal variations of the output current of a real current source, caused by the small-signal variations of the voltage across it.

Another concern in the design of the single-transistor current source is the generation of the gate voltage. In the following example, we will see how critical this issue is.

Example 6.1

The single-transistor current source of Figure 6.2 is to be designed for the output current of 100 μA. Assume V_{TH} = 0.7 V, k' = 200 $\mu A/V^2$, and λ = 0.01 V^{-1} for the transistor, and V_{Bias} = 1.2 V. Assuming that the channel-length modulation is negligible,

(a) Determine the required W/L for the transistor.

(b) If a power supply is used to provide the gate voltage (either directly or through a voltage division), and only ±1% voltage ripple is on V_{Bias}, how will the output current change?

Solution:

(a) W and L can be determined by the transistor's I_D-V_{GS} characteristics:

$$I_D = \frac{1}{2}k'\frac{W}{L}(V_{Bias} - V_{TH})^2$$

$$\frac{W}{L} = \frac{2 \times 100^{\mu A}}{200^{\mu A/V^2}(1.2^V - 0.7^V)^2} = 4$$

(b) Taking the ±1% ripple on V_{Bias} into account, we have:

$$V_{Bias(Max)} = 1.2+0.012 = 1.212 \text{ V}$$

$$\Rightarrow I_{D(Max)} = ½(4)(200\ \mu A/V^2)(1.212 \text{ V} - 0.7 \text{ V})^2 = 104.86\ \mu A$$

and

$$V_{Bias(min)} = 1.2 - 0.012 = 1.188 \text{ V}$$

$$\Rightarrow I_{D(min)} = ½(4)(200\ \mu A/V^2)(1.188 \text{ V} - 0.7 \text{ V})^2 = 95.26\ \mu A$$

which indicates almost ±5% variation in the output current.

To reduce the sensitivity of the transistor's output current to power supply variations, one can use the transistor's square law to generate a proper gate-source voltage as depicted in Figure 6.5. In this circuit, hereafter called the *reference branch*, the gate-source voltage, V_{GS}, is obtained by simultaneously solving

$$V_{GS} = V_{DD} - I_{REF}R_{REF},$$

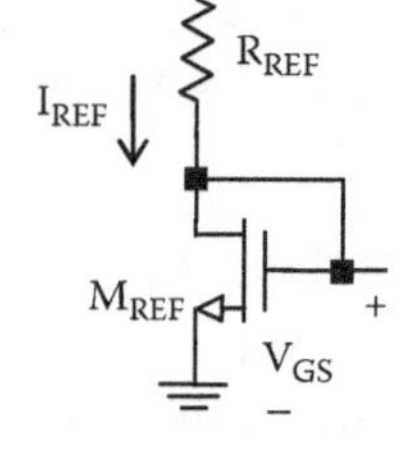

FIGURE 6.5
A reference branch to generate V_{GS}.

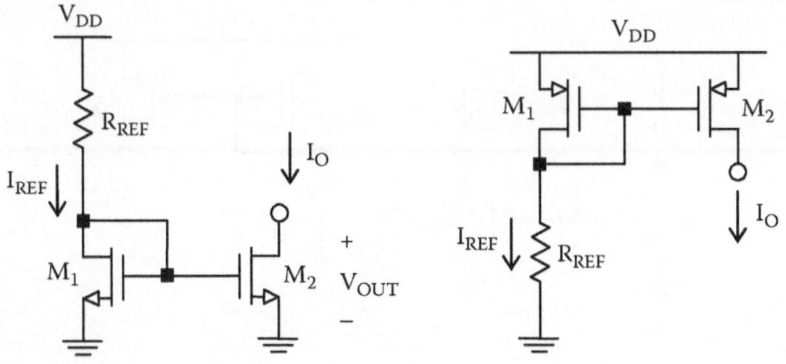

FIGURE 6.6
Simple current sources.

and

$$I_{REF} = \frac{1}{2} k'_N \frac{W}{L} (V_{GS} - V_{TH})^2.$$

It can be shown that in this circuit, variations in the supply voltage, V_{DD}, result in variations in the reference current (I_{REF}) and the generated V_{GS}, which are almost proportional to the supply variations and their square root, respectively. Now, if the prepared V_{GS} is applied to the transistor that acts as the current source (M_2), the so-called *simple current source* is formed (see Figure 6.6). To relate the output current and the reference current, one can write:

$$\frac{I_O}{I_{REF}} = \frac{(\beta_2/2)(V_{GS2} - V_{TH2})^2}{(\beta_1/2)(V_{GS1} - V_{TH1})^2}. \tag{6.3}$$

Assuming that M_1 and M_2 are physically identical, applying the same gate-source voltage will result in the same drain currents:

$$I_O = I_{REF}. \tag{6.4}$$

This approach of preparing the gate-source voltage for the current sourcing transistor (M_2) makes the output current less sensitive to supply voltage variations compared with the basic idea presented in Figure 6.4.

The basic concept of current copying in Figure 6.6 comes from both using identical transistors and the way they are connected, not how the reference current is generated. By eliminating R_{REF}, we will arrive at the so-called *simple current mirror*. As long as M_1 and M_2 are in saturation, the input current, i_{IN}, is copied and delivered to the output as i_O (see Figure 6.7).

Basically, a current mirror is a circuit that receives a current at the input and provides a copy of it at the output. This function can be theoretically

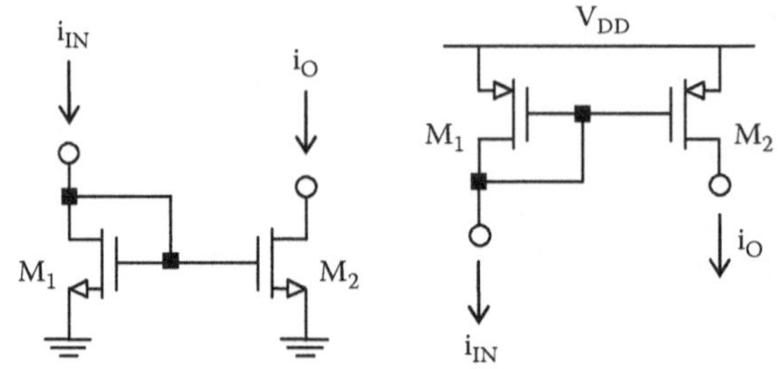

FIGURE 6.7
Simple current mirrors.

modeled by a dependent current source, as mentioned before. Current mirrors are used to copy both DC and AC currents. In the applications where a current mirror is mainly employed to take part in the AC function of a circuit, its input and output resistances become important. Because the input and output signals are currents, an ideal current mirror is expected to have zero input resistance and infinite output resistance. For the real current mirrors in Figure 6.7, we have:

$$R_i = 1/g_{m1}, \tag{6.5}$$

and

$$R_{out} = r_{o2}. \tag{6.6}$$

Example 6.2

Given W = 10 μm, L = 1 μm, $k' = 100\ \mu A/V^2$, $V_{TH} = 0.8$ V, and $\lambda = 0.01\ V^{-1}$ for the transistors, $V_{DD} = 5V$, and $R_{REF} = 100\ k\Omega$ in the simple current source depicted in Figure Ex. 6.2, find the output current, output resistance, and the minimum output voltage.

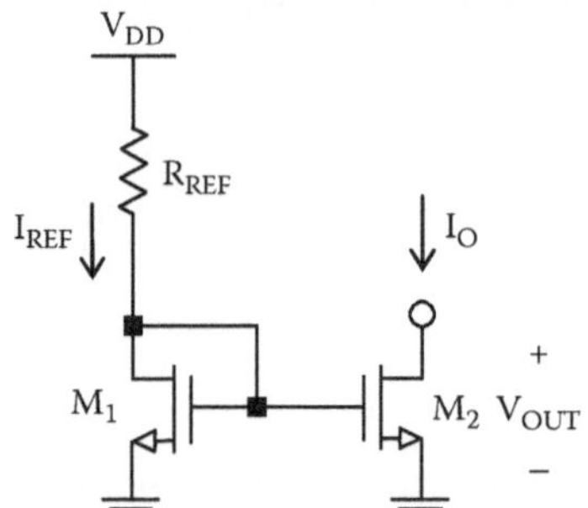

FIGURE EX. 6.2

Solution:
The parameter β is calculated as:

$$\beta = k'\frac{W}{L} = 100^{\mu A/V^2} \times \frac{10^{\mu m}}{1^{\mu m}} = 1^{mA/V^2}$$

The reference current, I_{REF}, is obtained using the KVL equation:

$$\text{KVL: } V_{DD} = I_{REF}R_{REF} + V_{GS1},$$

in which

$$I_{REF} = I_{D1}.$$

Thus:

$$V_{GS1} = V_{DD} - I_{D1}R_{REF},$$

which, if put in the transistor's characteristic equation, gives:

$$I_{D1} = \frac{\beta}{2}(V_{GS1} - V_{TH})^2$$

$$= \frac{\beta}{2}(V_{DD} - I_{D1}R_{REF} - V_{TH})^2$$

$$= \frac{1^{mA/V^2}}{2}(5^V - (I_{D1} \times 100^{k\Omega}) - 0.8^V)^2.$$

This equation can be rewritten as follows:

$$10^4 I_{D1}^2 - 842 I_{D1} + 17.64 = 0,$$

The roots of this quadratic equation are I_{D1} = 39.2 μA and 45 μA, which correspond to V_{GS1} = 1.08 V and 0.5 V, respectively. Only I_{D1} = 39.2 μA is acceptable for this circuit because the transistor needs a gate-source voltage greater than the threshold voltage to conduct current. Because M_1 and M_2 are matched, the output current is obtained:

$$I_O = I_{D1} = 39.2^{\mu A}.$$

Then, the output resistance is achieved:

$$R_{OUT} = r_{o2} = \frac{1}{\lambda I_{D2}} = \frac{1}{0.01^{V^{-1}} \times 39.2^{\mu A}} = 2.55^{M\Omega},$$

and the minimum output voltage is obtained:

$$V_{OUT}(\text{min}) = V_{DS2}(Sat.) = \sqrt{\frac{2I_{D2}}{\beta}} = \sqrt{\frac{2 \times 39.2^{\mu A}}{1^{mA/V^2}}} = 0.28^V.$$

There is another factor of nonideality in the operation of a simple current source/mirror. If the dependency of the drain current upon the drain-source voltage (channel-length modulation phenomenon) is taken into account, equation (6.3) should be rewritten as:

$$\frac{I_O}{I_{REF}} = \frac{(\beta_2/2)(V_{GS2} - V_{TH2})^2(1+\lambda_2 V_{DS2})}{(\beta_1/2)(V_{GS1} - V_{TH1})^2(1+\lambda_1 V_{DS1})} \tag{6.7}$$

for the current source of Figure 6.6. Unlike the other terms, the drain-source voltages for M_1 and M_2 are not necessarily equal. Because M_1 is diode-connected, V_{DS1} is forced to be equal to V_{GS1}, whereas V_{DS2} (V_{OUT}) is determined by the network connected to the drain of M_2. Thus, assuming that the transistors are physically identical, Equation (6.7) turns out to be:

$$\frac{I_O}{I_{REF}} = \frac{1+\lambda V_{OUT}}{1+\lambda V_{GS1}} \tag{6.8}$$

for the simple current source. It was explained that the output resistance represents the variations of the output current resulting from AC variations of the output voltage. Now, Equation (6.8) explains how the output current is affected by the DC component of the output voltage. For example, assuming that $\lambda = 0.01\ V^{-1}$, $V_{GS1} = 1$ V, and $V_{DS2} = 4$ V, this term introduces about 3% difference between the input current (I_{REF}) and the output current. It should be noted, however, that, in general, the output voltage in Equation (6.8) can be the whole voltage (including both DC and AC components).

One use of current sources is in differential amplifiers as the tail current source (see Figure 6.8). In these circuits, the reference branch consisting of M_3 and R_{REF} only provides M_4 with the DC gate voltage that it requires to generate the current I_{SS}. Thus, in AC analysis the gate of M_4 is grounded and the reference branch does not appear in the AC schematic of the amplifier at all. The internal resistance of the tail current source, R_{SS}, is then the resistance seen into the drain of M_4 while its gate is grounded, which is r_{o4}.

Although the output resistance of a simple current source is large enough to yield a large CMRR for the amplifier, one can use current sources with higher output resistance (e.g., the cascode current source introduced in the next section) as the tail current source to further increase the common-mode rejection ratio.

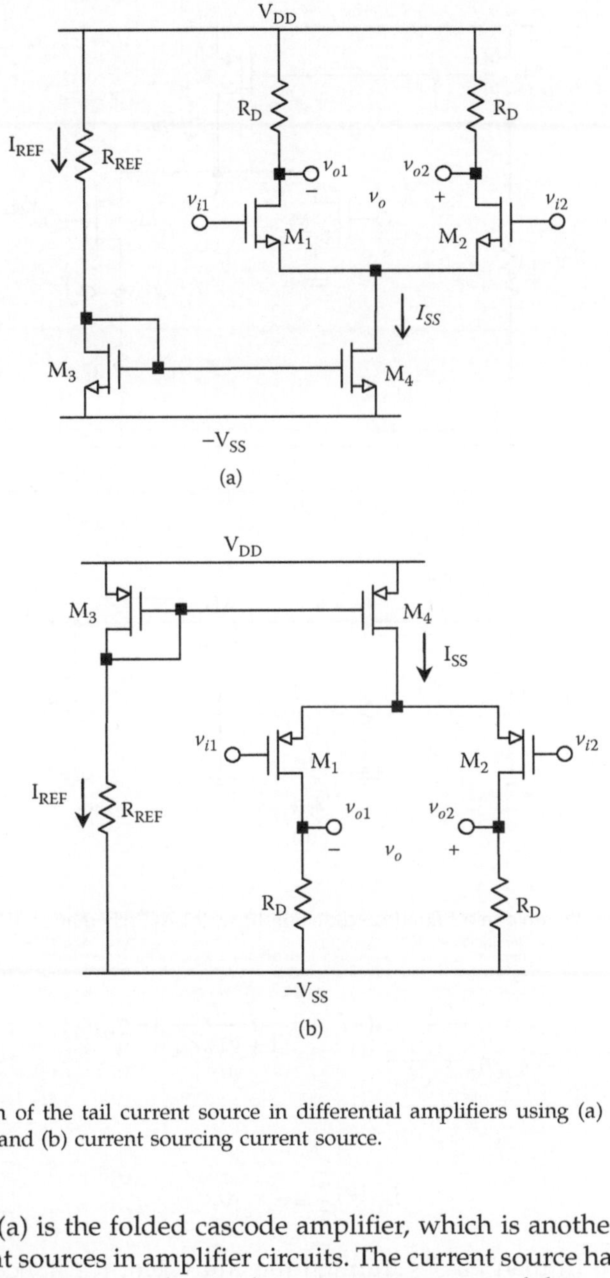

FIGURE 6.8
Implementation of the tail current source in differential amplifiers using (a) current sinking current source and (b) current sourcing current source.

Figure 6.9(a) is the folded cascode amplifier, which is another example of using current sources in amplifier circuits. The current source has a relatively large internal resistance, due to the output resistance of the current source, r_{o4}. As shown in the AC schematic of Figure 6.9(b), the AC current $-g_{m1}v_i$, which arrives at node A, is divided between the large output resistance of the current source, r_{o4}, and the small input resistance of the common-gate stage, $1/g_{m2}$. As a result, the majority of the current goes into M_2:

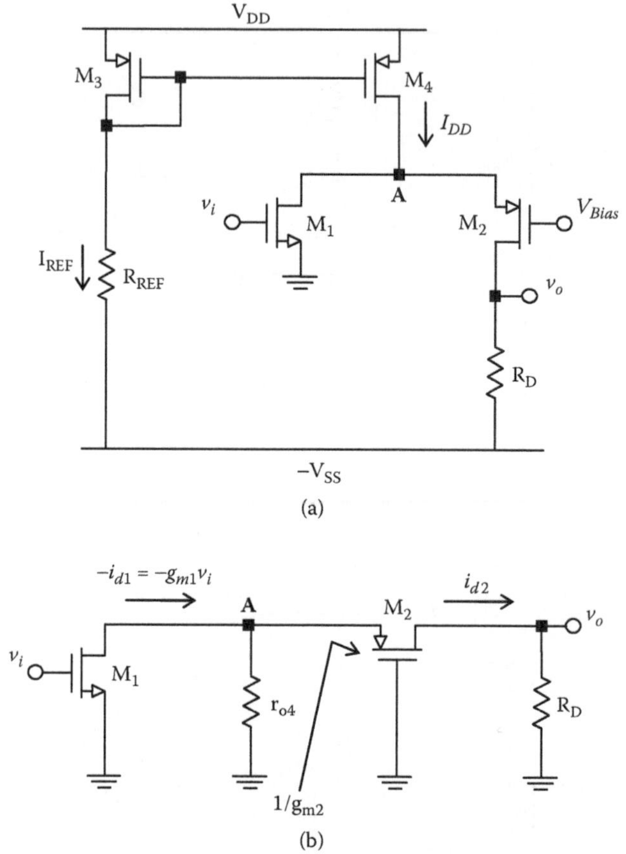

FIGURE 6.9
(a) Using a current source in a folded cascode amplifier, (b) AC schematic.

$$i_{s2} = \frac{r_{o4}}{r_{o4} + \dfrac{1}{g_{m2}}} \cdot (-i_{d1}) = \frac{g_{m2} r_{o4}}{1 + g_{m2} r_{o4}} \cdot (-g_{m1} v_i),$$

and thus,

$$i_{d2} = i_{s2} \cong -g_{m1} v_i.$$

The current i_{d2} then passes through the equivalent resistance seen at the output node, which is:

$$R_o = R_D \| [g_{m2} r_{o2} (r_{o1} \| r_{o4})] \cong R_D,$$

and makes the output voltage as:

$$v_o = i_{d2} R_D = (-g_{m1} v_i) R_D.$$

Therefore, the voltage gain of the amplifier is obtained:

$$A_V = \frac{v_o}{v_i} = -g_{m1} R_D.$$

To conclude, Figure 6.10 illustrates the i_{OUT}–v_{OUT} characteristics for a current source in the ideal case, and also when the minimum output voltage and the output resistance nonidealities are taken into account.

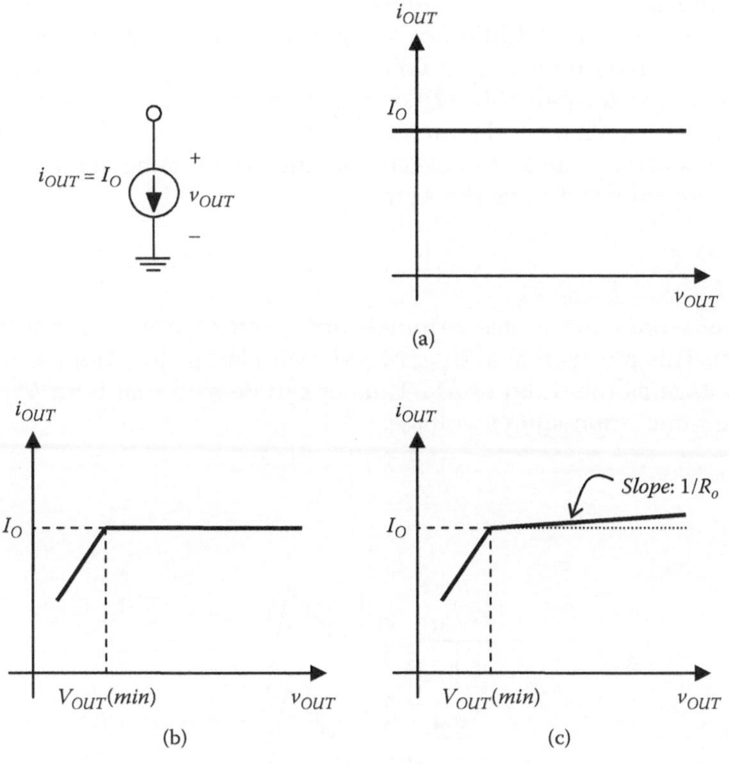

FIGURE 6.10
i_{OUT}–v_{OUT} characteristics of a current source (a) ideal, (b) considering the minimum output voltage, and (c) taking the output resistance into account.

6.3 Cascode Current Source/Mirror

To both increase the output resistance of a simple current source/mirror and make the output current less sensitive to the output voltage, the *cascode current source/mirror* is introduced.

The basic idea to increase the output resistance of the current source/mirror is to stack a second transistor on top of the original current sourcing transistor (see Figure 6.11). The transistor M_4 magnifies the resistance seen into the drain of M_2 (r_{o2}) to:

$$R_o = r_{o4}(1 + g_{m4}r_{o2}) \approx g_{m4}\, r_{o4}\, r_{o2}. \tag{6.9}$$

To generate V_{Bias1} and V_{Bias2}, a reference branch is added (see Figure 6.12), which functions in the same way that was explained for the simple current source. It is worth noting that the current copying function is still performed by M_1 and M_2.

To explain how the addition of M_3 and M_4 can reduce the sensitivity of the output current to the output voltage, let us assume that we have stacked one more transistor pair (M_3, M_4) on top of the basic current source/mirror (M_1, M_2), as visualized in Figure 6.13. If M_3 and M_4 are identical, because their drain currents are (either exactly or almost) the same, their gate-source voltages are expected to be the same:

$$V_{GS3} = V_{GS4}. \tag{6.10}$$

In other words, the source terminals of M_3 and M_4 are almost at the same potential. This means that adding M_3–M_4 pair clamps the drain of M_2 at the same voltage as the drain of M_1. Thus, it can be said that both M_1 and M_2 have the same drain-source voltage:

$$V_{DS1} = V_{DS2}. \tag{6.11}$$

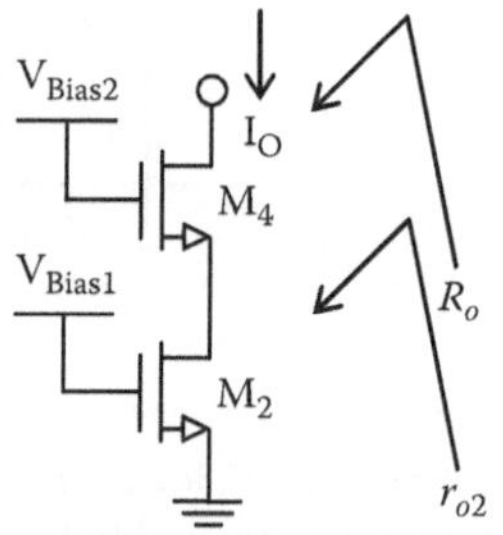

FIGURE 6.11
Increasing the output resistance by stacking a second transistor.

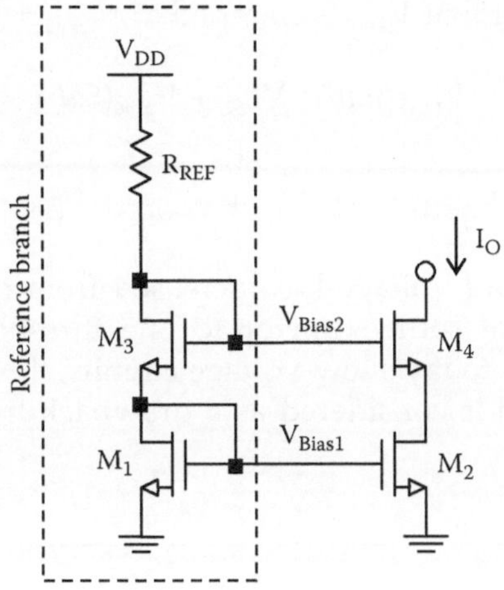

FIGURE 6.12
The cascode current source.

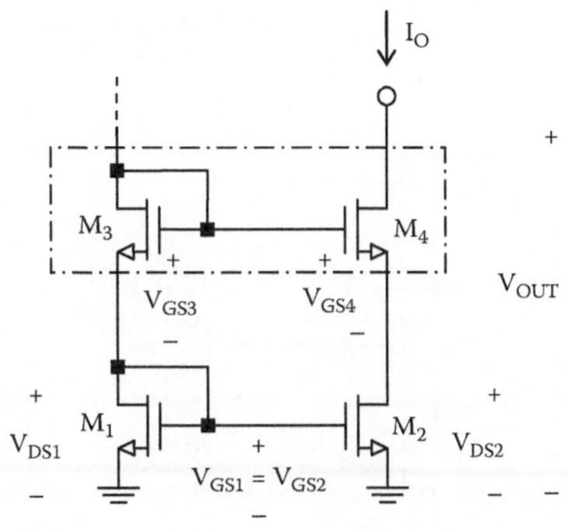

FIGURE 6.13
Reducing the sensitivity of the output current to the output voltage.

This makes the current copying function of a cascode current mirror much more accurate compared with a simple current mirror. In general, Equation (6.11) holds if transistors M_3 and M_4 are scaled with the same ratio as M_1 and M_2 are (i.e., $(W/L)_4/(W/L)_3 = (W/L)_2/(W/L)_1$).

Adding a pair of transistors significantly improves the quality of our current source/mirror, but it does not come without a cost. Unfortunately, the minimum output voltage required to keep both M_2 and M_4 in saturation has increased:

$$V_{OUT} = V_{DS2} + V_{DS4}, \tag{6.12}$$

Considering the fact that V_{DS2} is clamped at $V_{GS1,2} = V_{TH} + V_{OD1,2}$, we have:

$$V_{OUT}(min) = V_{DS2} + V_{DS4}(Sat.)$$

and

$$V_{OUT}(min) = V_{TH} + V_{OD1,2} + V_{OD4}. \tag{6.13}$$

The minimum output voltage has increased from one overdrive voltage for the simple current source/mirror to one threshold voltage plus two overdrive voltages. In today's low-voltage circuits, this is a significant waste of voltage swing, and is considered as a drawback for the cascode current source/mirror.

Example 6.3

The cascode current source depicted in Figure Ex. 6.3 is comprised of the same transistors used in Example 6.2. Given $V_{DD} = 5$ V, find a value for R_{REF} to have the same amount of output current as that of the simple current source of Example 6.2.

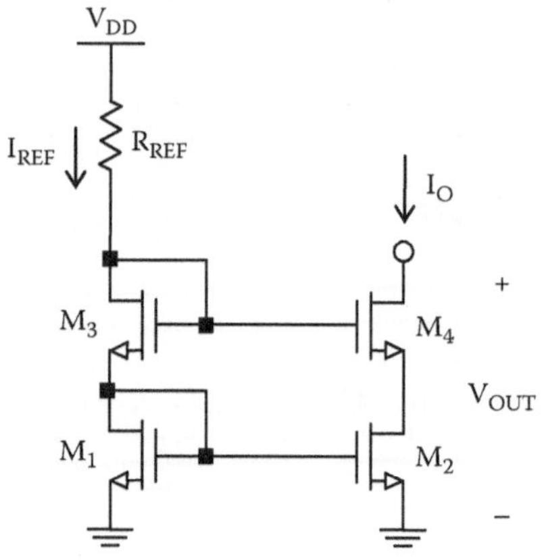

FIGURE EX. 6.3

Solution:
The parameter β for all the four transistors is:

$$\beta_{1-4} = k'\frac{W}{L} = 100^{\mu A/V^2} \times \frac{10^{\mu m}}{1^{\mu m}} = 1^{mA/V^2}$$

Knowing that $I_{D1\text{-}4} = I_{REF}$, one can write:

$$\text{KVL: } V_{DD} = I_{REF}R_{REF} + V_{GS3} + V_{GS1},$$

in which

$$I_{REF} = 39.2^{\mu A}$$

and

$$V_{GS3} = V_{GS1} = V_{GS}.$$

Therefore,

$$R_{REF} = \frac{V_{DD} - 2V_{GS}}{I_{REF}}.$$

Using the characteristic equation of the transistor, V_{GS} is obtained:

$$V_{GS} = V_{TH} + \sqrt{\frac{2I_D}{\beta}} = 0.8^V + \sqrt{\frac{2 \times 39.2^{\mu A}}{1^{mA/V^2}}} = 1.08^V.$$

Thus, the reference resistance, R_{REF}, is obtained:

$$R_{REF} = \frac{V_{DD} - 2V_{GS}}{I_{REF}} = \frac{5^V - 2 \times 1.08^V}{39.2^{\mu A}} \cong 72.5^{k\Omega}.$$

The output resistance of the transistors is the same as calculated in Example 6.2:

$$r_{o2-4} = \frac{1}{\lambda I_D} = \frac{1}{0.01^{V^{-1}} \times 39.2^{\mu A}} = 2.55^{M\Omega},$$

and their transconductance is:

$$g_{m1-4} = \sqrt{2\beta I_D} = \sqrt{2 \times 1^{mA/V^2} \times 39.2^{\mu A}} = 0.28^{mA/V}.$$

Therefore, the output resistance of the cascode current source is achieved:

$$R_o = g_{m4}\, r_{o4}\, r_{o2} = (0.28^{mA/V})(2.55^{M\Omega})(2.55^{M\Omega})$$

$$= 1.82^{G\Omega}.$$

Minimum output voltage for the cascode current source is

$$V_o(min) = V_{TH} + 2V_{OD}$$

$$= 0.8^V + (2 \times 0.28^V)$$

$$= 1.36^V.$$

6.4 Current Scaling

Sometimes, it is necessary to implement a current mirror with a current-copying factor other than unity. To do this using a current mirror, channel width/length ratios of the current-copying transistors should be sized differently:

$$\frac{i_O}{i_{IN}} = \frac{(\beta_2/2)(v_{GS} - V_{TH})^2}{(\beta_1/2)(v_{GS} - V_{TH})^2} = \frac{(W/L)_2}{(W/L)_1}. \tag{6.14}$$

Equation (6.14) holds for both simple and cascode current sources/mirrors. For M_3 and M_4 in a cascode current source/mirror, however, they do not have to be necessarily of the same size as M_1 and M_2 are, but because their drain currents are now scaled, it is recommended to scale them, too, to keep their gate-source voltages the same:

$$\frac{(W/L)_4}{(W/L)_3} = \frac{(W/L)_2}{(W/L)_1}. \tag{6.15}$$

6.5 Multi-Output Current Sources/Mirrors

Sometimes, having two or more current sources with matched currents is of critical importance. Designing multiple current sources in these cases has the drawback of extra power dissipation and area consumption by the additional reference branches. Furthermore, because of the unavoidable matching tolerances between the per-current-source reference branches, the generated currents are not expected to be identical. To overcome all these situations, it is suggested to design a *multi-output current source* consisting of one shared reference branch to generate the required bias voltages and of course as many current-sourcing transistors as needed. This idea is illustrated in Figure 6.14.

The idea of sharing the reference branch can also be combined with current scaling to implement multiple current sources with different output currents.

6.6 Bipolar Current Sources/Mirrors

The same idea that has been described so far to realize current sources/mirrors in CMOS technology can be generally used in bipolar technology to

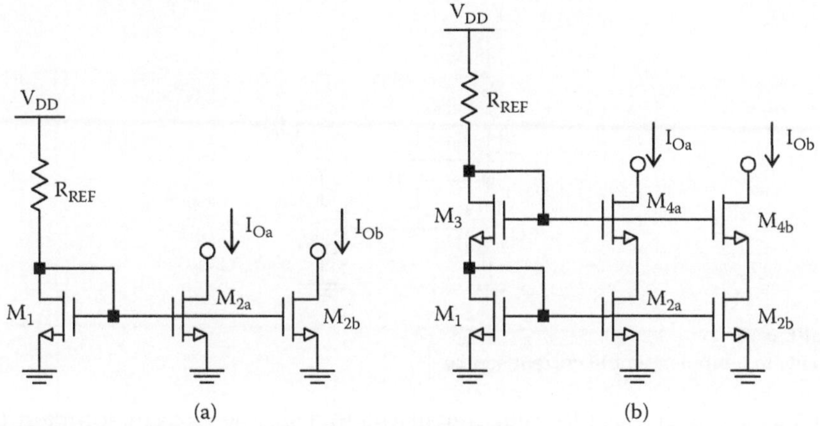

FIGURE 6.14
Multi-output current source: (a) simple current source, (b) cascode current source.

implement bipolar current sources/mirrors as well. Figure 6.15 is the bipolar version of simple and cascode current sources that can be basically analyzed in the same way their MOS counterparts were.

Bipolar current sources and their MOS counterparts have a few major differences, due to the intrinsic differences between bipolar and MOS transistors.

The minimum required output voltage for a bipolar current source/mirror to keep its transistors in the active mode is fixed and independent of the current level and transistor sizing. For example, the simple current source depicted in Figure 6.15(a) has a minimum output voltage of $V_{CE2}(Sat.)$, which is typically as small as 0.1–0.2 V no matter how much the output current level is.

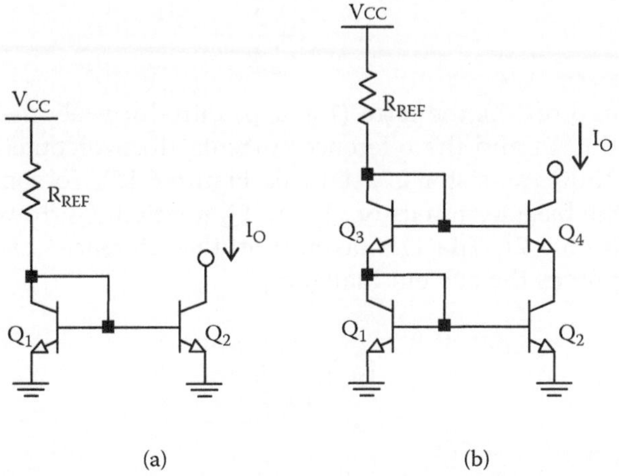

FIGURE 6.15
Bipolar current sources: (a) simple current source, (b) cascode current source.

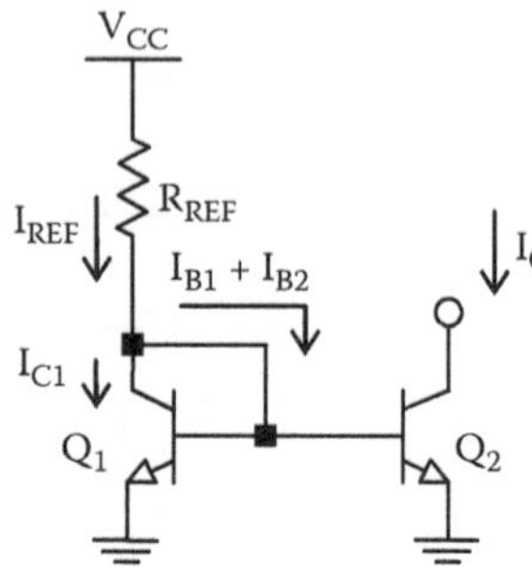

FIGURE 6.16
Currents in a bipolar simple current source.

The base currents for bipolar transistors in a simple current source-mirror are taken from the reference current (or the input current for current mirrors), which introduces a current mismatch between the two halves of the circuit. The current I_{C1} in the simple current source illustrated in Figure 6.16 is written as:

$$I_{C1} = I_{REF} - (I_{B1} + I_{B2}). \tag{6.16}$$

Assuming that Q_1 and Q_2 are identical, knowing that they have the same base-emitter voltage, and ignoring the possible base-length modulation effect, their collector currents are expected to be the same (and the base currents as well):

$$I_O (=I_{C2}) = I_{C1}. \tag{6.17}$$

Using Equation (6.16), Equation (6.17) can be rewritten as:

$$I_O = \frac{I_{REF}}{1+\frac{2}{\beta}}. \tag{6.18}$$

For a typical β of 100, the term (1 + 2/β) introduces about 2% mismatch between the output and the reference currents. To overcome this problem, sometimes a third transistor is used (see Figure 6.17). To source $2I_B$ to the node where the base terminals of Q_1 and Q_2 are tied, Q_3 draws $2I_B\beta/(\beta+1)$ from V_{CC}, and only $2I_B/(\beta+1)$ is taken from the reference current. This significantly improves the current matching:

$$I_O \cong \frac{I_{REF}}{1+\frac{2}{\beta^2}}. \tag{6.19}$$

The simplest way of current scaling in bipolar current sources/mirrors is to use parallel combinatin of transistors (by base-to-base, emitter-to-emitter, and collector-to-collector connecting them together) for either the reference

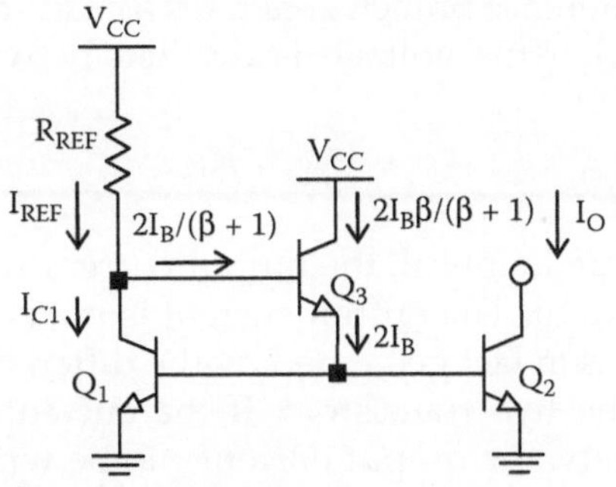

FIGURE 6.17
Reducing the current mismatch in a bipolar current source.

transistor or the current-sourcing transistor. In integrated circuits, emitter cross-sectional area, A_E, is the geometric parameter a designer can use to scale the current of a bipolar current source/mirror. This is because I_S for a bipolar transistor is proportional to A_E. As a result, any ratio between the emitter areas of Q_1 and Q_2 in Figure 6.15 can scale the current accordingly:

$$\frac{i_O}{i_{C1}} = \frac{A_{E2}}{A_{E1}} \tag{6.20}$$

In addition to the previously mentioned configurations that can be used both as a current source and as a current mirror, some other circuits act only as current sources. One of the well-known circuits of this type is the Widlar current source (see Figure 6.18).

The only difference between this circuit and the simple current source is R_E, the resistor that is added in the emitter of Q_2. To analyze the circuit, it

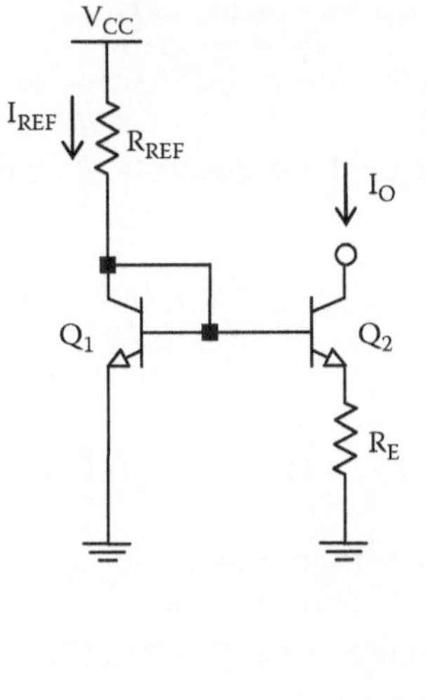

FIGURE 6.18
Widlar current source.

can be said that the reference branch generates a base voltage for the current-sourcing transistor, Q_2. This voltage is divided between the base-emitter junction of Q_2 and R_E:

$$V_{BE1} = V_{BE2} + I_E R_E. \tag{6.21}$$

Thus, if Q_1 and Q_2 are identical, the output current will always be smaller than the reference current. The output current is not a copy of the reference current anymore, but is in fact generated by the difference between the base-emitter voltages of the two transistors. If the current gain, β, is assumed much greater than unity, the output current can be written as:

$$I_O = (V_{BE1} - V_{BE2})/R_E. \tag{6.22}$$

Here is one of the cases where the base-emitter voltage should be precisely treated and cannot be approximated to $V_{BE}(ON)$. Thus, instead of taking $V_{BE1} = V_{BE2} = V_{BE}(ON)$, one should use the exact value of base-emitter voltages as functions of the associated collector currents. This way Equation (6.22) becomes:

$$I_O = \frac{V_T \ln \frac{I_{C1}}{I_{S1}} - V_T \ln \frac{I_{C2}}{I_{S2}}}{R_E}$$

$$= \frac{V_T \ln\left(\frac{I_{C1}}{I_{S1}} \cdot \frac{I_{S2}}{I_{C2}}\right)}{R_E}.$$

Because the transistors are assumed identical, $I_{S2} = I_{S1}$, thus:

$$I_O = \frac{V_T}{R_E} \ln \frac{I_{C1}}{I_{C2}}.$$

Neglecting the effect of the base currents, I_{C1} can be substituted by I_{REF}:

$$I_O = \frac{V_T}{R_E} \ln \frac{I_{REF}}{I_O}. \tag{6.23}$$

As we know,

$$I_{REF} = (V_{CC} - V_{BE1})/R_{REF}. \tag{6.24}$$

in which V_{BE1} can be approximated to $V_{BE}(ON)$ because it is subtracted from a much larger quantity, V_{CC}. Therefore, I_{REF} is treated as a known parameter, and Equation (6.22) becomes a single-variable nonlinear equation, which should be solved using numerical analysis methods or other techniques for nonlinear equations.

The output resistance of the Widlar current source is

$$R_o = r_{o2}(1 + g_{m2}R_E). \tag{6.25}$$

Although the addition of R_E apparently results in much higher output resistance than that of a simple current source, it is not as effective as the cascode technique. It is important to note, however, that for the Widlar current source, the minimum output voltage is still as low as that of a simple current source, whereas for the cascode current source the minimum output voltage is significantly larger.

6.7 Current Sources for Biasing and as Active Loads

In integrated circuits, biasing by using current sources is much more common than the other techniques studied in Chapter 3. This is because employing current sources for biasing enables the designer to generate, copy, and scale the required bias currents easily and in an area- and power-efficient fashion. Furthermore, sometimes the same current source that is used for biasing can act as a huge load resistance, called *active loads*, to yield a high gain. To describe the critical role of active loads in efficiently realizing high-quality monolithic amplifiers, let us start with trying to design a simple common-emitter voltage amplifier with a relatively high gain.

Suppose that in the amplifier in Figure 6.19, the transistor is somehow biased at a proper collector current, I_C. The amplifier's voltage gain can be written as:

$$A_V = \frac{v_{out}}{v_{in}} = -g_m R_C = \frac{-I_C R_C}{V_T} \tag{6.26}$$

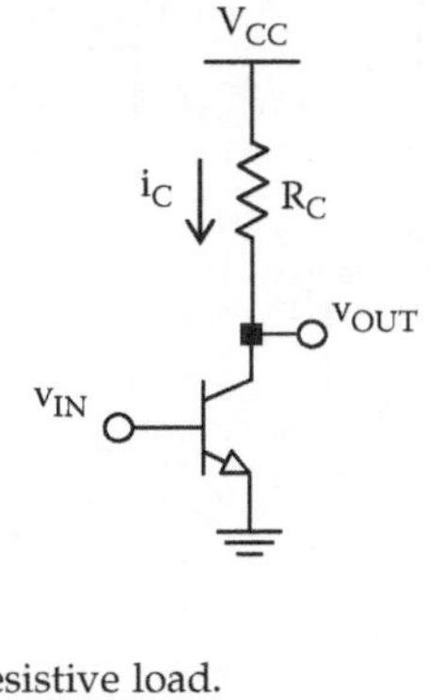

FIGURE 6.19
Common-emitter amplifier with resistive load.

At the first glance, it appears that by choosing a large enough R_C, a large voltage gain can be achieved; however, a limitation is imposed by DC conditions of the amplifier. The term I_CR_C in the numerator of Equation (6.24) is, in fact, the voltage drop across R_C. As we know, to keep the transistor in the active mode, I_CR_C cannot be larger than $V_{CC} - V_{CE}(Sat.)$. This means that even if an ideal transistor is used with $V_{CE}(Sat.)$ of 0 V, the amplifier with V_{CC} = 5 V cannot have a voltage gain of any larger than −200. To conclude, this limitation stems from the fact that the resistive load, which plays a role in the AC mode, is also present in the DC mode and contributes to the determination of the collector-emitter voltage of the transistor.

To overcome this problem, a transistor is used as the load. As depicted in Figure 6.20, looking into the drain/collector of an MOS/bipolar transistor, a large resistance is seen, but only in the AC mode. On the other hand, in the DC mode the source-drain/emitter-collector voltage is determined by the rest of the circuit that is connected to the drain/collector, independent of the currents and voltages in the AC mode.

The voltage V_{Bias} in Figure 6.20 puts the MOS/bipolar load transistor into saturation/active mode, which resembles the basic single-transistor current source. Figure 6.21 illustrates how current sources are used as active loads. In these circuits the drain/collector bias current for the amplifying transistor is set by the current source. It is worth noting that with this method of biasing, the gate-source/base-emitter DC voltage is determined according to the bias current of the transistor.

Now, with the active load, it can be claimed that no conflict exists between the DC and AC issues of the amplifier anymore. In DC mode, the supply voltage is divided between the amplifying and the load transistors (across their drain-source/collector-emitter), whereas in the AC mode, a very large gain can be achieved:

$$A_V = \frac{v_{OUT}}{v_{IN}} = -g_m.R_o, \tag{6.27}$$

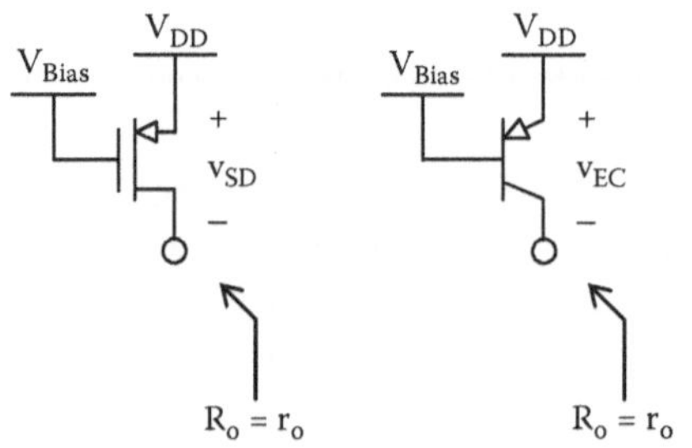

FIGURE 6.20
The transistor as an active load.

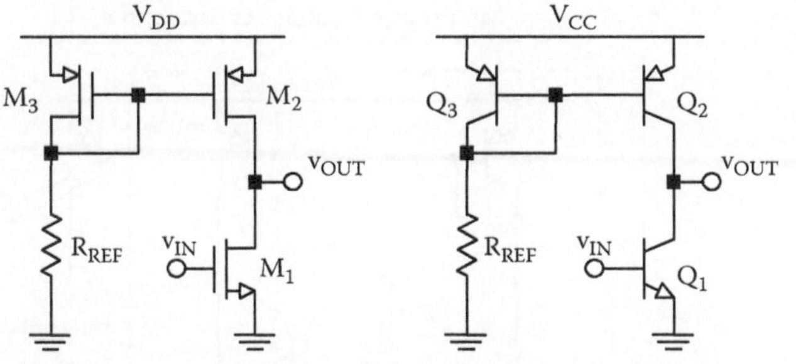

FIGURE 6.21
Common-source/common-emitter amplifier with active load.

where

$$R_o = r_{o1} \| r_{o2}. \tag{6.28}$$

To have an idea about how large a voltage gain can be achieved by an active load, let's analyze these amplifiers one level deeper and write the common-source and common-emitter gains ($A_{V,CS}$ and $A_{V,CE}$, respectively) in terms of their bias currents as:

$$A_{V,CS} = -\sqrt{2\beta_1 I_D}.\left(r_{o1} \| r_{o2}\right) = -\sqrt{2\beta_1 I_D}.\frac{1}{(\lambda_1+\lambda_2)I_D} = \frac{-\sqrt{2\beta_1}}{\lambda_1+\lambda_2}.\frac{1}{\sqrt{I_D}}, \tag{6.29}$$

and

$$A_{V,CE} = \frac{-I_C}{V_T}.\left(r_{o1} \| r_{o2}\right) = \frac{-I_C}{V_T}.\frac{V_{A1}.V_{A2}/I_C^2}{V_{A1}/I_C + V_{A2}/I_C} = \frac{-V_{A1}V_{A2}}{V_T(V_{A1}+V_{A2})}. \tag{6.30}$$

It is interesting to see that the CE voltage gain is independent of the bias current, and the CS gain increases as the bias current decreases. Assuming $\beta = 20\ mA/V^2$ and $\lambda_1 = \lambda_2 = 0.01\ V^{-1}$ for the MOS transistors, biased at I_D = 1 mA, 100 μA, and 10 μA, the CS gain would be –316, –1000, and –3160, respectively. Also, assuming $V_{A1} = V_{A2} = 100$ V for the bipolar transistors, the CE amplifier would have a gain of –2000.

It would be instructive to look at a current source/mirror combination, which has been used in some of the well-known commercial operational amplifiers (see Figure 6.22). (Q_1, Q_2, R_2) make a Widlar current source, and (Q_3, Q_4, Q_5) form a two-output current source. Both current sources share the same reference branch (Q_1, R_{REF}, Q_3), which saves power and makes the circuit smaller in size. The current source transistor, Q_4, performs as an active load for the transistor Q_7, which acts as a common-emitter amplifier stage.

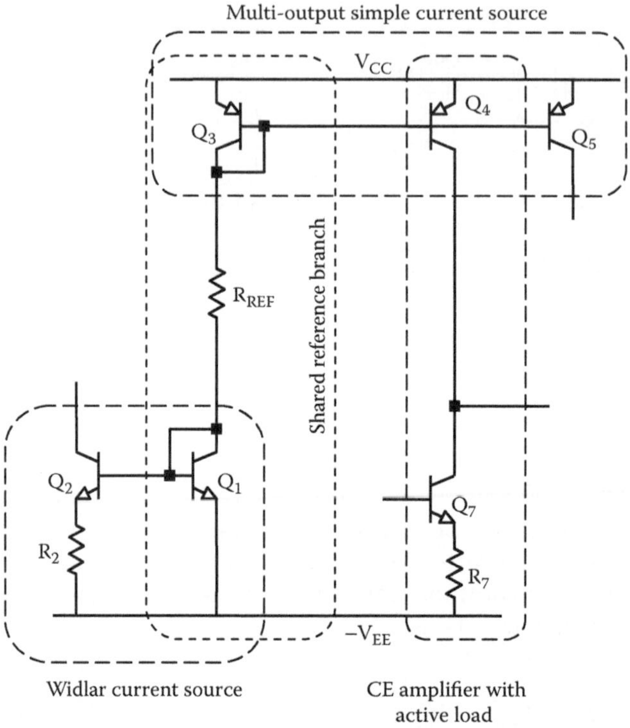

FIGURE 6.22
Combination of current source/mirror circuits.

Other examples of using current sources both for biasing and as active loads are given in Figure 6.23. Using an active load in a source/emitter follower (see Figure 6.23(a)), makes a good voltage buffer with a voltage gain of nearly 1. The circuit will also exhibit small attenuation at the output when driving the output current to the load. This is because the output resistance of the current source, r_{o2}, is usually much larger than the load resistance. The cascode amplifiers in Figure 6.23(b) can also have much larger voltage gains compared with their passive-load counterparts. Figure 6.23(c) is also a three-stage amplifier containing both bipolar and MOS transistors. The amplifying transistors are M_1 as a source follower, Q_2 as a common-base stage, and Q_4 in common-emitter configuration. Q_5, R_{REF}, and M_6 make the shared reference branch for two current sources: the bipolar current source (Q_5, Q_3), and the two-output MOS current source (M_6-M_8). The current-sourcing transistor Q_3 sets the DC collector current of Q_2 and serves it as an active load. M_1 and Q_2 are in a combination that looks like a hybrid differential amplifier, with M_7 as the tail current source considered for biasing. The DC drain current of M_1 is set to the difference between the collector current of Q_2 and the drain current of M_7. A proper bias current level for M_1 requires suitable current scaling in either the bipolar current source or the

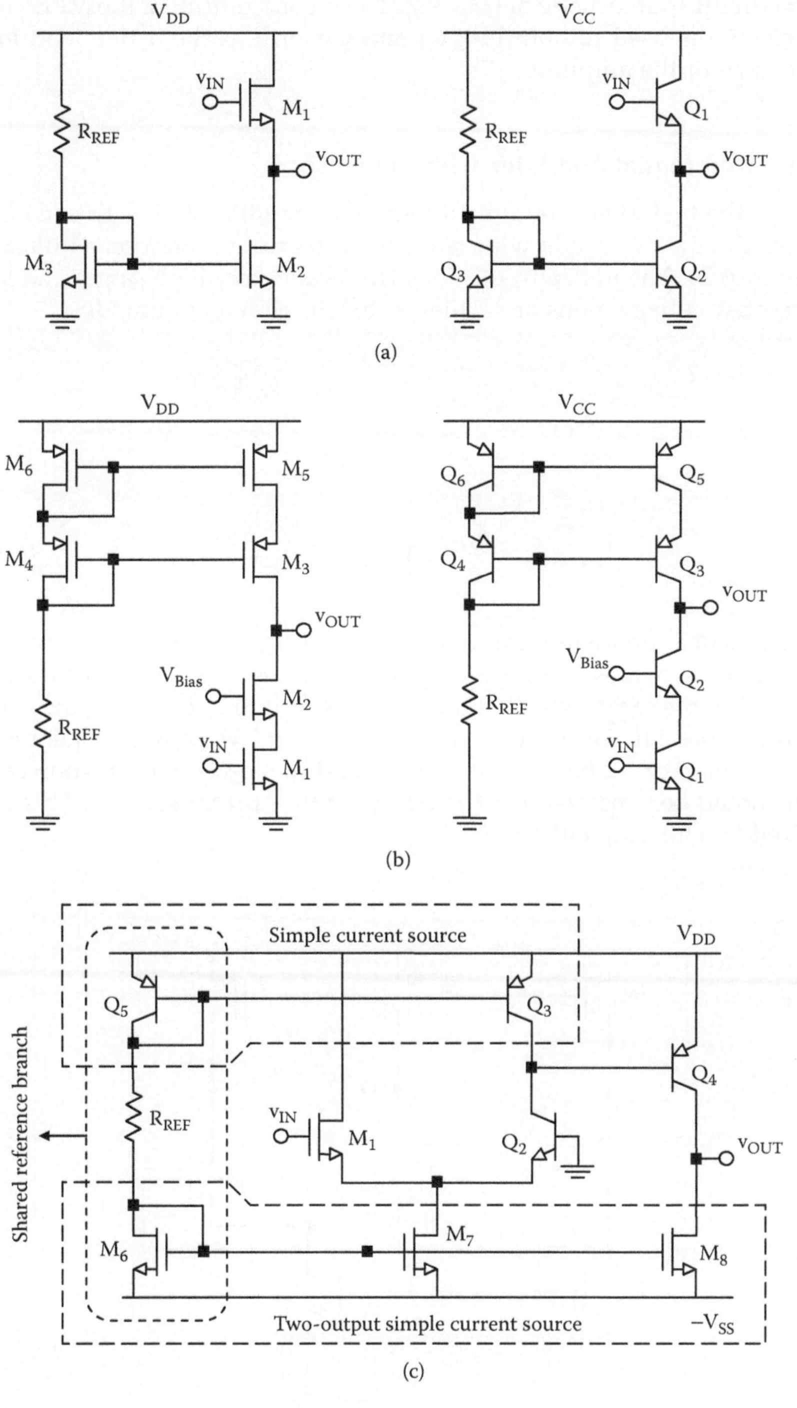

FIGURE 6.23
Examples of using current sources as active loads.

MOS current source made of (M_6, M_7). The second output of the MOS current source (M_8) is used to both bias Q_4 and perform as the active load for the third stage of the amplifier.

6.7.1 Differential Amplifier with Active Load

Perhaps the first idea to equip a differential amplifier with active loads is to replace the passive loads with two identical current sources, as illustrated in Figure 6.24. Theoretically, this idea works and very high single-ended and differential voltage gains are achieved for the differential mode:

$$A_{V,DM-SE} = \frac{v_{o2}}{v_i} = \frac{-v_{o1}}{v_i} = \frac{g_{m1}\left(r_{o1}\| r_{o3}\right)}{2} \quad (6.31(a))$$

and

$$A_{V,DM-DE} = \frac{v_o}{v_i} = 2A_{V,DM-SE} = g_{m1}\left(r_{o1}\| r_{o3}\right). \quad (6.31(b))$$

The practical problem with this idea is that the transistors M_1 and M_2 are biased from two sides: They are supposed to be biased at $I_{D5}/2$ from the source side, and, at the same time, their drains are forced to be receiving the currents that the upper current sourcing transistors M_3 and M_4 provide. Theoretically, a current ratio of 1:2 between the upper and the lower current sources is all what should be considered in the design of the current sources. This can be realized by choosing either:

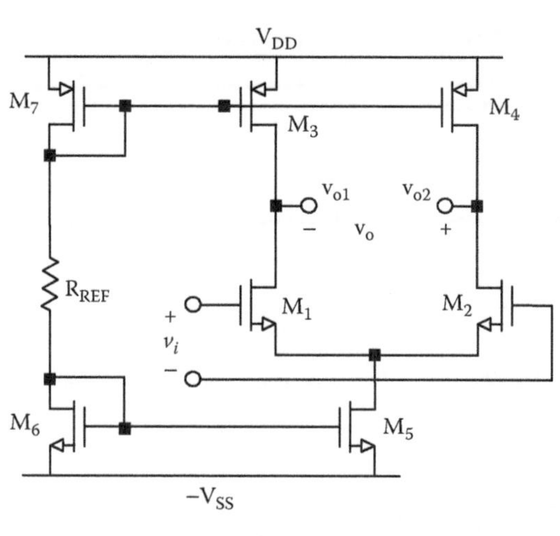

FIGURE 6.24
Replacing the passive loads of a differential amplifier with current sources.

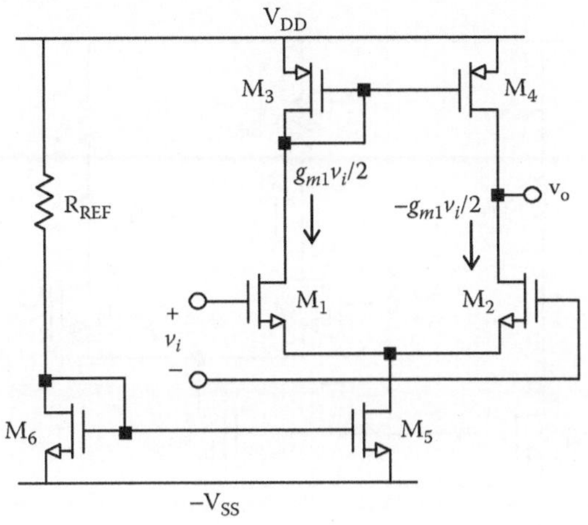

FIGURE 6.25
Differential amplifier with a current mirror as the active load.

$$(W/L)_5 = 2(W/L)_6,$$

or

$$(W/L)_3 = (W/L)_4 = (W/L)_7/2.$$

The unavoidable mismatches between the physical and electrical parameters of the transistors make this difficult to realize in practice. Although solutions are available for this issue, they require relatively complex circuitry, which is out of the scope of this book.

A very simple yet efficient way to have a differential amplifier with active load is to use a current mirror (see Figure 6.25).

In DC mode, the current mirror (M_3, M_4) only guarantees that the bias currents of M_3 and M_4 are equal, which is exactly what is dictated by the lower part of the circuit ($I_{D1} = I_{D2} = I_{D5}/2$). In AC mode, as a result of applying the input voltage, v_i, the AC drain currents $g_{m1}v_i/2$ and $-g_{m1}v_i/2$ are generated by M_1 and M_2, as before. The output resistance of the current mirror (i.e., the resistance seen into the drain of M_4) is r_{o4}, which is high enough to result in a good voltage gain. On the other side, the diode-connected transistor, M_3, determines the input resistance of the current mirror, which is as low as $1/g_{m3}$. Thus, a large voltage gain cannot be expected at the drain of M_1. Moreover, because the circuit is not symmetric in the AC mode anymore, all it can provide at the output is a single-ended output voltage. Despite the above explanation, the circuit *does not* behave like the differential amplifier with passive loads of $1/g_{m3}$ and r_{o4} illustrated in Figure 6.26.

In this circuit, the AC current of the right-hand half-circuit, $-g_{m1}v_i/2$, passes through r_{o4}, and generates the single-ended output at the drain of M_2, whereas the same amount of current at the left-hand half-circuit is left unused.

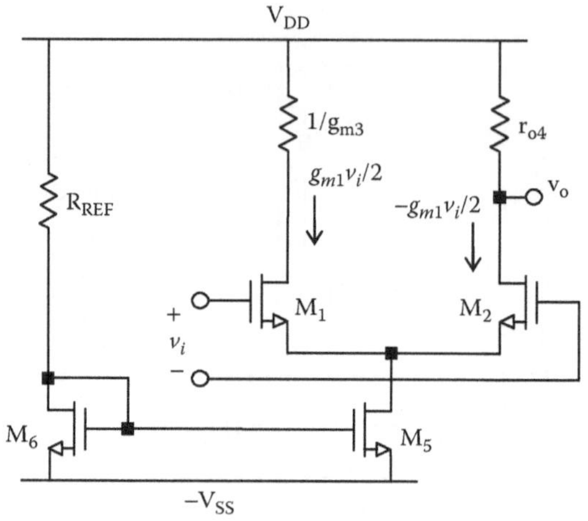

FIGURE 6.26
Incorrect schematic for the AC role of the current mirror as the active load.

This is in fact how a differential amplifier with passive loads works, but the current mirror is not modeled only by its input and output resistances. The main role of the current mirror is to source a current from the drain of M_4, which is a copy of whatever current is sunk from the drain of M_3, including both DC and AC components. Thus, the right replacement for the current mirror in the AC mode needs to have a dependent current source in addition to its input and output resistances as illustrated in Figure 6.27. This means that the current mirror saves the AC current of the left-hand

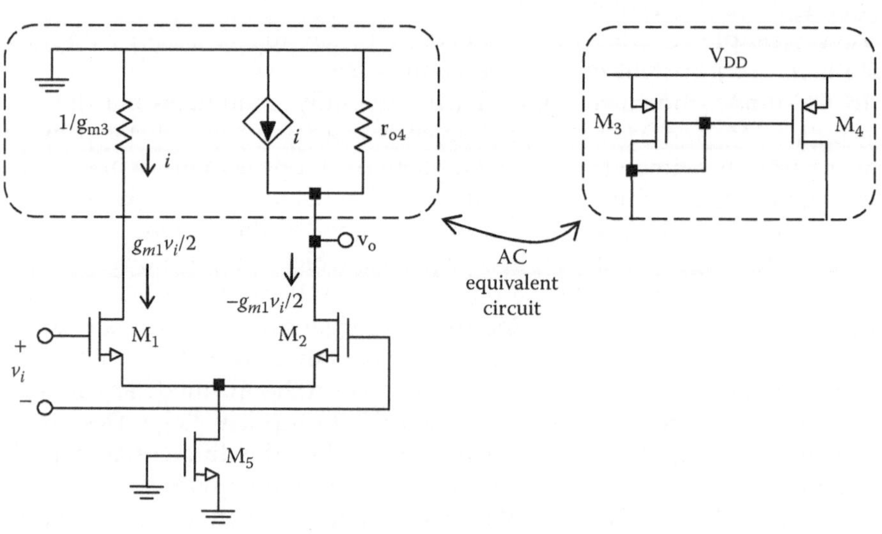

FIGURE 6.27
AC schematic of the amplifier of Figure 6.25.

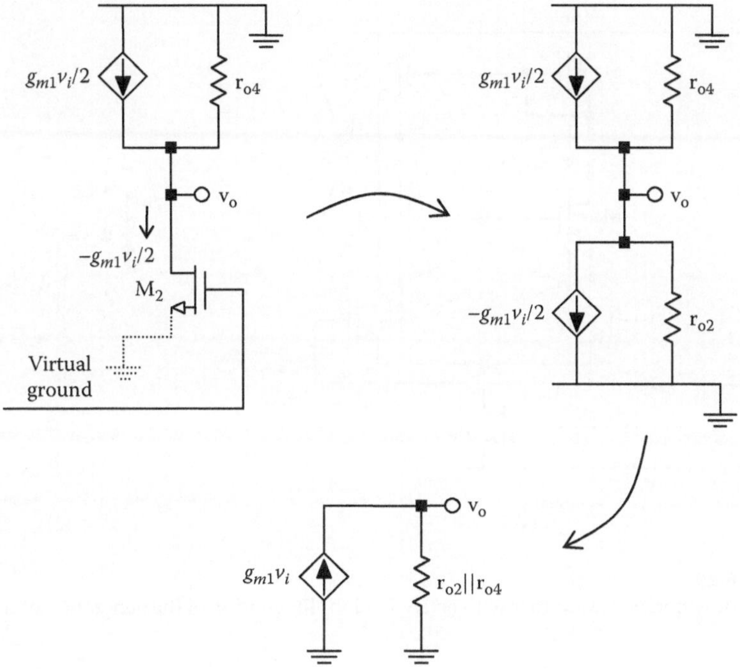

FIGURE 6.28
Generation of the differential-mode single-ended output voltage utilizing the AC currents of both half-circuits.

half-circuit, transfers it to the high-resistance output node, and converts it to AC output voltage in the same way that the AC current of the right-hand half-circuit is. This is illustrated in Figure 6.28. In other words, even though the output of this circuit is single-ended, the output voltage and therefore the voltage gain are the same as those of the differential output of a passive-load differential amplifier:

$$v_o = (g_{m1}v_i)\left(r_{o2} \| r_{o4}\right) \quad \Rightarrow \quad A_V = \frac{v_o}{v_i} = g_{m1}\left(r_{o2} \| r_{o4}\right). \tag{6.32}$$

Current mirrors can be used in any other differential amplifier as the active load to both increase gain and convert a differential input signal to a single-ended output signal. As an example, Figure 6.29(a) is a differential cascode amplifier with a simple current mirror as the active load. Similar to the basic differential amplifier, the AC components of the drain currents of the transistors M_1 and M_2 are written as:

$$i_{ac} = i_{d1} = -i_{d2} = g_{m1}v_i/2. \tag{6.33}$$

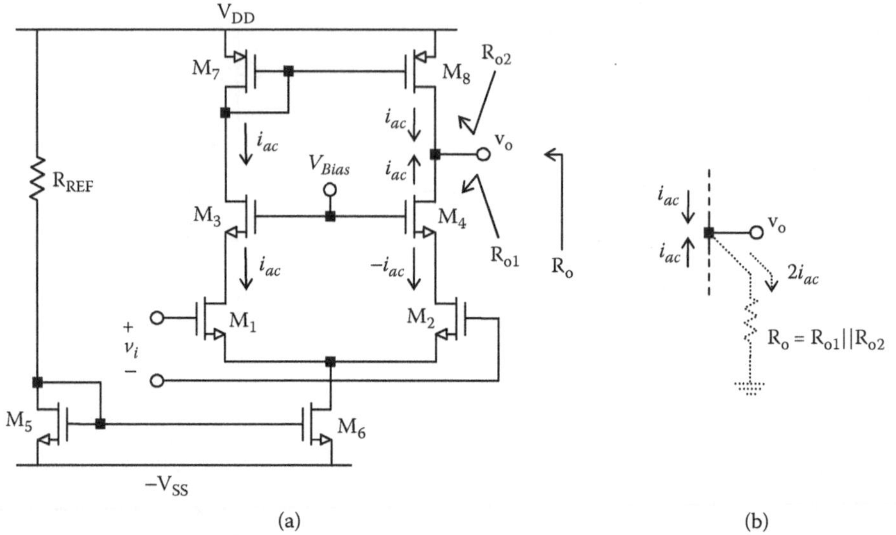

FIGURE 6.29
(a) Cascode differential amplifier with active load (b) Illustration of the generation of the output voltage.

As illustrated in Figure 6.29(b), two i_{ac} current components arrive the output node, and pass through the output resistance:

$$R_o = R_{o1} \| R_{o2}, \tag{6.34(a))}$$

where

$$R_{o1} = (g_{m4} r_{o4}).r_{o2} \tag{6.34(b))}$$

and

$$R_{o2} = r_{o8}. \tag{6.34(c))}$$

It should be noted that because the circuit is studied in differential mode, the source of M_2 is virtually grounded when finding the resistance R_{o1}. It is obvious that in this circuit R_{o2} is much smaller than R_{o1}, and thus dominates the output resistance of the amplifier:

$$R_o = R_{o1} \| R_{o2} \cong R_{o2} = r_{o8}. \tag{6.34(d))}$$

Based on the illustration in Figure 6.29(b), the output voltage of the amplifier can be written as:

$$v_o = 2i_{ac}.R_o. \tag{6.35}$$

Substituting i_{ac} from Equation (6.33), one can write:

$$v_o = 2(g_{m1}v_i/2).R_o = (g_{m1}R_o)v_i. \tag{6.36}$$

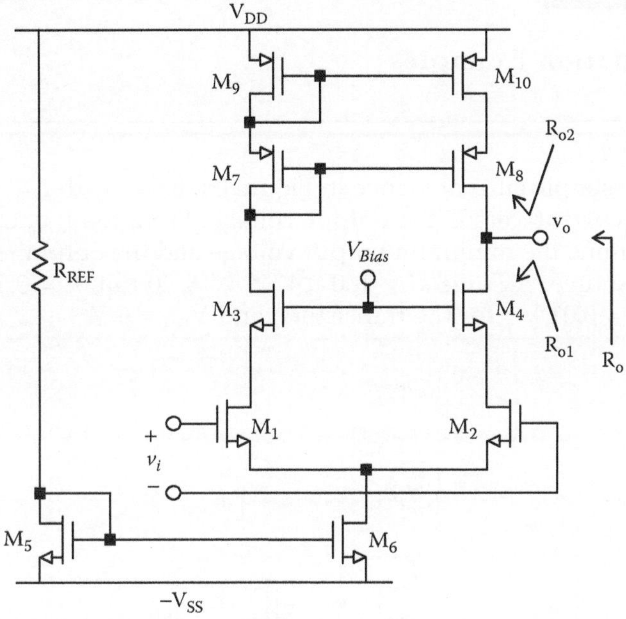

FIGURE 6.30
Cascode differential amplifier with a cascode current mirror as the active load.

and the voltage gain of the amplifier is given as:

$$A_V = \frac{v_o}{v_i} = g_{m1}R_o \cong g_{m1}r_{o8} \, . \tag{6.37}$$

To take advantage of the high output resistance of the lower half of the circuit (i.e., R_{o1}) to achieve a higher voltage gain, it is suggested that a cascode current mirror be used as the active load (see Figure 6.30). This way, the upper half of the amplifier will exhibit much larger output resistance:

$$R_{o2} = (g_{m8}r_{o8}).r_{o10}, \tag{6.38(a)}$$

the output resistance increases to:

$$R_o = R_{o1}\|R_{o2} = [(g_{m4}r_{o4}).r_{o2}] \,\|\, [(g_{m8}r_{o8}).r_{o10}], \tag{6.38(b)}$$

and the voltage gain becomes:

$$A_V = \frac{v_o}{v_i} = g_{m1}R_o = g_{m1}\left[(g_{m4}r_{o4}r_{o2})\big\|(g_{m8}r_{o8}r_{o10})\right]. \tag{6.39}$$

which is indeed a very large gain. Assuming $g_{m1\text{-}8} = 1\ mA/V$, $r_{o1\text{-}4} = 2\ \text{M}\Omega$, and $r_{o7\text{-}10} = 1\text{M}\Omega$, the output resistance is 800 MΩ, and the gain can go as high as 800,000.

6.8 Simulation Examples

Example 6.4

Simulate the simple current source in Figure Ex. 6.4(a) with $I_{REF} = 1\ mA$, and plot output current versus the output voltage. From the i_o-v_o characteristic curve, determine the minimum output voltage and the output resistance for the current source. Assume $k' = 100\ \mu A/V$, W = 20 μm, L = 0.5 μm, V_{TH} = 0.7 V, and λ = 0.01 V^{-1} for the transistors, and V_{DD} = 5 V.

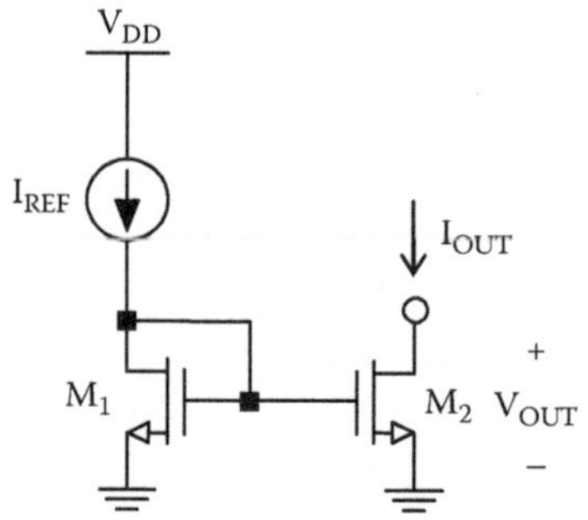

FIGURE EX. 6.4(a)

Solution:

```
VDC  VCC 0 5V
Vout out 0 5v
Iref VCC in 1ma
M1 in  in 0 0 NMOSmodel W=20u L=0.5u
M2 out in 0 0 NMOSmodel W=20u L=0.5u
.MODEL NMOSmodel NMOS(VTO=0.7 KP=100U LAMBDA=0.01)
.LIB NOM.LIB
.DC Vout 0 5v 1mv
.PROBE
.OP
.END
```

Figure Ex. 6.4(b) is the simulated i_{OUT}-v_{OUT} characteristic curve for the current source. The knee, from which the desired operation of the current source starts, is located at around 0.7 V. This is $V_{OUT}(min.)$ for the current source, which is expected to be equal to $V_{DS}(Sat.)$ for transistor M_2. Looking at the part of the simulation output file that contains biasing conditions and the bias-dependent AC parameters of the transistors (given next) it is verified that $V_{DS}(Sat.)$ for transistor M_2 is 0.7 V.

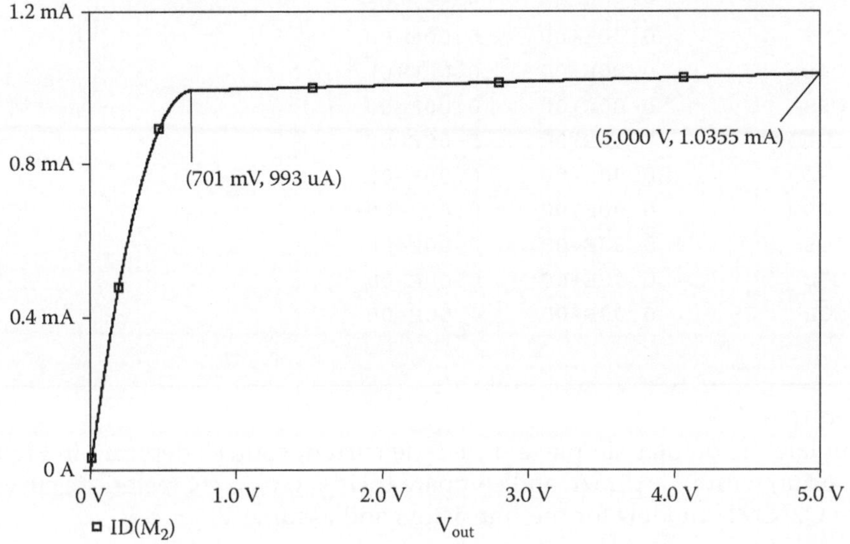

FIGURE EX. 6.4(b)

A second point on the characteristic curve, allows one to find the output resistance of the simulated current source as the inverse of the slope of the curve:

$$R_o = \frac{5.000^V - 0.701}{1.0355^{mA} - 0.993^{mA}} \cong 101^{k\Omega}$$

which is in total agreement with the calculated output resistance:

$$R_o = r_o = \frac{1}{\lambda I_D} = 100\,k\Omega$$

```
****OPERATING POINT INFORMATION  TEMPERATURE=27.000 DEG C***

**** MOSFETS
NAME          M1            M2
MODEL         NMOSmodel     NMOSmodel
ID            1.00E-03      1.04E-03
VGS           1.40E+00      1.40E+00
VDS           1.40E+00      5.00E+00
VBS           0.00E+00      0.00E+00
VTH           7.00E-01      7.00E-01
VDSAT         7.02E-01      7.02E-01
GM            2.85E-03      2.95E-03
```

```
GDS        9.86E-06    9.86E-06
GMB        0.00E+00    0.00E+00
CBD        0.00E+00    0.00E+00
CBS        0.00E+00    0.00E+00
CGSOV      0.00E+00    0.00E+00
CGDOV      0.00E+00    0.00E+00
CGBOV      0.00E+00    0.00E+00
CGS        0.00E+00    0.00E+00
CGD        0.00E+00    0.00E+00
CGB        0.00E+00    0.00E+00
```

Example 6.5

Simulate the bipolar simple and cascode current sources depicted in Figure Ex. 6.5(a) with I_{REF} = 1mA, and compare their i_{OUT}-v_{OUT} characteristic curves. Use Q2N3904 models for the transistors and assume V_{CC} = 5 V.

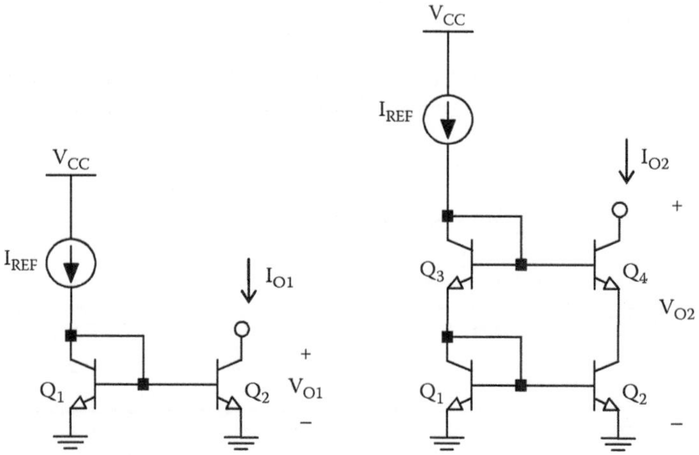

FIGURE EX. 6.5(a)

Solution:

```
VDC   VCC 0 5V
Vout  out 0 5v
*Simple Current Source/Mirror
Iin1 VCC in1 1ma
Q11  in1  in1 0 Q2N3904
Q12  out in1 0 Q2N3904
*Cascode Current Source/Mirror
```

```
Iin2 VCC in2 1ma
Q21  a a 0 Q2N3904
Q22  b a 0 Q2N3904
Q23  in2 in2 a Q2N3904
Q24  out in2 b Q2N3904
.LIB NOM.LIB
.DC Vout 0 5V 0.001
.PROBE
.OP
.END
```

The upper part of Figure Ex. 6.5(b) depicts the simulated characteristic curves for both current sources. The minimum output voltage is obtained

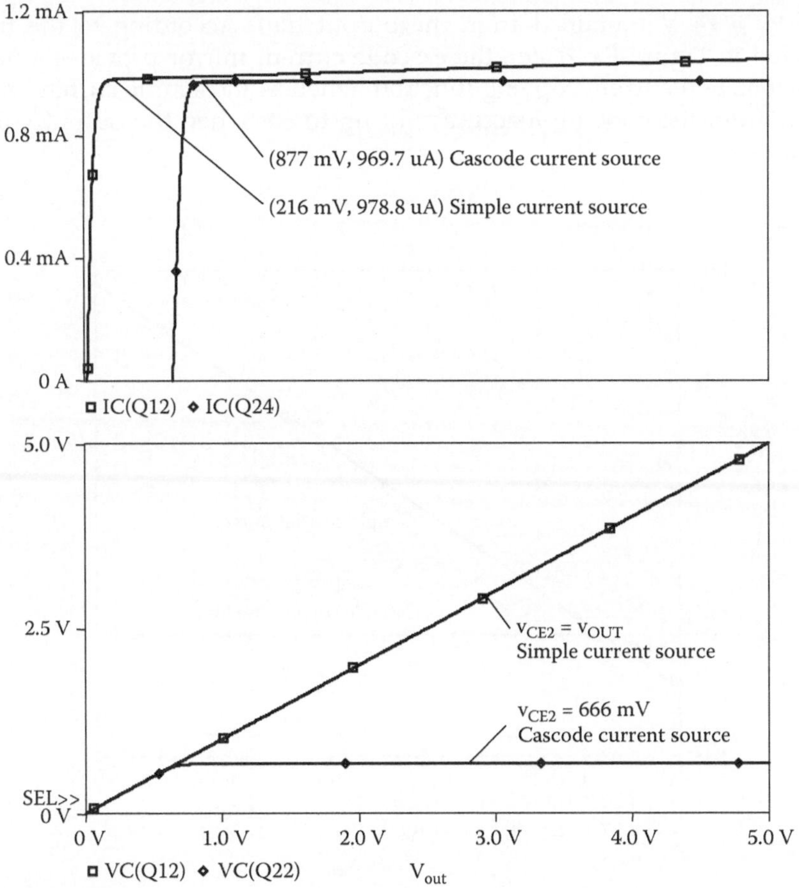

FIGURE EX. 6.5(b)

0.216 V and 0.877 V for the simple and cascode current mirrors, respectively, which agree with the expected values from analysis.

The lower part of Figure Ex. 6.5(b) demonstrates the function of the stacked pair (Q_3, Q_4) in the cascode current source. It clamps the collector-emitter voltage of Q_2 to the base-emitter voltage of Q_1, whereas V_{CE2} in the simple current source follows the output voltage.

To see how the difference between the collector-emitter voltages of the current-copying transistors Q_1 and Q_2 affects the current matching of the simple and cascode current mirrors, Figure Ex. 6.5(c) compares the term

$$\frac{1+\dfrac{V_{CE2}}{V_A}}{1+\dfrac{V_{CE1}}{V_A}}$$

with V_A = 74 V (obtained from the output file). According to the traces depicted in Figure Ex. 6.5(c), the cascode current mirror provides a nearly 100% matched current copying function, whereas the simple current mirror suffers from the copying inaccuracy of up to 6% when the output voltage increases to 5 V.

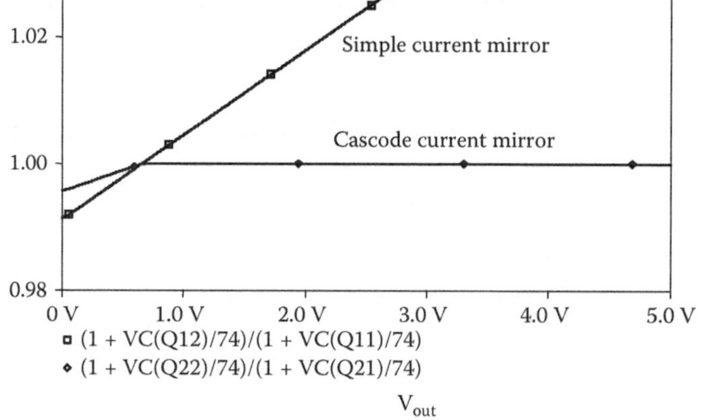

FIGURE EX. 6.5(c)

6.9 Problems

6.1 Find the output current, output resistance, and the minimum output voltage for each of the current sources depicted in Figure Prob. 6.1. Assume $|V_{BE}(ON)| = 0.7$ V, $|V_{CE}(Sat.)| = 0.1$ V, $V_A = 80$ V, and $\beta = 100$.

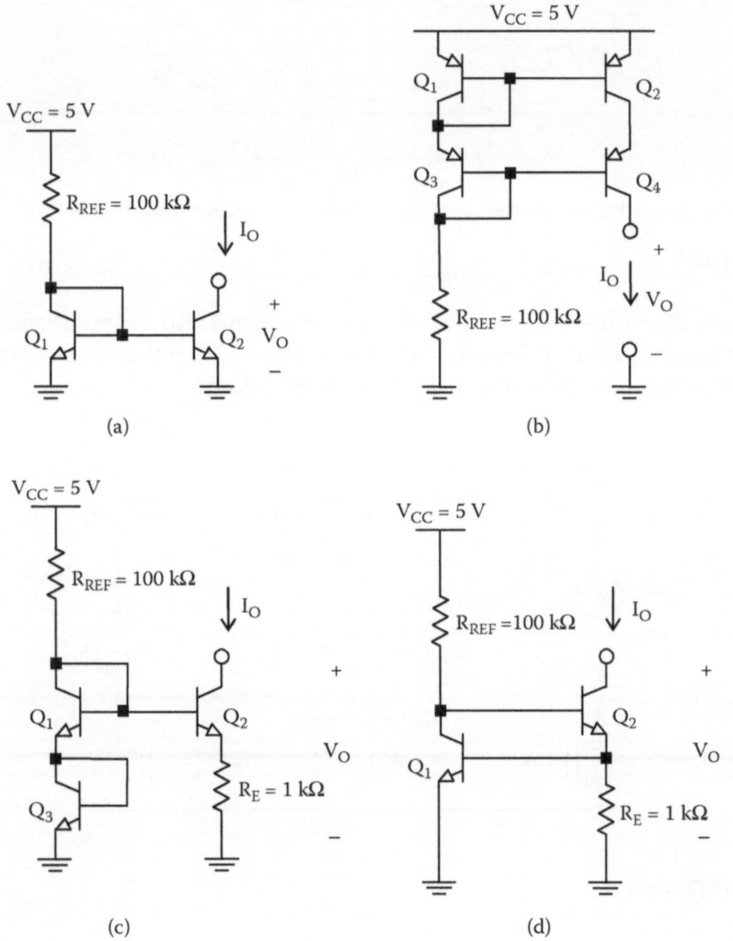

FIGURE PROB. 6.1

6.2 For the current sources of Problem 6.1, change the supply voltage from 5 V to 6 V, and determine their relative supply dependency factor defined as:

$$\Delta I_O\% = \frac{I_O\big|_{V_{CC}=6^V} - I_O\big|_{V_{CC}=5^V}}{I_O\big|_{V_{CC}=5^V}} \times 100\%$$

6.3 Given $V_{DD} = 3.3$ V, $|V_{TH}| = 0.7$ V, $k' = 100\ \mu A/V^2$, $R_{REF} = 250\ k\Omega$, and $(W/L)_{REF} = 20$ for the multi-output current source shown in Figure Prob. 6.3, find proper values for $(W/L)_1$ and $(W/L)_2$ so that $I_{O1} = 50\ \mu A$ and $I_{O2} = 100\ \mu A$.

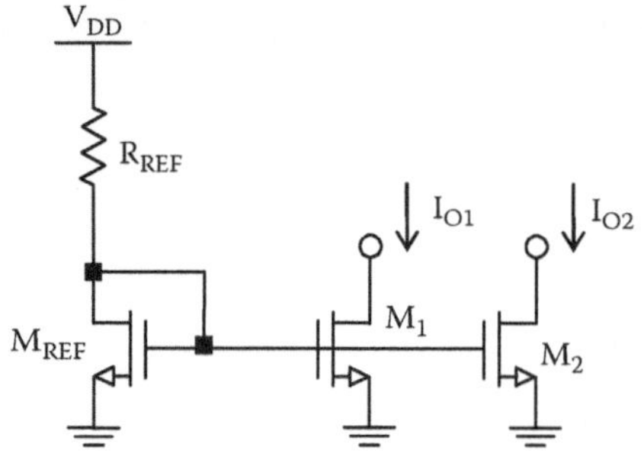

FIGURE PROB. 6.3

6.4 Find the voltage gain $A_V = v_o/v_i$, and the input and output resistances of the amplifier in Figure Prob. 6.4. It is assumed that the transistors (M_1, M_2), (M_3, M_4), and (M_5, M_6) are matched.

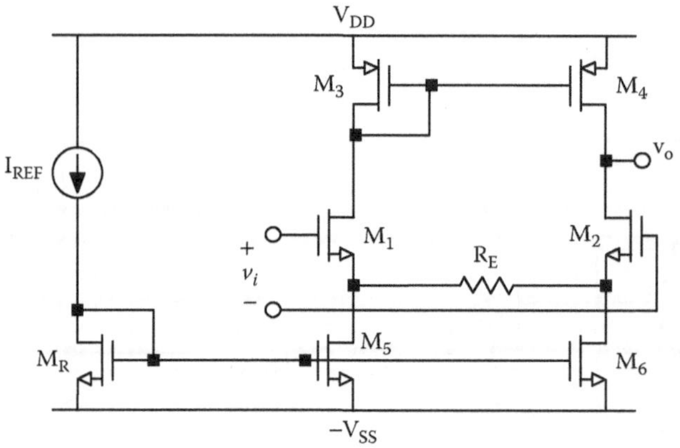

FIGURE PROB. 6.4

6.5 Figure Prob. 6.5 is a differential folded cascode amplifier with active load. Find the output resistance and voltage gain $A_V = v_o/v_i$.

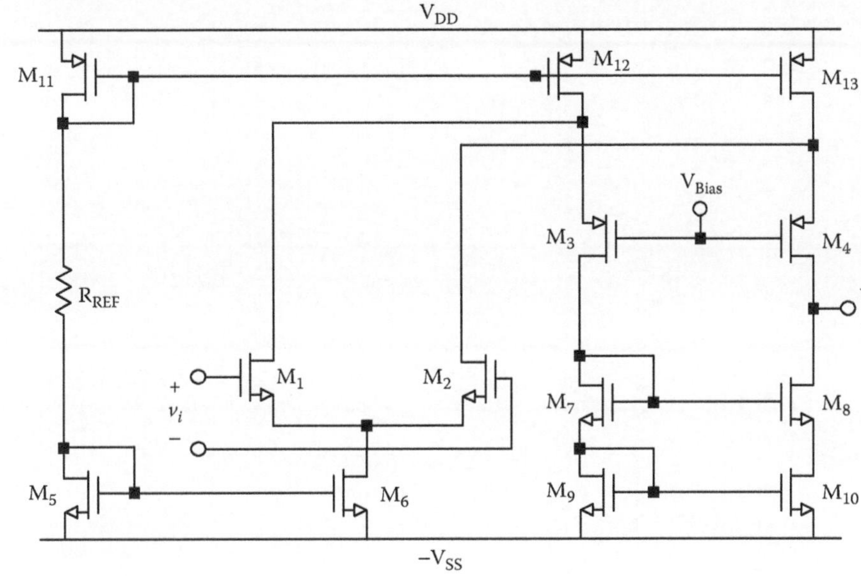

FIGURE PROB. 6.5

6.6 Figure Prob. 6.6 shows a differential amplifier comprised of the Darlington pair with current source (studied in Chapter 5) as the primitive half circuit. Assuming that all the transistors are identical with $\beta = 100$ and $V_A = 100\text{V}$, and that $V_{CC} = V_{EE} = 5\text{V}$, $R_{REF} = 930\text{k}\Omega$, and $R_C = 400\text{k}\Omega$,

(a) Find the input and output resistances in common-mode and differential-mode, the voltage gain v_o/v_i, and the CMRR.

(b) Remove M_5 and M_6 and repeat part (a).

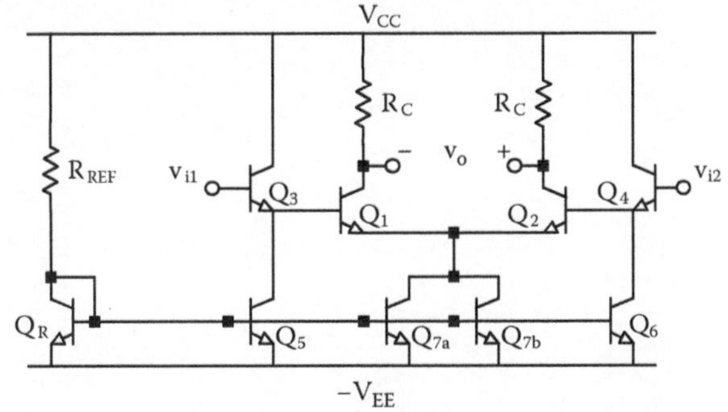

FIGURE PROB. 6.6

Chapter 7

Analysis of Amplifiers at Low Frequencies

7.1 Introduction

So far, we have learned to analyze an amplifier in DC mode to find its bias conditions and in AC mode to determine its functional specifications such as gain, and input and output resistances. When performing AC analysis, it has always been assumed that the signal frequency is so high that all the external capacitors (i.e., coupling and bypass) can be replaced by short circuits, and all the transistors' internal capacitances (introduced in the next chapter) are treated as open circuits. In general, during the AC analysis, we always assume that the signal has a moderate frequency. This chapter deals with the amplifiers' behavior when the frequency approaches and even passes the low extreme of the operational frequency band. Furthermore, the impact of the amplifier's low-frequency behavior on its time response will also be studied.

7.2 Basic Concepts in Frequency Domain

Figure 7.1 illustrates a typical plot for an amplifier's gain in the frequency domain. At *midband frequencies*, where the amplifier is usually supposed to operate, the gain is A_0. As we approach the extremes, the gain starts to roll off. As long as the amplifier exhibits at least 70% of the midband gain ($A_0\sqrt{2}/2$ to be exact), it is said that the amplifier operates within the midband frequency range. As depicted in Figure 7.1, the low and high cutoff frequencies, f_{CL} and f_{CH},* are indeed the edges of the amplifier's frequency band (midband frequency range), where the results of the previously described AC analysis are valid. As seen before, the amplifier's behavior in AC mode (at midband frequencies) is assumed to be frequency invariant. Beyond and

* Also referred to as –3dB cutoff frequencies $f_{-3dB,L}$ and $f_{-3dB,H}$.

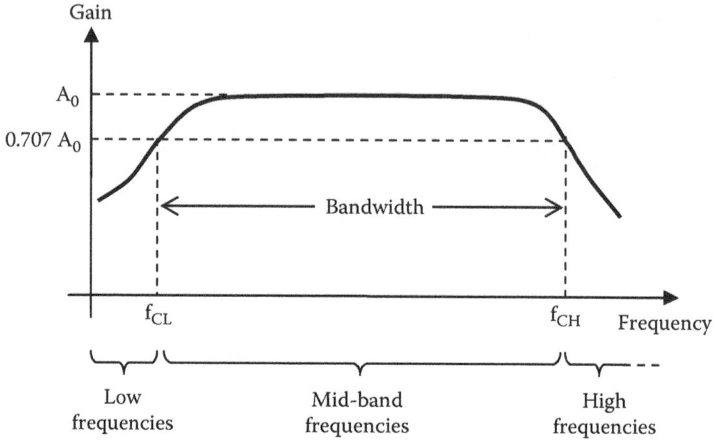

FIGURE 7.1
Typical plot for the gain of an amplifier vs. frequency.

even around the cutoff frequencies, where the amplifier's functional specifications (including its gain) tend to change with frequency, the amplifier should be analyzed using a more sophisticated method to obtain more precise results.

Usually, the amplifier's frequency dependent behavior at low frequencies is caused by the (external*) coupling and bypass capacitors, and the amplifier's bandwidth is limited at high frequencies by the transistors' (internal) parasitic capacitances. Starting from DC (f = 0 Hz), all the capacitors in the circuit have infinite impedances and are treated as open circuits. As the frequency increases, the impedances of all capacitors start to decrease. Because the values of the external capacitors are larger than the internal capacitor by orders of magnitude, their impedance approaches finite values earlier (in frequency) than that of the internal capacitors, as frequency increases. Each capacitor has a *region of activity* in the frequency domain, which starts from "where it comes out of the open circuit state" to "where it enters the short circuit state." It is obvious that the *open circuit* and *short circuit* concepts discussed here are relative and are determined in a comparison with the adjacent impedances. For example, a capacitor that is in parallel with a 1-$k\Omega$ resistor cannot be assumed as an open circuit if its impedance becomes lower than 10 $k\Omega$. In addition, a capacitor in series with a 1-$k\Omega$ resistor is assumed as a short circuit as soon as its impedance goes under 0.1 $k\Omega$.

Each capacitor, including both the internal and the external ones, contributes to the amplifier's transfer function by introducing a *zero* and a *pole*. At low frequencies, each capacitor's region of activity starts with a zero and ends with a pole. This means that starting from DC, whenever we pass through

* It should be noted that in advanced frequency domain analysis and design, there are external capacitors with very small capacitances that determine the high-frequency behavior of amplifiers.

a region of activity of an external capacitor on the frequency axis the gain increases. This will continue to the point where *the last external capacitor shorts.* Beyond this frequency, the gain remains nearly constant. This is the amplifier's low cutoff frequency, which is in fact where the midband frequencies begin. The gain remains nearly constant up to the point where *the first internal capacitor comes out of the open-circuit state.* This is the end of the midband frequency range and is referred to as the amplifier's high cutoff frequency.

As in any other linear time-invariant (LTI) system, an amplifier's linear behavior in the frequency domain is formulated by its transfer function. The transfer function that models an amplifier's frequency-dependent behavior at low and midband frequencies is called the *low-frequency transfer function,* which is in general written as:

$$A(s) = \frac{A_0 \prod_{i=1}^{N} (s + z_i)}{\prod_{i=1}^{N} (s + p_i)} \tag{7.1}$$

where A_0 is the amplifier's midband gain, N is the number of the (external) capacitors, and z_i and p_i are the zero and pole for the i-th capacitor.

Before dealing with the frequency analysis methods that lead to finding an amplifier's transfer function, let us learn how to plot the amplifier's frequency response. This provides us with a helpful insight with which to study the amplifiers' behavior in the frequency domain.

7.3 Plotting an Amplifier's Response at Low Frequencies

Being a function of the complex frequency $s = \sigma + j\omega$, each transfer function is a function of both σ and $j\omega$. The transient response of a system is represented by the terms that are a function of σ, and the terms that are a function of $j\omega$ represent the steady-state behavior. To make the frequency-domain analysis as simple as possible, in most cases the complex frequency s is replaced only with $j\omega$. In other words, $A(j\omega)$ will be studied instead of $A(s)$. This means that the analysis is reduced to the system's steady-state response for a single-frequency input (i.e., a sine wave with a frequency of ω). However, the analysis is performed for periodic sinusoidal inputs, but the results can be easily used to obtain the system's steady-state response to any periodic input. Although σ is set to zero, the transfer function is, in general, still a complex (and not purely imaginary) function, and obviously a function of signal frequency ω. A variety of approaches are available to plot such a function in the frequency domain. If the transfer function's imaginary part

(Im{$A(j\omega)$}) is sketched versus its real part (Re{$A(j\omega)$}), the plot is called the *Nyquist plot*. Plotting the magnitude of the transfer function ($|A(j\omega)|$) as a function of its phase ($< A(j\omega)$) is referred to as the *Nichols plot*. In frequency-domain analysis of the amplifiers it is usually preferred to use *Bode plot*, which in fact consists of two plots: the magnitude of the transfer function ($|A(j\omega)|$) and its phase ($< A(j\omega)$) both as functions of the frequency ω or f [10]. As we know, ω is the angular frequency with dimension Radiant/Second (Rad/Sec), whereas f the is the frequency with dimension Hertz (Hz) that is related to ω as:

$$f = \omega/2\pi. \tag{7.2}$$

Example 7.1

Draw the magnitude and phase plots for an amplifier with transfer function:

$$A(s) = \frac{10s}{s+10}.$$

Solution:

First, s is replaced with $j\omega$:

$$A(j\omega) = \frac{10j\omega}{j\omega + 10}$$

Then, the magnitude and phase are derived:

$$|A(j\omega)| = \frac{10\omega}{\sqrt{(\omega^2 + 10^2)}}$$

and

$$\angle A(j\omega) = 90° - \tan^{-1}\left(\frac{\omega}{10}\right).$$

The plots of these functions versus ω are presented in Figure Ex. 7.1.

The plots in Figure Ex. 7.1 look fine, but as soon as we have zeros or poles at higher frequencies, we need to compress the low-frequency parts of the curves to be able to draw the high-frequency parts. This means that on such a plot one cannot concentrate at both low- and high-frequency parts of the

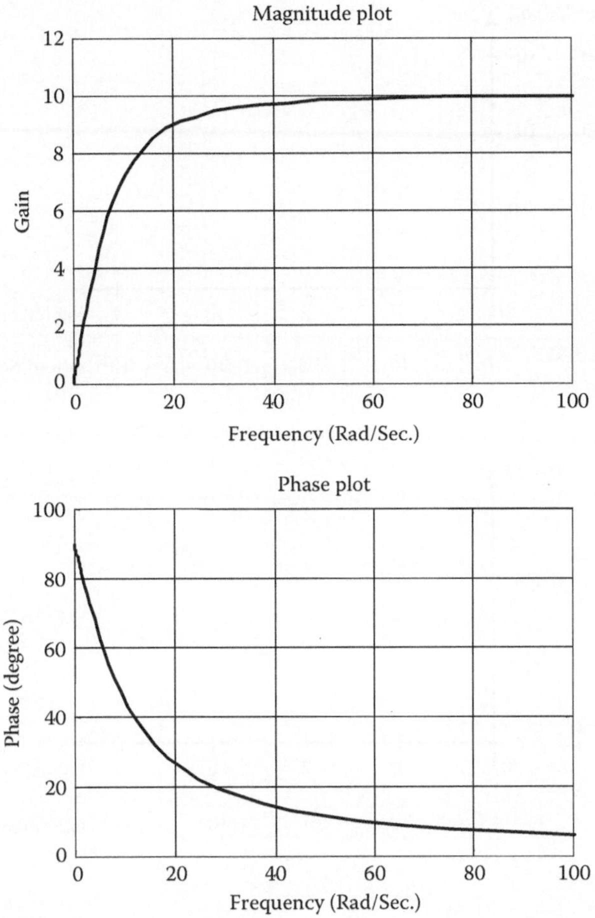

FIGURE EX. 7.1

curve at the same time. To overcome this problem, the frequency axis is usually logarithmically scaled. This way, the low-frequency region on the frequency axis is stretched, and as the frequency increases, the higher-frequency regions are condensed more and more. For example, the interval from 0 to 1 on the log(ω) axis is assigned to only 9 Hz, whereas the interval from 3 to 4 is considered for 9 kHz. Each division on the frequency axis is called a *decade*. In addition, because of the possibility of gain variations over a very wide range as the frequency changes, sometimes up to a couple of orders of magnitude, it is preferred to scale the gain axis logarithmically, too. In such occasions, representing the gain in decibels (dB) as

$$|A(j\omega)|_{dB} = 20\ \log_{10}|A(j\omega)|$$

is also a choice.

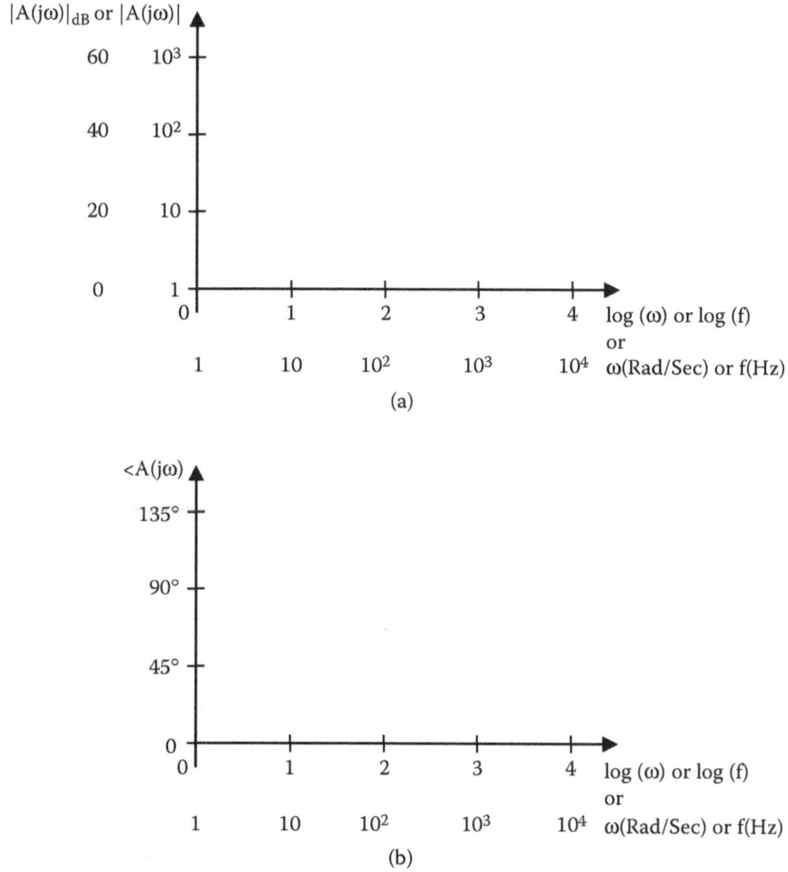

FIGURE 7.2
Scaling the axes for (a) magnitude and (b) phase Bode plots.

To conclude, magnitude and phase Bode plots are usually drawn with logarithmic frequency scale, and the gain is also drawn either on a logarithmic axis or in dB on a linear axis, as illustrated in Figure 7.2.

Redrawing the plots of Figure Ex. 7.1 based on the scales in Figure 7.2, the curves of Figure 7.3 will be resulted.

As clearly indicated, the plots of Figure 7.3 can be easily approximated by line segments or asymptotes. In fact, one can simply draw approximate magnitude and phase plots using some line segments, which will be precise enough for rough analysis and design of amplifiers.

A variety of methods are available for drawing Bode plots for linear systems in general using line segments, which are also referred to as *asymptotic plots*. The approaches introduced in this book for plotting Bode plots are indeed optimized *specifically for regular electronic amplifiers*.

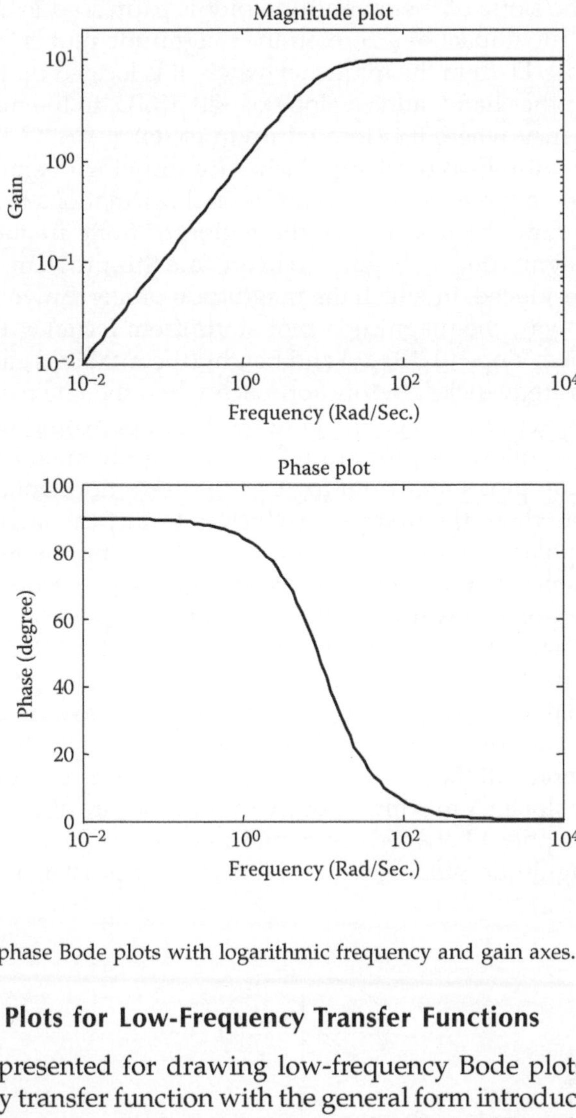

FIGURE 7.3
Magnitude and phase Bode plots with logarithmic frequency and gain axes.

7.3.1 Bode Plots for Low-Frequency Transfer Functions

The method presented for drawing low-frequency Bode plots starts with a low-frequency transfer function with the general form introduced in Equation (7.1). The gain or magnitude axis is assumed to be in dB, and the frequency axis is scaled for log(ω). First, the absolute value of the midband gain is calculated in dB ($|A_0|_{dB}$). Then, each zero and pole of the transfer function are located on the frequency axis by "o" and "x", respectively. The zeros, which are at 0 Rad/Sec, are assumed to be located at $-\infty$ on the frequency axis.

The *magnitude plot* is a continuous piecewise line that indicates the variations of the magnitude of the transfer function (in dB) versus frequency (on a logarithmically scaled axis). In the typical approaches for drawing the magnitude plot, the line usually starts at $log(\omega) = -\infty$ and goes toward higher frequencies. At first, the line starts with a certain slope, and as it passes the frequencies where the zeros and poles are located, the slope is adjusted accordingly. The slope of the magnitude plot is expressed in dB/Decade or simply dB/D. The impact of a zero on the magnitude plot is the addition of a slope of +20 dB/D. from the frequency where it is located up to $log(\omega) = +\infty$. A pole, on the other hand, adds a slope of –20 dB/D. to the magnitude plot from the frequency where it is located up to $log(\omega) = +\infty$.

The problem with the typical approaches for drawing magnitude plots for *low-frequency transfer functions* is the additional calculations that one should do to find out how the line arrives the region of finite frequencies, where drawing the magnitude plot begins. To overcome this problem, the following approach is introduced, in which the magnitude plot is drawn in the reverse direction. This way, the magnitude plot starts from $log(\omega) = +\infty$ where it is a line with known slope (0 dB/D.) and height (the midband gain in dB), and comes to lower frequencies. In this approach when the line passes each zero or pole, the impact of that zero or pole on the slope would end, as clearly expressed in the following procedure for drawing the magnitude plot.

The magnitude plot starts from $\log(\omega) = +\infty$ as a horizontal line at $|A_0|_{dB}$ up to the point where the first pole (which is in fact the largest among all the zeros and poles) is encountered. Passing this point, the line, which has so far had a slope of zero, would break and the slope becomes +20 dB/D. The line keeps going toward $-\infty$ and passing each pole or zero the slope adjusts accordingly. If it is a pole, the slope increases by +20 dB/D. and if it is a zero, –20 dB/D. is added to the slope. If, at a certain point, more than one zero or pole exists, then each zero or pole will have its own effect on the slope. This continues until the line passes the last zero (which is in fact the smallest among all the zeros and poles), then moves toward $\log(\omega) = -\infty$ with the same slope. A maximum error of 3 dB occurs at *corner frequencies* (either zeros or poles) between the actual magnitude curve and the piecewise-linear magnitude plot drawn based on this approach, as illustrated in Figure 7.4.

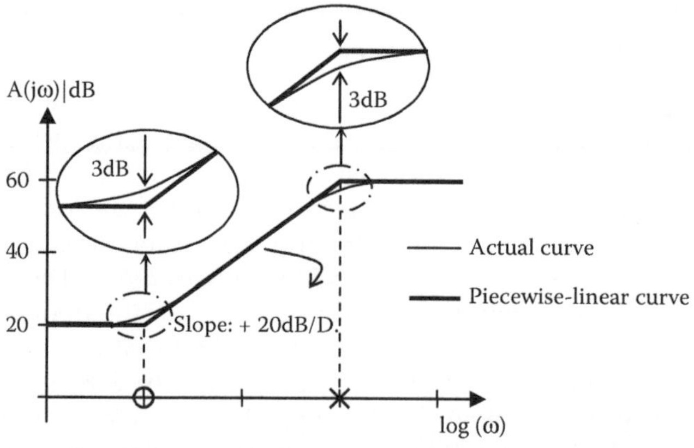

FIGURE 7.4
The actual and the asymptotic piecewise-linear magnitude curves at corner frequencies.

Example 7.2

Using the method described previously, draw the magnitude plot for an amplifier with transfer function:

$$A(s) = \frac{10(s+1)}{s+10}.$$

Solution:

The amplifier's midband gain is 10, which will be 20 dB (= 20 $log_{10}(10)$). It has a pole at 10 Rad/Sec and a zero at 1 Rad/Sec, which will be located at 1 and 0 on log(ω) axis, respectively (see Figure Ex. 7.2). Magnitude curve starts from log(ω) = +∞ at 20dB with slope 0. After passing log(ω) = 1, the slope becomes +20 dB/D. as the result of passing a pole. At log(ω) = 0, the slope becomes 0 because a zero is passed, and because no other corner frequency occurs afterward, the magnitude keeps going with the same slope until it arrives log(ω) = −∞.

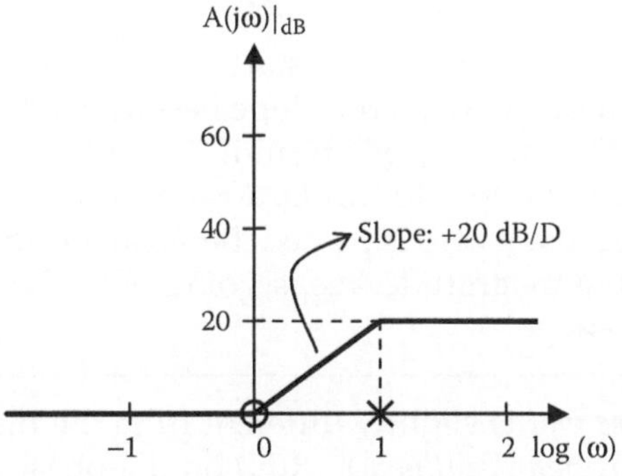

FIGURE EX. 7.2

Example 7.3

Repeat the previous example for transfer function:

$$A(s) = \frac{10^3 s(s+0.1)}{(s+10)(s+100)}.$$

Solution:

The amplifier's midband gain is 10^3, which will be 60dB (=20 $log_{10}(10^3)$). It has two zeros at 0 and 0.1 Rad/Sec and two poles at 10 and 100 Rad/Sec, which will be located at −∞, −1, 1, and 2 on log(ω) axis, respectively (see

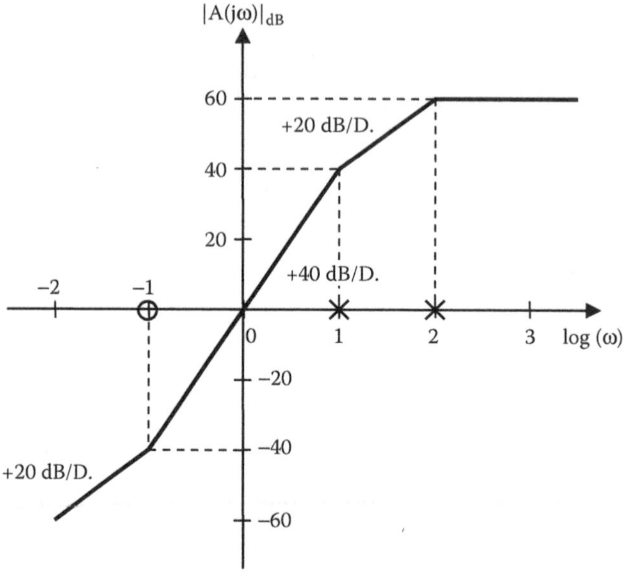

FIGURE EX. 7.3

Figure Ex. 7.3). The magnitude curve starts from $\log(\omega) = +\infty$ at 60 dB with slope 0. After passing $\log(\omega) = 2$, the slope becomes +20 dB/D. as the result of passing a pole. The curve goes on until it passes the pole at $\log(\omega) = 1$, which changes the slope to +40 dB/D. At $\log(\omega) = -1$, the slope becomes +20 dB/D. again because a zero is passed. Because no other corner frequency occurs afterward, the magnitude keeps going with the same slope until it arrives at $\log(\omega) = -\infty$.

Drawing the *phase plot* is slightly different from the magnitude plot. Each zero or pole has a *region of influence (ROI)* on the phase curve, which starts one decade before the zero or pole location and ends one decade after that. For example, for a pole at 1 kRad/Sec, the ROI is from 100 Rad/Sec up to 10 kRad/Sec Within its ROI, a zero/pole adds +/−45°/D. to the slope of the phase curve.* To draw the phase plot, first the ROIs of all the zeros and poles are marked on the frequency axis. Then, the phase curve starts from $\log(\omega) = +\infty$ at 0° with slope 0 toward lower frequencies. Once the phase curve enters an ROI, its slope increases or decreases by 45°/D. depending on whether that the ROI belongs to a zero or a pole, respectively. Upon exiting the ROI, the slope changes accordingly. Within the intervals where more than one ROIs overlap, the curve's slope is determined by the algebraic summation

* Throughout this book, all the zeros and poles are assumed to be located in the left-hand half-plane (LHP) of the complex frequency plane. The effect of a zero located in the right-hand half-plane (RHP) on the phase plot is similar to that of an LHP pole. In addition, an RHP pole is treated like an LHP zero when drawing the phase plot.

of the influences. In general, if within a frequency interval ROIs of N_Z zeros and N_P poles overlap, the slope is determined as:

$$\text{Slope} = (N_Z - N_P) \times 45°/\text{D}. \tag{7.3}$$

After the piecewise-linear phase plot exits the last ROI, it moves toward $\log(\omega) = -\infty$ with slope zero. If the phase plot has been drawn based on the preceding rules, the final phase should have become +90° times the number of zeros at 0 Rad/Sec.

Example 7.4

Draw the phase plot for an amplifier with transfer function:

$$A(s) = \frac{10(s+1)}{s+1000}.$$

Solution:

The amplifier's zero and pole are located at 0 and 3 on log(ω) axis with ROIs from −1 to 1 and from 2 to 4, respectively. Based on the aforementioned rules, the phase plot is drawn as depicted in Figure Ex. 7.4.

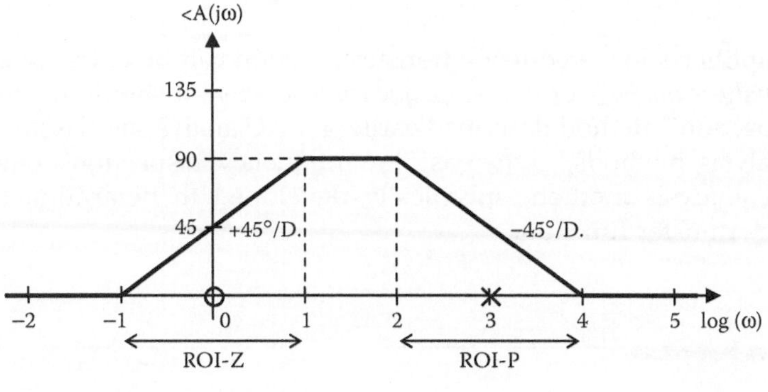

FIGURE EX. 7.4

Example 7.5

Draw the phase plot for an amplifier with transfer function:

$$A(s) = \frac{10s(s+1)}{(s+10^3)(s+10^4)}.$$

Solution:
This example, in addition to what was seen in the previous example, includes a zero at 0 Rad/Sec and two poles with overlapping ROIs. The phase plot is drawn as depicted in Figure Ex. 7.5.

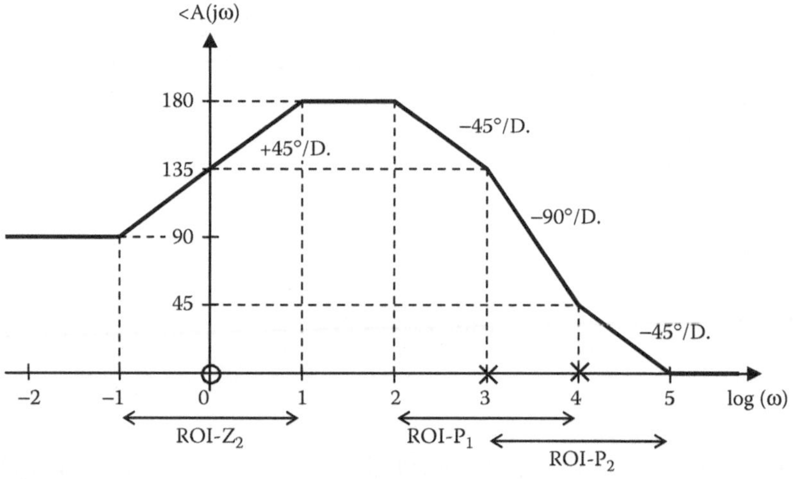

FIGURE EX. 7.5

An amplifier's low-frequency transfer function can be determined either by *AC analysis methods* or by *low-frequency inspection*. It should be noted that the "inspection" method described earlier for AC analysis is classified under "AC analysis methods," whereas "low-frequency inspection" is a totally different analysis method, specifically developed to determine the low-frequency transfer function.

7.4 Low-Frequency Analysis by AC Analysis Methods

The classic approach to analyze an amplifier at low frequencies is similar to the AC analysis methods explained earlier except the external capacitors are not replaced by short circuits. Instead, each capacitor is represented by its impedance in *complex frequency*, or *s-domain*. Then, the amplifier is simply analyzed using the previously described AC analysis methods for its gain or other AC specifications.

Example 7.6 describes how this approach is applied to amplifiers at low frequencies.

Example 7.6

For the amplifier in Figure Ex. 7.6(a),

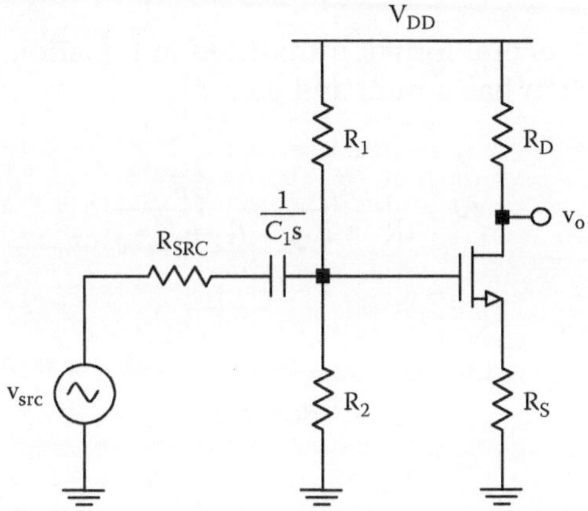

FIGURE EX. 7.6(a)

(a) Find the low-frequency transfer function.

(b) Assuming that $R_G = R_1 \| R_2 = 100\ k\Omega$, $R_{SRC} = 1\ k\Omega$, $R_S = 100\ \Omega$, $R_D = 10\ k\Omega$, $C_1 = 1\mu F$, and $g_m = 10\ mA/V$, draw the amplifier's Bode plot.

Solution:

(a) First, we draw the amplifier's AC schematic with capacitor C_1 replaced with its impedance $1/C_1 s$ (see Figure Ex. 7.6(b)).

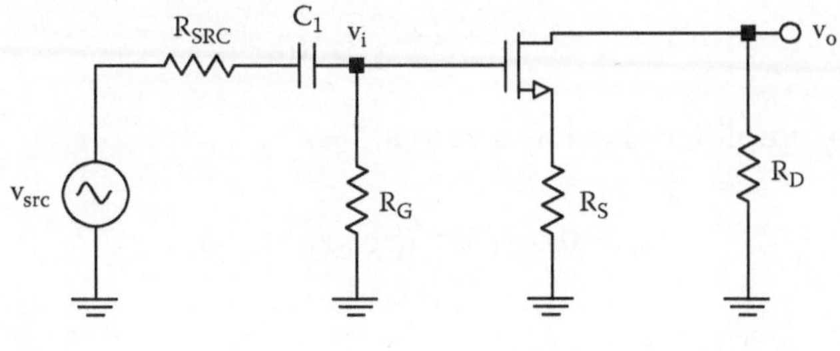

FIGURE EX. 7.6(b)

Then, applying the inspection method for AC analysis, the amplifier's gain is written as:

$$A_{VS}(s) = \frac{v_o(s)}{v_{src}(s)} = \frac{v_i}{v_{src}} \cdot \frac{v_o}{v_i} = \frac{R_G}{R_G + \left(R_{SRC} + \dfrac{1}{C_1 s}\right)} \cdot \frac{-R_D}{R_S + 1/g_m}$$

$$= \frac{R_G}{R_G + R_{SRC}} \cdot \frac{-R_D}{R_S + 1/g_m} \cdot \frac{s}{s + \frac{1}{C_1(R_G + R_{SRC})}}$$

Based on the general format introduced in Equation (7.1), the above transfer function has a midband gain of

$$A_0 = \frac{R_G}{R_G + R_{SRC}} \cdot \frac{-R_D}{R_S + 1/g_m},$$

a zero at

$$z_1 = 0 Rad/Sec.,$$

and a pole at

$$P_1 = \frac{1}{C_1(R_G + R_{SRC})}.$$

(b) Substituting the numeric values, the midband gain is

$$A_0 = \frac{100k}{101k} \cdot \frac{-10k}{0.2k} \cong -50$$

or

$$20 log_{10}|-50| = 34\ dB,$$

and the amplifier also has a zero at

$$z_1 = 0 Rad/Sec.\ (log_{10}|z_1| = -\infty),$$

and a pole at

$$p_1 = \frac{1}{1\mu \times 101k} \cong 1k Rad/Sec.\ (log_{10}|p_1| = 3).$$

The magnitude and phase plots for this transfer function are drawn in Figure Ex. 7.6(c). It should be added that the phase plot only exhibits the

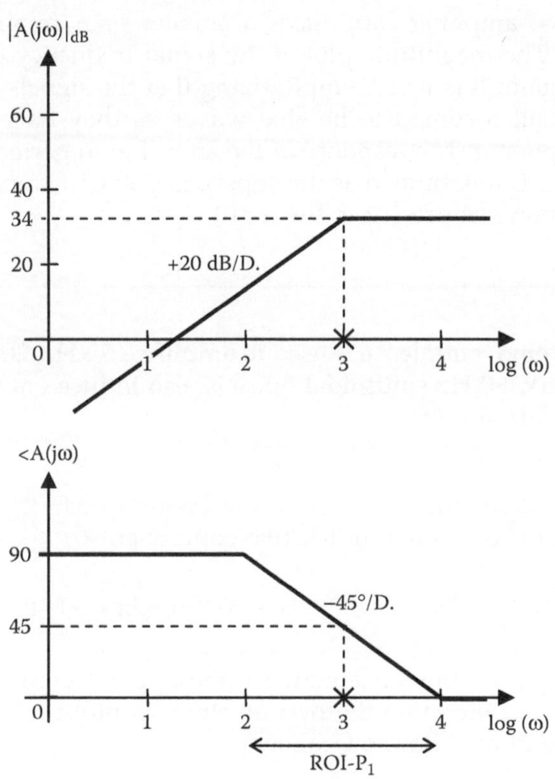

FIGURE EX. 7.6(c)

phase contributions of the zero and the pole. Because the midband gain is negative, the phase plot should be shifted down by 180° to illustrate the total phase shift of the amplifier.

Now that we have learned how to determine a low-frequency transfer function and how to draw its Bode plot, let us see how the magnitude and phase plots can be used to analyze an amplifier at low and midband frequencies. Considering the amplifier of Example 7.6 as an example, and keeping in mind that the magnitude and phase plots approximate the real magnitude and phase characteristic curves, it can be said that the circuit amplifies the signals that are above 10^4 Rad/Sec (1.6 kHz) with a gain of −50 and no phase difference. If a signal with a frequency between 10^3 Rad/Sec and 10^4 Rad/Sec is applied, it will be amplified with the same gain, but there will be a phase difference between the input and the output. The phase plot at the signal frequency determines the amount of the phase difference. As the signal frequency goes below 10^3 Rad/Sec, in addition to a phase difference between

45° and 90°, the amplifier introduces a smaller gain compared with the midband gain. The magnitude plot at the signal frequency determines the amount of the gain. It is worth emphasizing that the signals that are talked about here are all assumed to be sine waves as they have only a single frequency component. The response of the amplifier to periodic nonsinusoidal input signals is determined as the superposition of its reponses to all the frequency components present in the input signal.

Example 7.7

The amplifier from Example 7.6 is used to amplify a 5-kHz, 20-mV sine wave signal. If a 30-mV, 60-Hz sinusoidal noise is also induced at the input port, determine the output voltage.

Solution:

The input signal is composed of the two components:

$$v_{src}(t) = 20^{mV}\text{Sin}(2\pi\times 5^{kHz}t) + 30^{mV}\text{Sin}(2\pi\times 60^{Hz}t + \phi),$$

where ϕ is the phase difference between the applied input signal and the induced noise. To determine the output, the two input frequency components should be located on the frequency axis:

$$\omega_1 = 2\pi \times 5^{Hz} = 31.4^{kRad/Sec} \Rightarrow \log(\omega_1) = 4.5$$

and

$$\omega_2 = 2\pi \times 60^{Hz} = 377^{Rad/Sec} \Rightarrow \log(\omega_2) = 2.6$$

Then the gains and phases associated with both frequencies are read from the amplifier's Bode plot (see Figure Ex. 7.7). Evidently the 5-kHz signal is amplified with a gain of 34 dB (–50) and no phase difference exists. The 60-Hz noise, however, is amplified by a gain of 26 dB (–20) and has a phase shift of 63° (1.1 Rad.). Thus, the output signal can be written as:

$$v_o(t) = -1^{V}\text{Sin}(2\pi \times 5^{kHz}t) - 0.6^{V}\text{Sin}(2\pi \times 60^{Hz}t + 1.1 + \phi).$$

The frequency response of an amplifier at low and midband frequencies is usually similar to a high-pass filter. The cutoff frequency is determined by the circuit elements, such as the external capacitors, some of the resistors, and sometimes the transistor(s). Usually, the resistances and the transistor parameters are set to meet the DC and AC requirements (e.g., DC bias point,

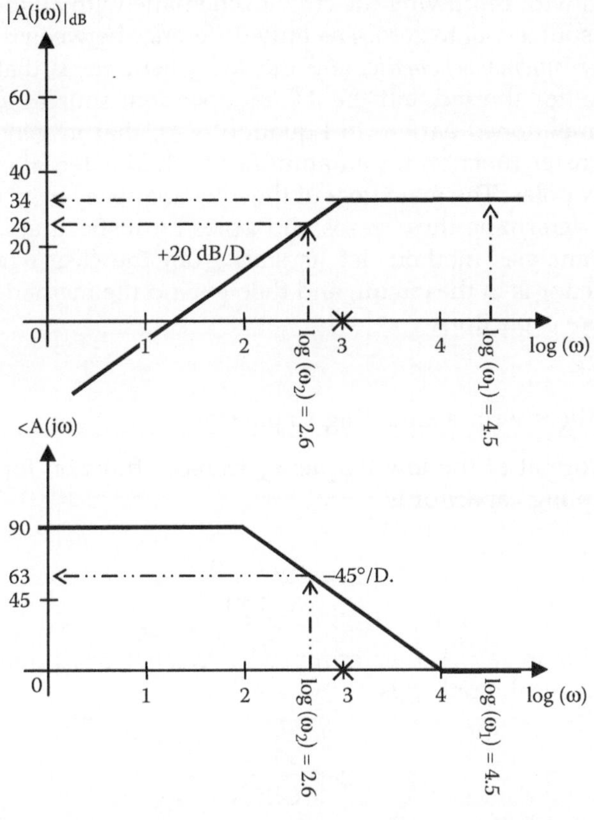

FIGURE EX. 7.7

gain, and input and output resistances). So, *the external capacitance(s) will often be the only parameter(s) that a designer has for setting the low cutoff frequency after the circuit is already designed for DC and AC specifications.*

7.5 Low-Frequency Analysis by Inspection

As long as one external capacitor is in the amplifier, the previously described AC analysis methods return relatively simple transfer functions. As soon as the number of the capacitors increases, however, it becomes increasingly difficult to figure out which parameters have more significant contributions to the amplifier's frequency domain behavior. The *low-frequency inspection* method is a straightforward approach to determine all the zeros and poles individually, without manipulating big fractional transfer functions. The analysis starts with redrawing the circuit schematic with all the DC and AC independent sources set to zero. The only difference between this schematic, hereafter *low-frequency schematic*, and the AC schematic, is that the external capacitors are not shorted and the AC independent sources are also set to zero. It was mentioned earlier, in Equation (7.1), that in general, the low-frequency transfer function for an amplifier with N external capacitors has N zeros and N poles. The main goal of the low-frequency inspection method is to directly determine these zeros and poles by inspecting the circuit. To simplify the analysis method, let us start from the case where only one external capacitor is in the circuit, and then extend the method to the amplifiers with more capacitors.

7.5.1 Amplifiers with a Coupling Capacitor

The general format of the low-frequency transfer function for an amplifier with one coupling capacitor is

$$A(s) = \frac{A_0(s+z)}{(s+p)}, \tag{7.4}$$

where the midband gain A_0 is determined by AC analysis, the zero z is always at 0 Rad/Sec, and the pole p is:

$$p = \frac{1}{CR^S}, \tag{7.5}$$

in which, C is the capacitance of the coupling capacitor, and R^S is the equivalent resistance seen by the capacitor C.

Taking a second look at the transfer function obtained for the amplifier studied in Example 7.6, it is clear that one can easily obtain exactly the same transfer function using the low-frequency inspection method.

Example 7.8

For the common-base amplifier in Figure Ex. 7.8(a),

(a) Find the low-frequency transfer function.
(b) Assuming that $R_B = R_1 \| R_2 = 10\ k\Omega$, $R_S = 100\ \Omega$, $R_E = 100\ \Omega$, $R_C = 10\ k\Omega$, $\beta = 100$, and $r_e = 25\ \Omega$, find C_1 to set the amplifier's low cutoff frequency at 100 Hz.

Solution:

(a) By inspection, the amplifier's midband gain is written as:

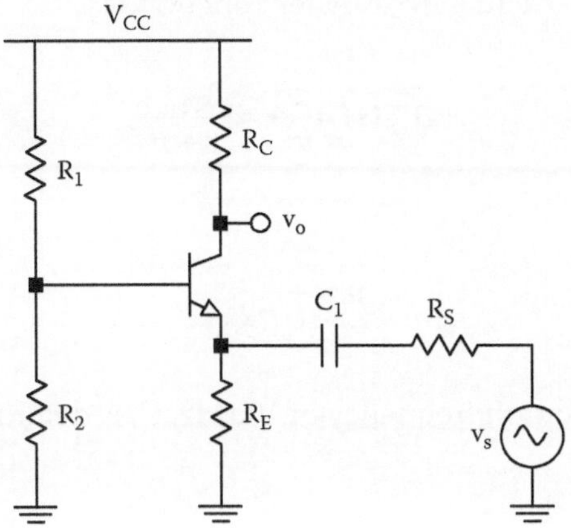

FIGURE EX. 7.8(a)

$$A_{VSO} = \frac{v_i}{v_s} \cdot \frac{v_o}{v_i} = \frac{R_E \left\| \left(r_e + \frac{R_B}{\beta} \right) \right.}{\left\{ R_E \left\| \left(r_e + \frac{R_B}{\beta} \right) \right. \right\} + R_S} \cdot \frac{R_C}{r_e + \frac{R_B}{\beta}}.$$

Now, to apply the low-frequency inspection method, the amplifier's low-frequency schematic is drawn as illustrated in Figure Ex. 7.8(b). The equivalent resistance seen by capacitor C_1, $R^S_{C_1}$ is determined:

$$R^S_{C_1} = R_S + \left(R_E \left\| \left(r_e + \frac{R_B}{\beta} \right) \right. \right)$$

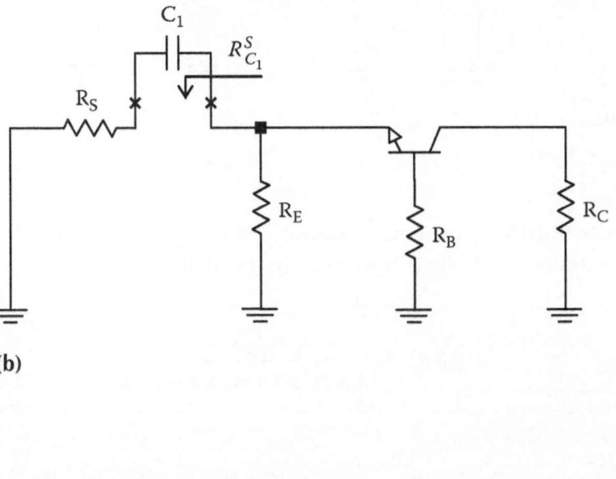

FIGURE EX. 7.8(b)

Then, the low-frequency transfer function is:

$$A_{VS}(s) = \frac{v_o}{v_s} = \frac{A_{VS0}.s}{s+p},$$

where

$$p = \frac{1}{C_1.R_{C_1}^S}.$$

(b) To have a low cutoff frequency of 100 Hz, C_1 is determined as follows:

$$p = \frac{1}{C_1.R_{C_1}^S} \Rightarrow C_1 = \frac{1}{p.R_{C_1}^S} = \frac{1}{(2\pi \times 100)\times\left\{0.1^k + \left[0.1^k \middle\| \left(0.025^k + \frac{10^k}{100}\right)\right]\right\}} = 10.2\mu F.$$

Example 7.9

Find the low-frequency transfer function for the two-stage amplifier in Figure Ex. 7.9(a).

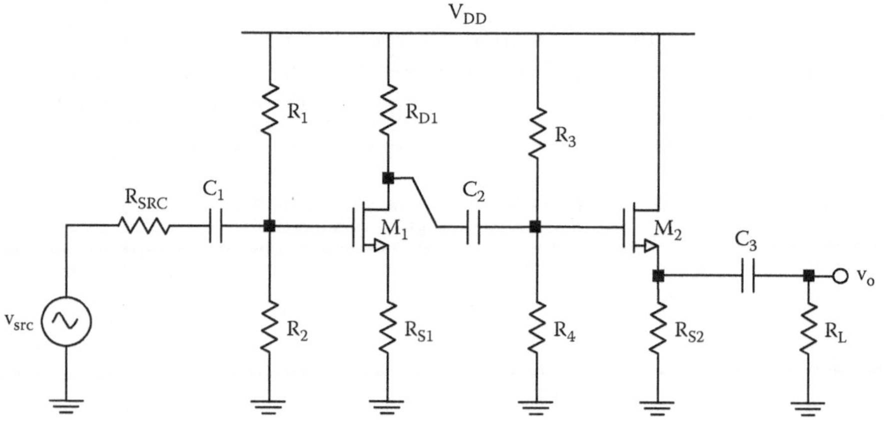

FIGURE EX. 7.9(a)

Solution:

Because of having three coupling capacitors the low-frequency transfer function has three zeros at 0 Rad/Sec and three poles:

$$A_{VS}(s) = \frac{A_{VS0}s^3}{(s+p_1)(s+p_2)(s+p_3)}.$$

Using the AC schematic of the amplifier in Figure Ex. 7.9(b), the midband gain, A_{VS0}, is obtained by inspection as:

$$A_{VS0} = \frac{v_i}{v_{src}} \cdot \frac{v_{o1}}{v_i} \cdot \frac{v_o}{v_{o1}}$$

$$= \frac{R_A}{R_A + R_{SRC}} \cdot \frac{-R_B}{R_{S1} + \dfrac{1}{g_{m1}}} \cdot \frac{R_C}{R_C + \dfrac{1}{g_{m2}}} \cdot$$

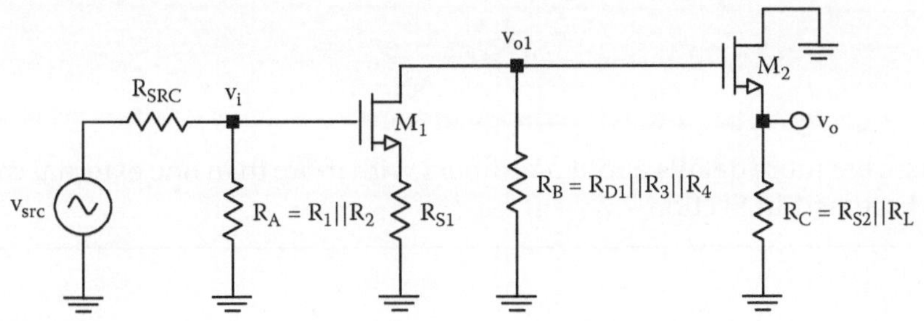

FIGURE EX. 7.9(b)

Figure Ex. 7.9(c) illustrates the resistances that each of the capacitors would see:

$$R^S_{C1} = R_{SRC} + R_A$$

$$R^S_{C2} = \left[R_{D1} \| r_{o1}(1 + g_{m1}R_{S1}) \right] + R_{G2}$$

$$R^S_{C3} = \left(R_{S2} \middle\| \frac{1}{g_{m2}} \right) + R_L.$$

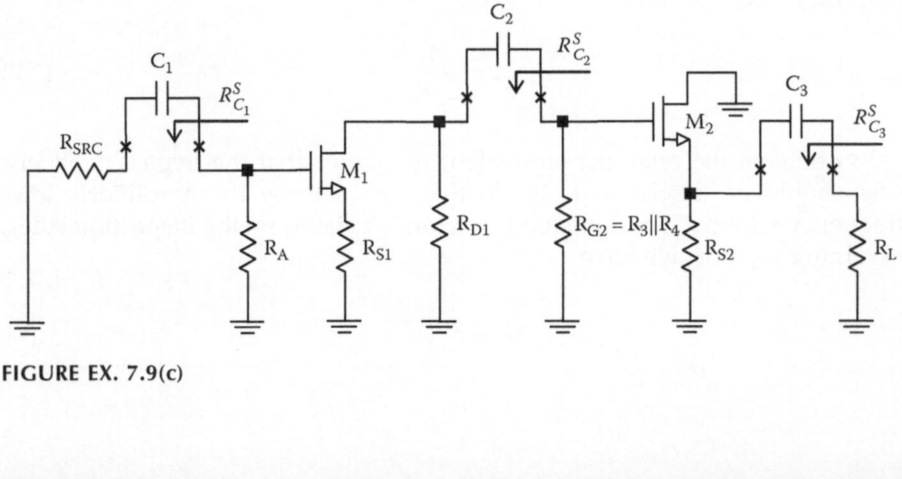

FIGURE EX. 7.9(c)

The low-frequency poles p_1, p_2, and p_3 are then obtained:

$$p_1 = \frac{1}{C_1 R_{C_1}^S},$$

$$p_2 = \frac{1}{C_2 R_{C_2}^S},$$

and

$$p_3 = \frac{1}{C_3 R_{C_3}^S}.$$

There are more details about amplifiers with more than one external capacitor discussed in Section 7.7.

7.5.2 Amplifiers with a Bypass Capacitor

Similar to a coupling capacitor, a bypass capacitor also causes a zero and a pole in the low-frequency transfer function. Although its pole is determined similar to that of a coupling capacitor (Equation (7.5)), a bypass capacitor's zero is obtained as:

$$z = \frac{1}{CR_{BP}}, \tag{7.6}$$

in which, C is the capacitance of the bypass capacitor, and R_{BP} is the resistance bypassed by the capacitor C.

As an example, a common-source amplifier with C_1 as a bypass capacitor is depicted in Figure 7.5. For this circuit, the bypassed resistor is R_S, and the amplifier's zero becomes:

$$z = \frac{1}{C_1 R_S}. \tag{7.7}$$

To calculate the pole, the equivalent resistance that the bypass capacitor sees should be determined. To do this, let us draw the amplifier's low-frequency schematic as depicted in Figure 7.6. Based on the inspection rules, assuming $R_D << r_o$ we have

$$R_{C_1}^S = R_S \left\| \frac{1}{g_m} \right.,$$

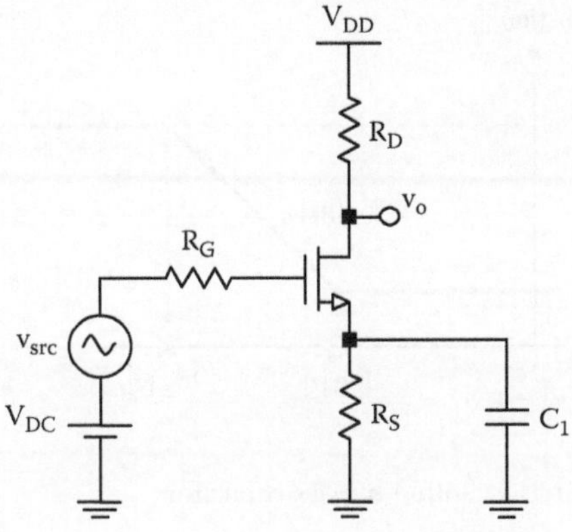

FIGURE 7.5
An amplifier with a bypass capacitor.

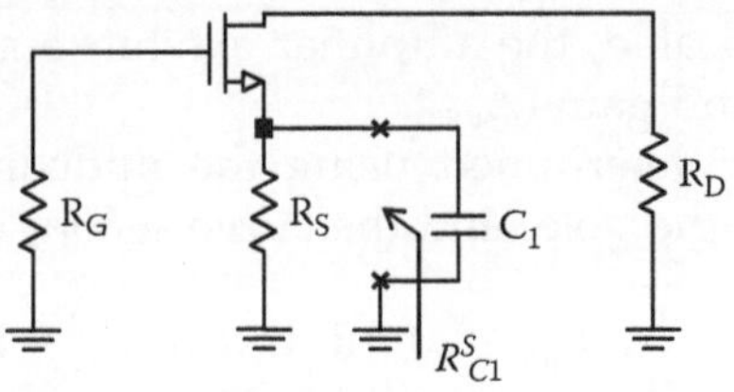

FIGURE 7.6
Low-frequency schematic of the amplifier of Figure 7.5.

and the pole is

$$p = \frac{1}{C_1 R_{C_1}^S} = \frac{1}{C_1 \cdot \left(R_S \left\| \frac{1}{g_m} \right. \right)}.$$

The midband gain of the amplifier can also be determined by inspection as:

$$A_{VSO} = -g_m R_D.$$

Now, the low-frequency transfer function can be written as:

$$A_{VS}(s) = \frac{v_o}{v_s} = \frac{A_{VSO} \cdot (s+z)}{s+p} = (-g_m R_D) \frac{s + \dfrac{1}{C_1 R_S}}{s + \dfrac{1}{C_1 \left(R_S \left\| \dfrac{1}{g_m} \right. \right)}}. \qquad (7.8)$$

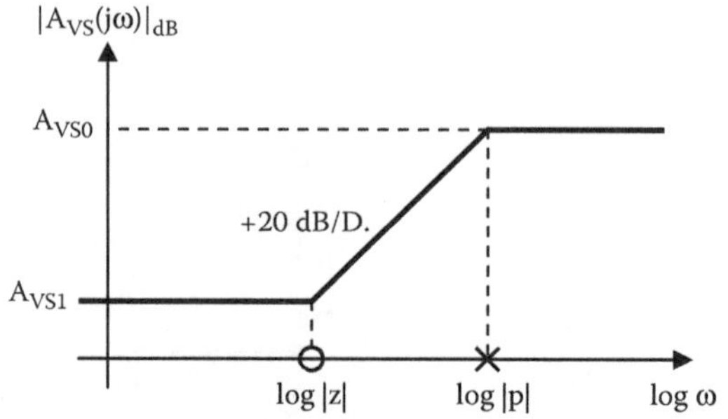

FIGURE 7.7
Magnitude plot for an amplifier with a bypass capacitor.

It will be instructive if the magnitude plot for this transfer function, which is drawn in Figure 7.7, is interpreted based on the behavior of the circuit in the frequency domain. Starting from DC (f = 0 Hz or ω = 0 Rad/Sec) up to where the "zero" is located, the amplifier exhibits a small gain (A_{VS1}) compared with the midband gain (A_{VS0}).

A_{VS1} can be simply determined using the midband gain, the distance between the zero and the pole, and the curve's slope in that interval:

$$|A_{VS1}|_{dB} = |A_{VS0}|_{dB} - 20\text{ dB/D} \times (\log|p| - \log|z|),$$

which leads to

$$A_{VS1} = A_{VS0}\frac{z}{p} = \frac{(-g_m R_D)\left(R_S \left\| \frac{1}{g_m}\right.\right)}{R_S} = \frac{-g_m R_D}{R_S + \dfrac{1}{g_m}}.$$

Taking a second look at the amplifier, this is in fact the gain when the bypass capacitor does not function (as if it is not connected at all). This confirms the previously described concept that an external capacitor is equivalent to an open circuit before the frequency reaches its "zero." As the frequency exceeds the "zero," the bypass capacitor comes out of open-circuit and provides a path, in parallel with R_S, for the AC current to flow from the transistor's source lead to the ground. In other words, within the frequency interval between the zero and the pole the capacitor partially bypasses R_S. The higher the frequency, the lower the capacitor's impedance, the stronger the bypassing function, and as the result, the higher the gain will become. The gain keeps growing until the frequency reaches the "pole." It was described earlier that an external capacitor's pole is the frequency where it enters the short circuit state. For frequencies greater than the pole, the capacitor is approximated to a short circuit, and the gain stops growing. For this amplifier, this is where the midband frequency range starts.

Example 7.10

Assuming that $R_S = 10\ k\Omega$, $R_D = 100\ k\Omega$, $C_1 = 1\ \mu F$, and $g_m = 10\ mA/V$ for the amplifier in Figure 7.5,

(a) Find the low-frequency transfer function.
(b) Draw the Bode plot.

Solution:

(a) Because no current is flowing through R_G, the source signal appears at the gate and there will be no attenuation at the input. Thus, the midband gain is simply determined by inspection as:

$$A_{VSO} = -g_m R_D = -10.$$

Because of having one external capacitor, the transfer function has one zero and one pole. The zero is the inverse of the product of the bypass capacitor, C_1, and the bypassed resistance, which is R_S:

$$z = \frac{1}{C_1 R_S} = \frac{1}{10^{-6} \times 10^4} = 100 Rad/Sec. = 2\pi \times 16 Hz$$

To calculate the pole, the equivalent resistance that the bypass capacitor sees should be determined. Based on the inspection rules we have:

$$R_{C_1}^S = R_S \left\| \frac{1}{g_m} = 10^k \right\| 0.1^k = 100\Omega$$

and the pole is:

$$p = \frac{1}{C_1 . R_{C_1}^S} = \frac{1}{10^{-6} \times 100} = 10k Rad/Sec. = 2\pi \times 1.6kHz.$$

Finally, the low-frequency transfer function can be written as:

$$A_{VS}(s) = \frac{v_o}{v_s} = \frac{A_{VSO}.(s+z)}{s+p} = \frac{-10(s+100)}{s+10^4}$$

(b) To draw the Bode plot, we should convert the gain into dB and locate the zero and the pole on $log(\omega)$ axis:

$$|A_{VS0}|_{dB} = 20log_{10}|-10| = 20\ dB,$$

$$log_{10}|z_1| = 2,$$

and

$$log_{10}|p_1| = 4.$$

The magnitude and phase plots for this transfer function are drawn in Figure Ex. 7.10. On the magnitude plot, the curve starts from log(ω) = +∞ at 20dB with slope 0. After passing the pole at log(ω) = 4, the slope becomes +20 dB/D. At log(ω) = 2, the slope becomes 0 because a zero is passed. Passing the 2-decade interval on the horizontal axis in between the pole and the zero with 20-dB/D. slope, 40 dB is spanned on the vertical axis. This means that the plot will be at –20 dB after it passes the zero. Because no other corner frequency occurs afterward, the magnitude keeps going horizontally until it arrives log(ω) = –∞. On the phase plot, the ROI for the zero is from 1–3, and that of the pole is extended from 3 up to 5. As the result,

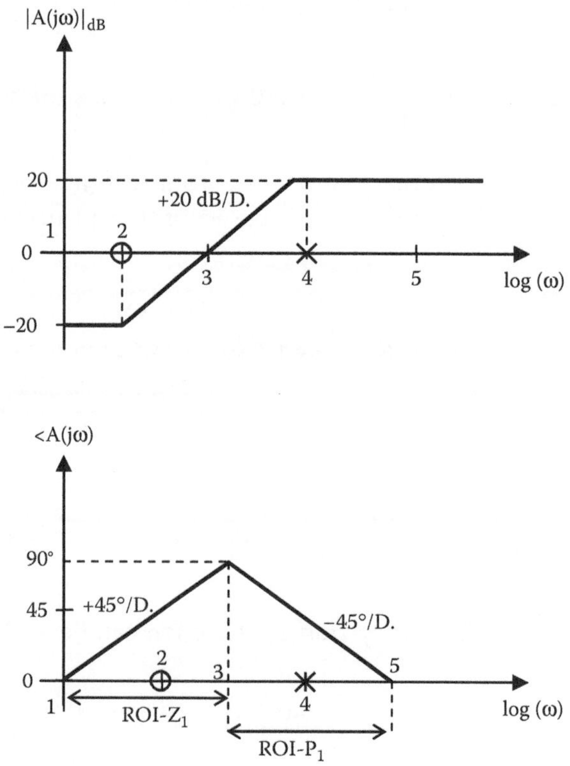

FIGURE EX. 7.10

from 1–3 the phase curve is under the influence of a zero (slope: +45°/D.) and from 3–5 a pole determines the slope (slope: –45°/D.).

7.6 Time Response

Before going into the details of the role of low-frequency zeros and poles in an amplifier's response in the time domain, it would be helpful to have a correspondence between the frequency components in a waveform and its time-domain shape. From the basics of signals and systems, *Fourier series expansion* is a method that can express any periodic signal in terms of sinusoidal components. As an example, a square wave with unit peak amplitude, zero DC component, and frequency f_0 can be expanded in terms of its fundamental (n = 1) and harmonic (n ≥ 2) frequency components as:

$$x(t) = \frac{4}{\pi}\sum_{n=1}^{+\infty}\frac{1}{n}\sin(2\pi n f_0 t).$$

Figure 7.8 illustrates this series with only the fundamental frequency component, and also the first 2, 5, 10, 20, and 100 harmonic terms.

It is instructive to pay attention to the features of the waveform as the number of the terms grows. Even with the first term (i.e., the fundamental component), it can be said that at least the signal frequency is preserved. The more terms taken into account, the closer the result will be to an ideal square wave. Sharp edges in the waveform begin to appear when the higher frequency harmonics are added. Although the first 100 frequency components have been able to make the relatively sharp edges in the waveform, the next infinite number of harmonics are still needed to recover the corners and also the mathematically sharp (slope: ±∞) edges. In general, it can be said that the *fast variations in a waveform is represented by high-frequency components*. In other words, if a waveform with fast variations (e.g., the sharp edges and corners in a square wave) is low-pass filtered, it will be smoothened such that neither the ultra-high-slope edges nor the corners will be seen in it anymore. This is because there will be no frequency component in the signal representing the waveform's fast variations as before.

On the other hand, Figure 7.9 illustrates the role of the low-frequency components in making a square wave. As can be seen, the last waveform in Figure 7.8 is taken as the reference (nearly a square wave). Eliminating first the fundamental frequency component and then the first 10 components

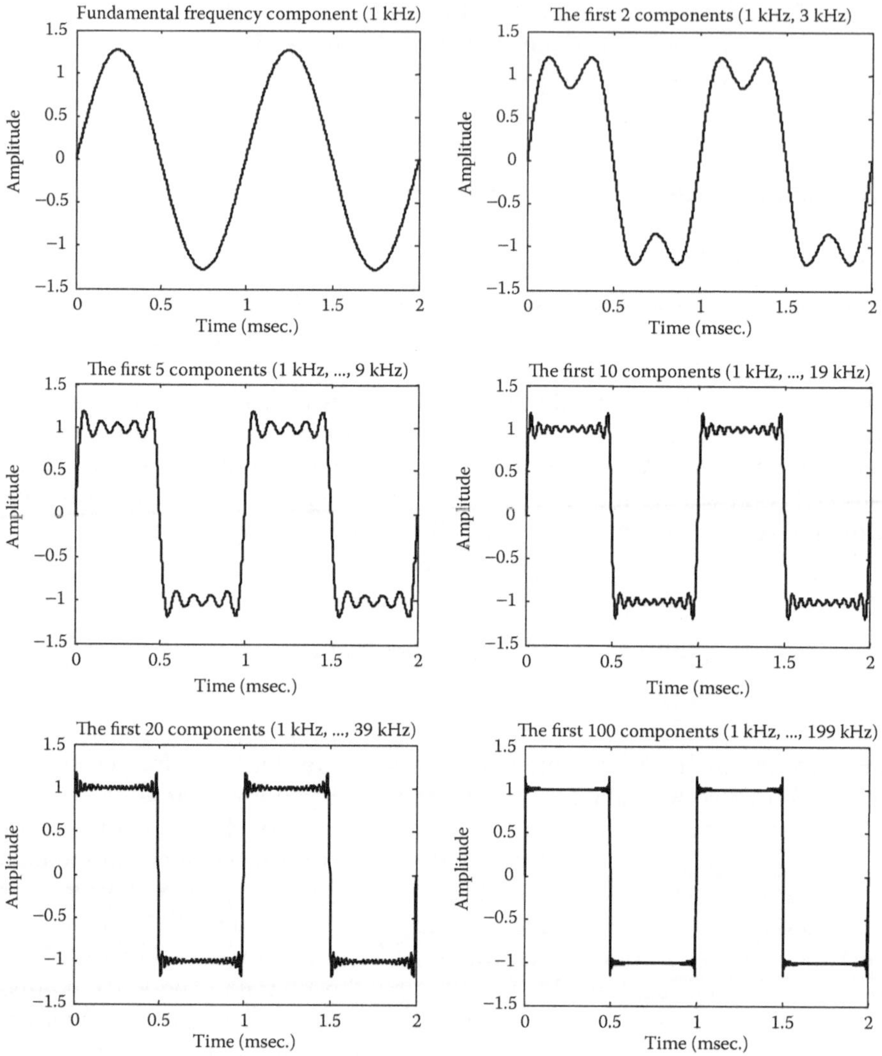

FIGURE 7.8
Composing a square wave using its fundamental and harmonic frequency components.

demonstrates how the low-frequency terms contribute to making the flat parts of the waveform. Therefore, it can be said that the *parts of a waveform that are flat or very-slowly varying with time are represented by low-frequency components*.

After learning to analyze an amplifier's low-frequency response, now it is time to study how the external capacitors affect the amplifier's time response. Let us start with the case where the amplifier has only one coupling capacitor. It was demonstrated that the low-frequency transfer function for such an amplifier is, in general,

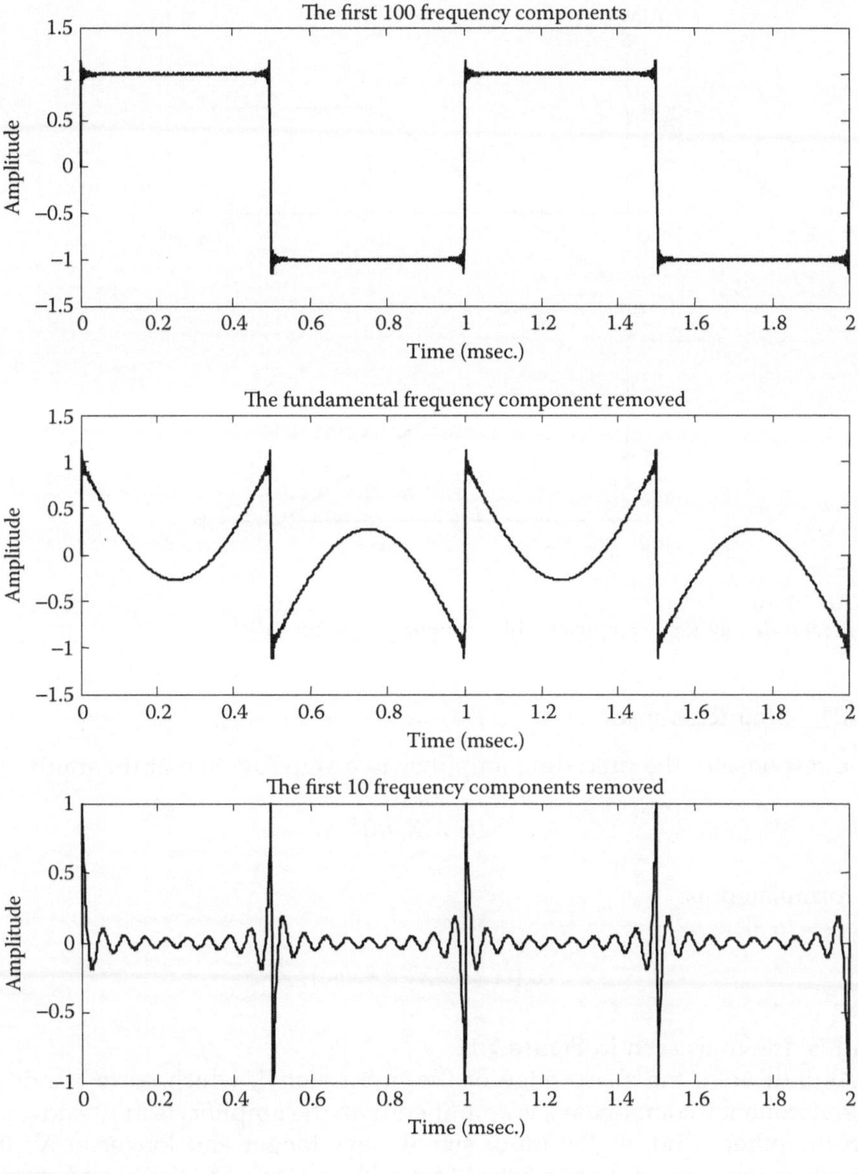

FIGURE 7.9
Illustration of the role of the low-frequency components in making a square wave.

$$A(s) = \frac{x_o}{x_s} = \frac{A_o s}{s + p}.$$

where x_s and x_o are the input and the output signals, respectively. The general Bode plot for such an amplifier is depicted in Figure 7.10 for reference and convenience.

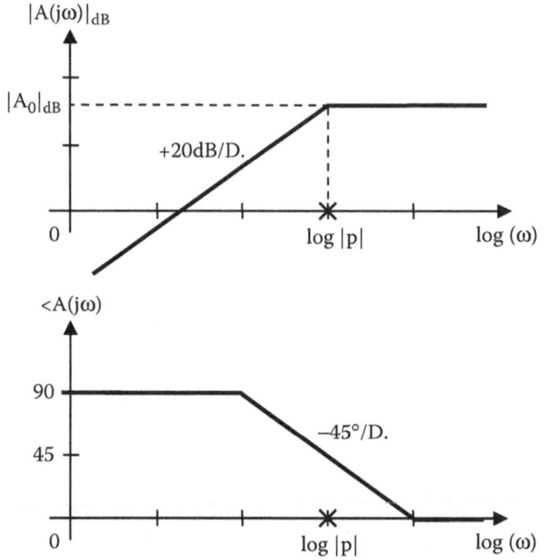

FIGURE 7.10
General bode plot for an amplifier with a coupling capacitor.

7.6.1 Step Response

The response of the preceding amplifier to a step function at the input

$$x_s(t) = X_I.u(t)$$

is formulated as:

$$x_o(t) = A_0 X_I.e^{-p.t} u(t),$$

and is drawn as seen in Figure 7.11.

As indicated, the sharp edge of the input signal, which corresponds to high frequency components, is amplified with the amplifier's midband gain. On the other hand, as the input signal stays longer and longer at X_I, the amplifier is increasingly convinced that it looks like a DC signal and should be rejected. After around five time constants, 5τ, it can be said that the output signal has retuned to zero as if no signal is applied at the input anymore.

7.6.2 Response to a Square Wave

Taking a square wave as a train of alternatively occurring positive and negative step functions, the amplifier's response to a square wave is a composition of its responses to positive and negative step functions, as illustrated

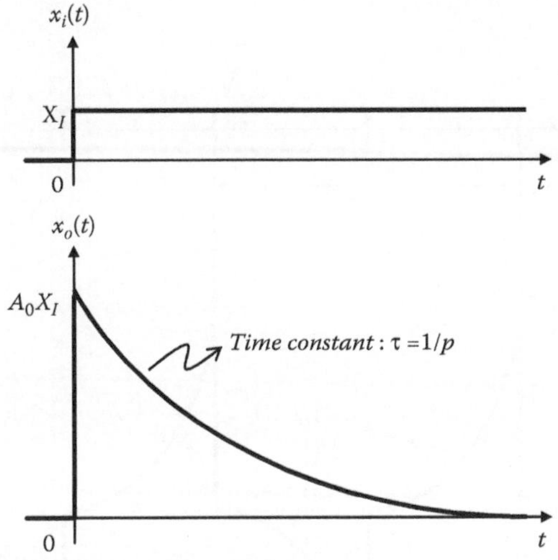

FIGURE 7.11
Step response of an amplifier with a coupling capacitor.

in Figure 7.12. As seen, each half-cycle of the input square wave in Figure 7.12 is long enough to allow the associated exponential decay to return to where it started from. Two extremes for the signal frequency, $f_S = 1/T_S$, compared with the amplifier's pole, deserve to be studied in detail.

7.6.2.1 $2\pi f_S << p$

When the signal frequency is much smaller than the amplifier's pole, the exponential decay at the output has enough time to settle down to its steady-state value and stay at that level for a while before the next step occurs at the input. In this case, the circuit's time response will be as depicted in Figure 7.13. When comparing the responses in Figure 7.12 and Figure 7.13, the reader should be aware of the fact that in both figures the amplifier's pole is the same, but the time axis in Figure 7.13 is compressed a lot to be able to plot the waveforms that now have much longer periods than the ones in Figure 7.12.

As the output waveform indicates, the amplifier's response to the flat parts of the input signal becomes zero very quickly. On the other hand, the edges of the input signal are well preserved and amplified with the midband gain. This is indeed a high-pass filtering function in the frequency domain, which noting the fact that the signal frequency in this case is below the low cutoff frequency, completely agrees with high-pass characteristic in Figure 7.10.

In addition to the high-pass filtering action, it can also be said, from the waveforms in Figure 7.13, that the amplifier also takes the derivative of the signal. The fact is that in this case, where the signal frequency is assumed to be much lower (at least 10 times, i.e., 1 decade on the logarithmic frequency axis) than the amplifier's pole, the amplifier is analyzed at the region that

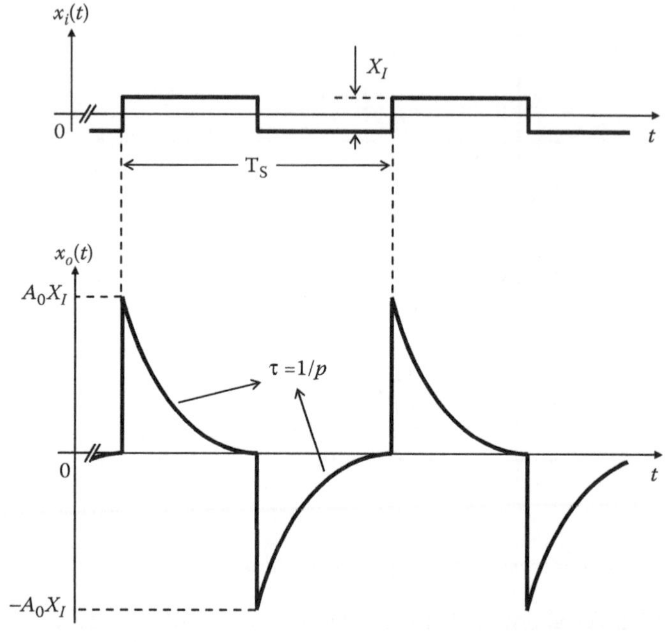

FIGURE 7.12
Response of an amplifier with a coupling capacitor to a square wave at the input.

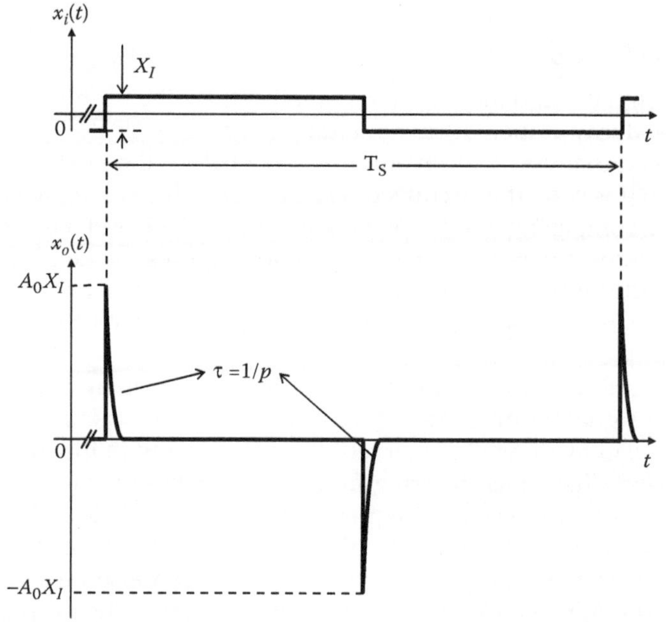

FIGURE 7.13
Response of an amplifier with a coupling capacitor to a square wave when the signal frequency is much lower than the amplifier's pole.

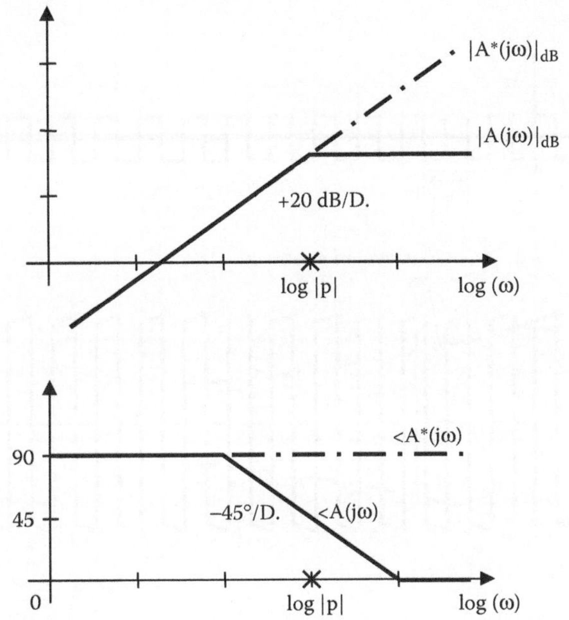

FIGURE 7.14
Approximating the amplifier with a derivative circuit at very low frequencies.

the magnitude plot has a slope of +20 dB/D. and the phase is constantly at +90°. Thus, it can be said that it does not matter to the input signal if the amplifier's transfer function is replaced with

$$A^{*}(s) = A_0 s,$$

which is in fact a transfer function that represents both amplification and derivation functions. This is illustrated in Figure 7.14.

7.6.2.2 $2\pi f_s >> p$

The other extreme is where the signal frequency is much higher than the amplifier's pole. In this case, again assuming the input square wave as a train of step funcitons, the next step occurs at the input before the output finds enough time to exhibit an exponential decay in response to the previous step. The circuit's time response is depicted in Figure 7.15. Now, it can be said that the wave shape of the input square wave is preserved during the amplification process. It should be noted that in Figure 7.15, however, the input signal frequency is intentionally not chosen much higher than the amplifier's pole only to exaggerate the exponential decays. If the signal frequency is chosen much higher than the pole, the decays will be negligible compared with the output peak-to-peak amplitude, and the output waveform will look like a perfect square wave.

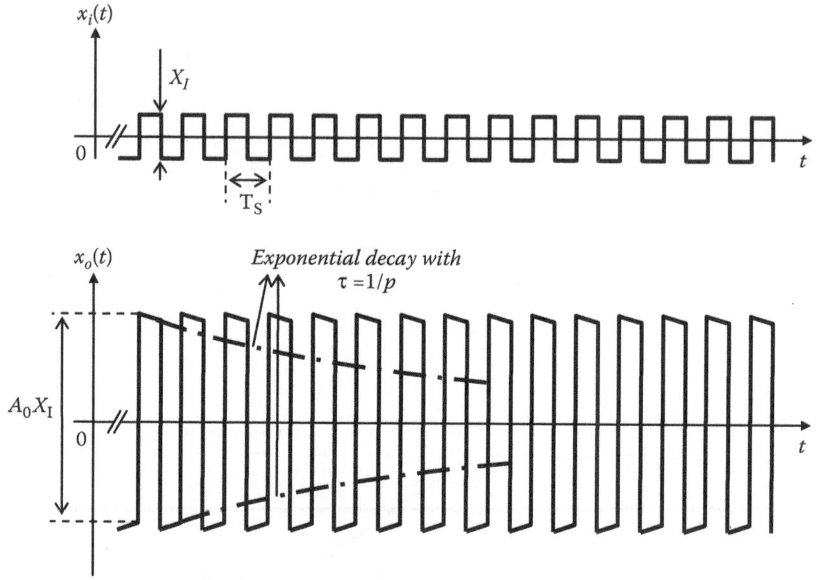

FIGURE 7.15
Response of an amplifier with a coupling capacitor to a square wave when the signal frequency is much higher than the amplifier's pole.

In the case where the signal frequency is assumed to be much higher (at least 10 times, i.e., 1 decade on the logarithmic frequency axis) than the amplifier's pole, the amplifier is analyzed at the region where the magnitude plot is flat and the phase is constantly at 0°, which is the ideal frequency response for an amplifier. Thus, it can be said that it does not matter to the signal if the amplifier's transfer function is replaced with

$$A^{*}(s) = A_0,$$

which is, in fact, a transfer function that represents only the amplification function, as illustrated in Figure 7.16.

7.6.3 Response to a Sine Wave

When a sine wave is applied to an amplifier, its amplitude and phase will be affected by the transfer function's magnitude and phase at the signal frequency. Based on the general Bode plot in Figure 7.10, if *the signal frequency is much lower than the pole,* in addition to exhibiting a gain smaller than the midband gain, the amplifier will also introduce +90° (π/2) phase shift to the signal:

$$x_i(t) = X_I Sin(\omega_S t) \Rightarrow x_o(t) = |A(j\omega_S)|.X_I.Sin(\omega_S t + \pi/2) = |A(j\omega_S)|.X_I.Cos(\omega_S t).$$

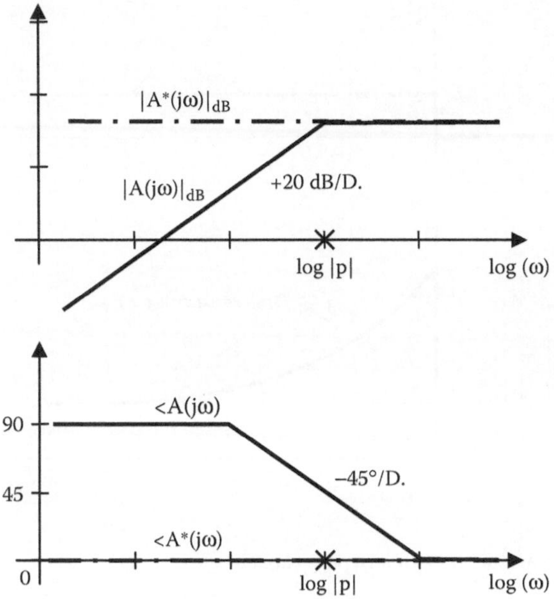

FIGURE 7.16
Exhibiting perfect amplification at midband frequencies.

Applying a very-low-frequency sine function as the amplifier's input and getting a cosine function at the output means that the amplifier is taking the derivative of the applied signal, which is the same operation previously described for very-low-frequency square wave inputs.

In case *the signal frequency is much higher than the pole,* the amplifier amplifies the input signal with the midband gain and there will be no phase difference between the input and the output caused by the low-frequency zeros or poles.

7.6.4 The Bypass Capacitor Case

The interested reader can show that the response of an amplifier with a bypass capacitor to either a step function or a square wave is much similar to what was described for the coupling capacitor case. The only exception is that even if enough time is available for the exponential decay to settle down, the steady-state output will not be zero. For an amplifier with a bypass capacitor, where the low-frequency transfer function has the general form of:

$$A(s) = \frac{x_o}{x_s} = \frac{A_o(s+z)}{s+p},$$

it can be shown that the step response is generally similar to the one in Figure 7.17. The gain $A_1 = A_0(z/p)$ is in fact the very-low-frequency gain described earlier in Figure 7.7.

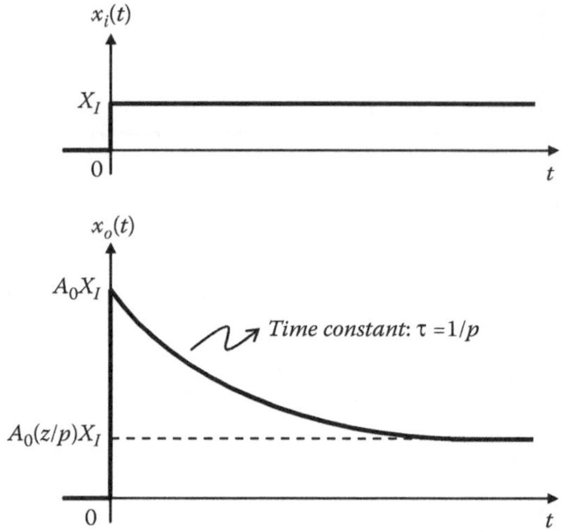

FIGURE 7.17
Step response of an amplifier with a bypass capacitor.

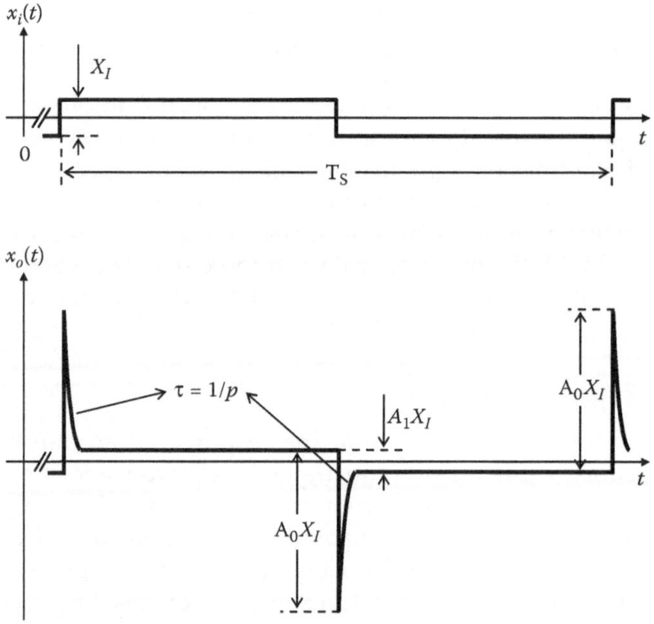

FIGURE 7.18
Response of an amplifier with a bypass capacitor to a square wave when the signal frequency is much lower than the amplifier's pole.

As illustrated in Figure 7.18, the amplifier's response to a square wave input where the signal frequency is much lower than the pole frequency is composed of alternatively positive and negative step responses of the same type as presented in Figure 7.17. In this case, because the half cycle of the input square wave is much longer than the transient regime of the amplifier's response, the initial jump at the output and its exponential decay look like an impulse. If the frequency of the input square wave becomes much higher than the pole, the amplifier's response will be similar to the one in Figure 7.15.

7.7 More Than One External Capacitor

To simplify the analysis methods that have been explained so far, amplifiers having only one external capacitor were considered. Now, let us generalize the low-frequency analysis to the cases where we have more than one external capacitor.

7.7.1 Transfer Function

Each external capacitor is represented in the low-frequency transfer function by one zero and one pole. In general, the low-frequency transfer function for an amplifier with N external capacitors is

$$A(s) = \frac{x_o}{x_i} = \frac{A_0 \prod_{i=1}^{N} (s + z_i)}{\prod_{i=1}^{N} (s + p_i)}, \tag{7.9}$$

in which A_0 is the midband gain, and z_i and p_i are the zero and the pole for the *i-th* external capacitor. *The zeros* for coupling capacitors are all at 0 Rad/Sec, and the bypass capacitors' zeros are determined by the bypassed resistance and the bypass capacitance as explained before. *The pole* for the *i-th* external capacitor, which is similarly determined for both coupling and bypass capacitors, is

$$p_i = \frac{1}{C_i R^S_{eq,i}},$$

in which $R^S_{eq,i}$ is the equivalent resistance that the *i-th* capacitor sees, *assuming that all the other capacitors are already shorted.*

Example 7.11

A common-emitter amplifier is depicted in Figure Ex. 7.11(a). Assuming that $R_B = R_1 \| R_2 = 10\ k\Omega$, $R_S = 100\ \Omega$, $R_E = 200\ \Omega$, $R_C = 10\ k\Omega$, $R_L = 10\ k\Omega$, $\beta = 100$, $r_e = 25\ \Omega$, and $C_1 = C_2 = C_3 = 1\ \mu F$, find the low-frequency transfer function.

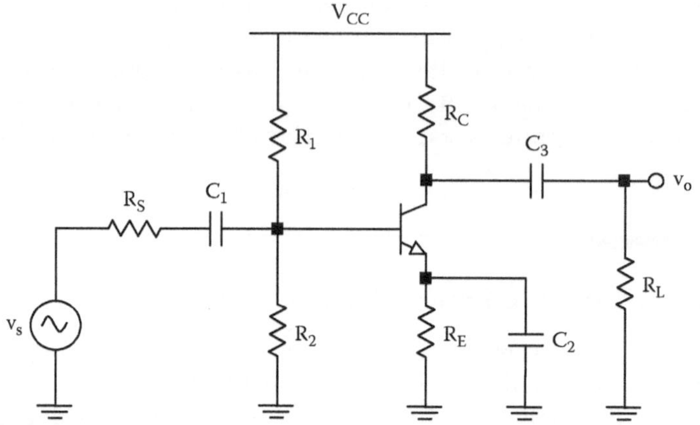

FIGURE EX. 7.11(a)

Solution:

Having three external capacitors, the transfer function will have three zeros and three poles:

$$A_{VS}(s) = \frac{v_o}{v_s} = \frac{A_{VS0}\prod_{i=1}^{3}(s+z_i)}{\prod_{j=1}^{3}(s+p_j)}.$$

Before dealing with zeros and poles, let us determine the midband gain, which is:

$$A_{VS0} = \frac{R_i}{R_i + R_S} \cdot \frac{-R_C \| R_L}{r_e},$$

where

$$R_i = R_1 \| R_2 \| \beta r_e = R_B \| r_\pi = 10^{k\Omega} \| 2.5^{k\Omega} = 2^{k\Omega}.$$

Therefore, the gain will be:

$$A_{VS0} = \frac{2k}{2k + 0.1k} \cdot \frac{-10k \| 10k}{0.025k} = -190.$$

The zeros for C_1 and C_3, which are both coupling capacitors, are at 0 *Rad/Sec*:

$$z_1 = z_3 = 0 \ Rad/Sec,$$

and for the bypass capacitor, C_2, the zero is determined by its capacitance and the resistance that it bypasses:

$$z_2 = \frac{1}{C_2 . R_E} = \frac{1}{10^{-6} \times 200} = 5kRad / Sec. = 2\pi \times 796Hz$$

To determine the poles, the circuit's low-frequency schematic should be drawn as depicted in Figure Ex. 7.11(b). Then the equivalent resistance that each capacitor sees can be found when the other capacitors are shorted (see Figure Ex. 7.11(c)–(e)).

The equivalent resistances seen by the capacitors C_1–C_3 are then numerically obtained as follows:

$$R_{C_1}^S = R_S + \left(R_B \| r_\pi\right) = 0.1^k + \left(10^k \| 2.5^k\right) = 2.1k\Omega,$$

$$R_{C_2}^S = R_E \left\| \left(r_e + \frac{R_B \| R_S}{\beta} \right) \right. = 0.2^k \left\| \left(0.025^k + \frac{10^k \| 0.1^k}{100} \right) \right. = 0.2^k \| 0.026^k = 0.023k\Omega,$$

and

$$R_{C_3}^S = R_C + R_L = 10^k + 10^k = 20k\Omega.$$

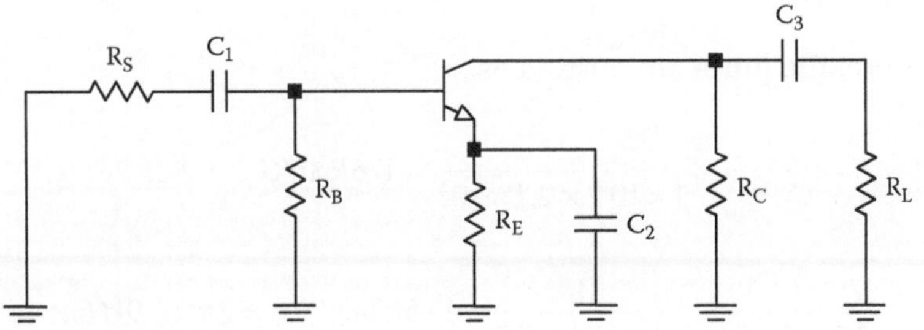

FIGURE EX. 7.11(b)

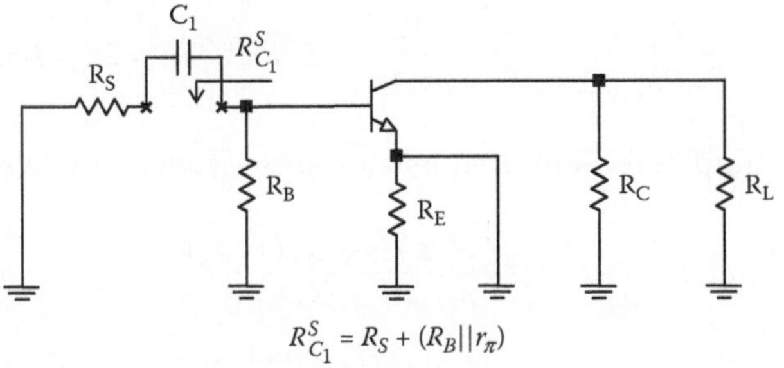

FIGURE EX. 7.11(c)

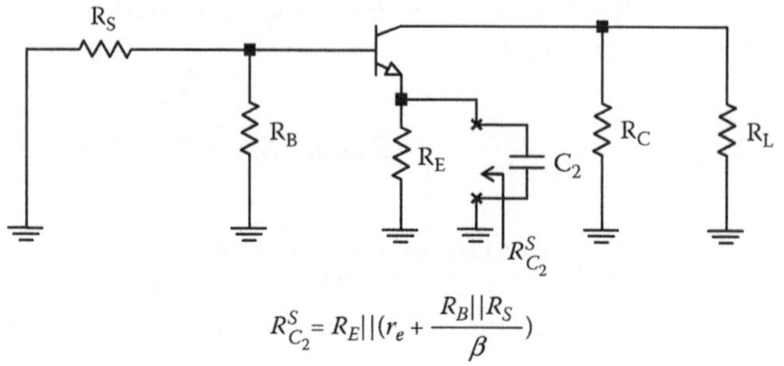

FIGURE EX. 7.11(d)

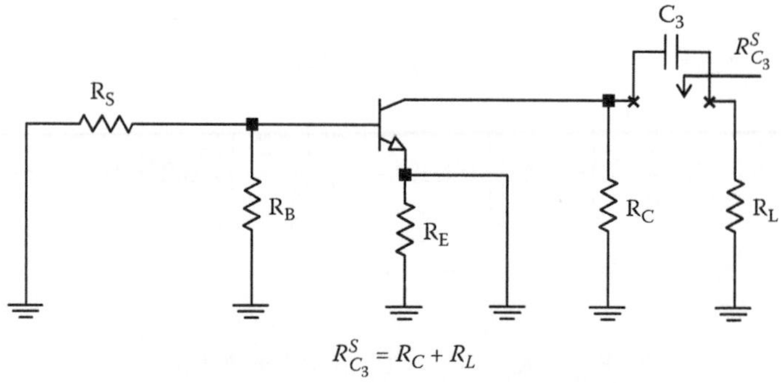

FIGURE EX. 7.11(e)

Therefore, the poles are written as:

$$p_1 = \frac{1}{C_1.R^S_{C_1}} = \frac{1}{1\times10^{-6}\times2.1\times10^3} = 476Rad/Sec. = 2\pi\times76Hz,$$

$$p_2 = \frac{1}{C_2.R^S_{C_2}} = \frac{1}{1\times10^{-6}\times23} = 43.5kRad/Sec. = 2\pi\times6.9kHz,$$

and

$$p_3 = \frac{1}{C_3.R^S_{C_3}} = \frac{1}{1\times10^{-6}\times2\times10^4} = 50Rad/Sec. = 2\pi\times8Hz.$$

Finally, the amplifier's low-frequency transfer function is written as:

$$A_{VS}(s) = \frac{v_o}{v_s} = \frac{A_{VS0}.(s+z_1).(s+z_2).(s+z_3)}{(s+p_1).(s+p_2).(s+p_3)}$$

$$= \frac{-190s^2(s+5^{kRad/Sec.})}{(s+50^{Rad/Sec.}).(s+476^{Rad/Sec.}).(s+43.5^{kRad/Sec.})}$$

$$= \frac{-190s^2(s+2\pi\times796^{Hz})}{(s+2\pi\times8^{Hz}).(s+2\pi\times76^{Hz}).(s+2\pi\times6.9^{kHz})}.$$

7.7.2 Low Cutoff Frequency

As observed in Example 7.11, if all the external capacitors have the same capacitance, the last pole belongs to the capacitor that sees the smallest equivalent resistance. It was explained earlier that, *conceptually* the low cutoff frequency is where the largest low-frequency pole occurs (i.e., where the last external capacitor can be assumed as a short circuit). By *definition*, however, the low cutoff frequency for an amplifier with the low-frequency transfer function, $A(s)$, and the midband gain, A_0, is the frequency at which the gain reduces to 0.707 A_0:

$$|A(j\omega_{CL})| = 0.707A_0, \tag{7.10}$$

or in decibels:

$$|A(j\omega_{CL})|_{dB} = |A_0|_{dB} - 3\ dB. \tag{7.11}$$

In other words, when the magnitude axis is in dB, the low cutoff frequency is the frequency at which the gain reduces to 3 dB below the midband gain, and thus the name *–3-dB cutoff frequency*. This is illustrated in Figure 7.19.

Remembering the fact that the difference between the actual Bode magnitude plot and the piecewise-linear asymptotic plot is exactly 3 dB at where the pole is located (Figure 7.4), it can be said that the amplifier's (–3 dB) low cutoff frequency is where the last low-frequency pole is located. However, this definition holds only if the other zeros and poles are much smaller (at least 10 times) than the largest pole:

$$\omega_{CL} = 2\pi f_{CL} = max(p_i) \qquad i = 1, \ldots, N. \tag{7.12}$$

A more accurate formula for the low cutoff frequency, which takes the contributions of all the low-frequency zeros and poles into account, is:

$$\omega_{CL} = 2\pi f_{CL} = \sqrt{\sum_{i=1}^{N} p_i^2 - 2\sum_{j=1}^{N} z_j^2}\ . \tag{7.13}$$

This is a general formula that usually gives a good estimation for the low cutoff frequency even if the largest pole is not significantly larger than the other corner frequencies.

Based on Equation (7.13), for an amplifier with more than one external capacitor if a pole gets closer to the largest pole, it will push the low cutoff frequency to higher frequencies. One of the simulation examples in the next section is considered to compare the accuracy of the above formulas for the low cutoff frequency.

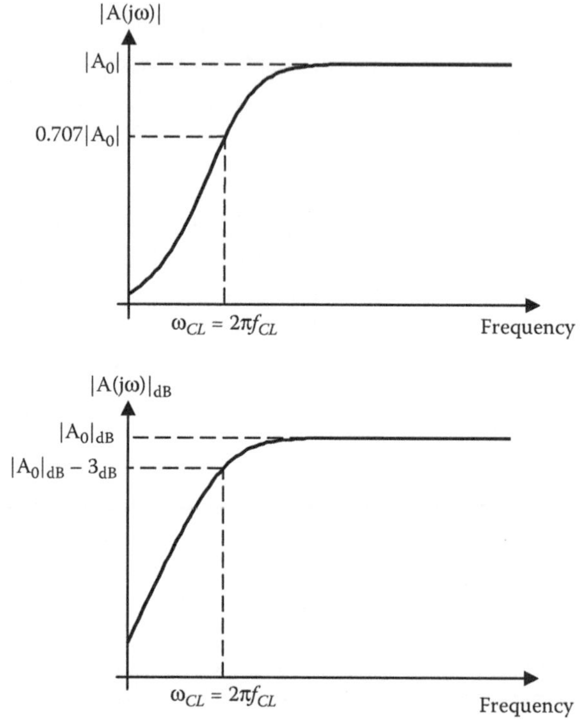

FIGURE 7.19
Definition of the low cutoff frequency.

7.8 Simulation Examples

These simulation examples are provided to enhance the reader's insight and understanding about the topics discussed throughout this chapter. In this section simulations are performed using the SPICE program unless otherwise stated.

Example 7.12

Low-frequency transfer function of an amplifier is given as

$$A(s) = \frac{10^3 s(s+0.1)}{(s+10)(s+100)}.$$

Draw the magnitude and phase plots using MATLAB and compare them with the associated asymptotic plots.

Solution:

```
k=1e3;
z=[0
   -0.1];
p=[-10
   100];
[num den]=zp2tf(z, p, k);
w=logspace(-3, 5, 100);
[mag ph]=bode(num, den, w);
subplot 211
loglog(w, mag)
xlabel('Frequency (Rad./Sec.)')
ylabel('Magnitude')
subplot 212
semilogx(w, ph)
xlabel('Frequency (Rad./Sec.)')
ylabel('Phase (Degree)')
```

The magnitude and phase plots resulted from simulation are depicted in Figure Ex. 7.12(a).

The asymptotic magnitude plot for this transfer function has been already drawn in Example 7.3, and is given again in Figure Ex. 7.12(b) for comparison convenience. The asymptotic phase plot starts from +90° at $\log\omega = -\infty$ because of having a zero at 0 Rad./Sec. It then increases to up to 180° while passing through the ROI of the zero at 0.1 Rad./Sec, and returns to 0 as a result of the poles at 10 and 100 Rad./Sec.

Example 7.13

An amplifier has a midband gain of $A_0 = 100$, and zeros at $z_1 = 0$ Rad./Sec. and $z_2 = 10$ Rad./Sec., and poles at $p_1 = 100$ Rad./Sec. and $p_2 = 1$ kRad./Sec.

(a) Using MATLAB, draw the magnitude and phase plots for $p_1 = 100$, 300, 500, 700, and 1000 Rad./Sec.

(b) Using the plots obtained in part (a), determine the low cutoff frequency for each of the magnitude plots.

(c) Plot the low cutoff frequencies obtained in part (b) as a function of the location of p_1. Compare these values with the values that Equation (7.12) and Equation (7.13) return for this example.

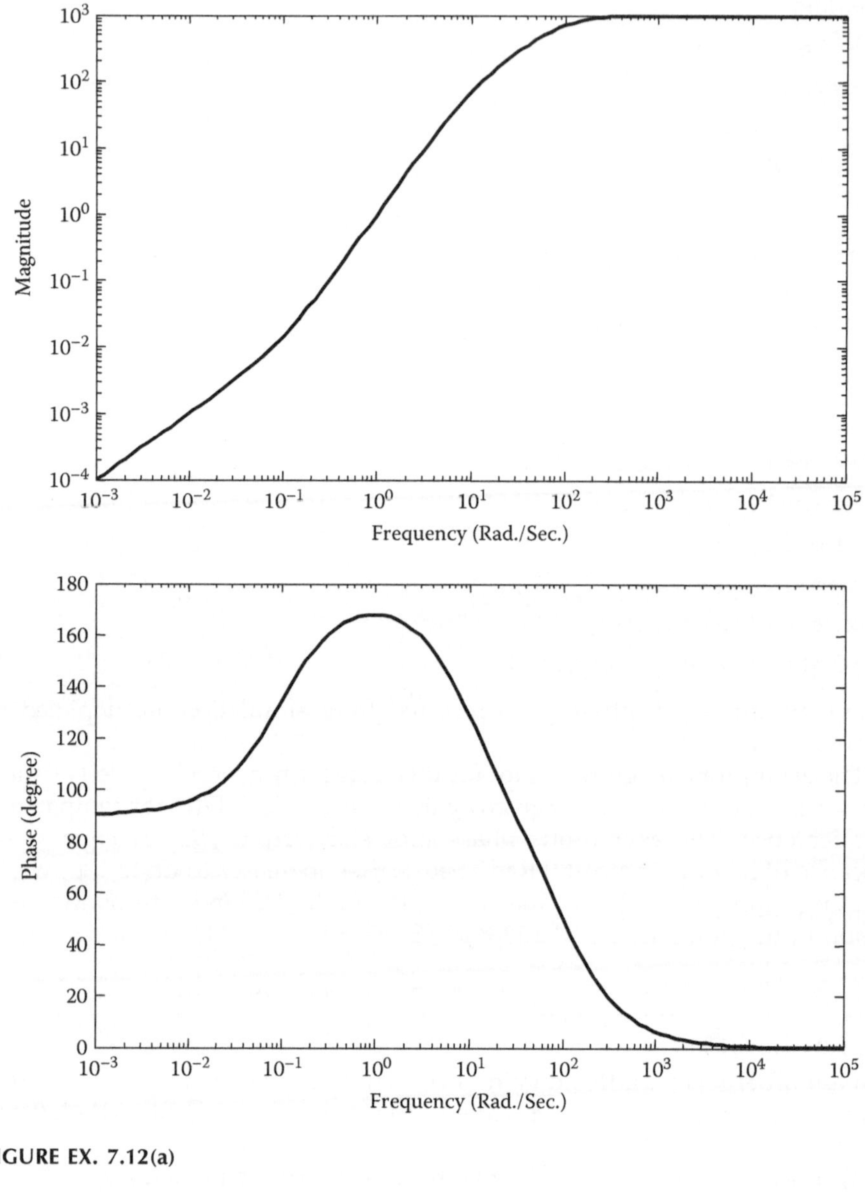

FIGURE EX. 7.12(a)

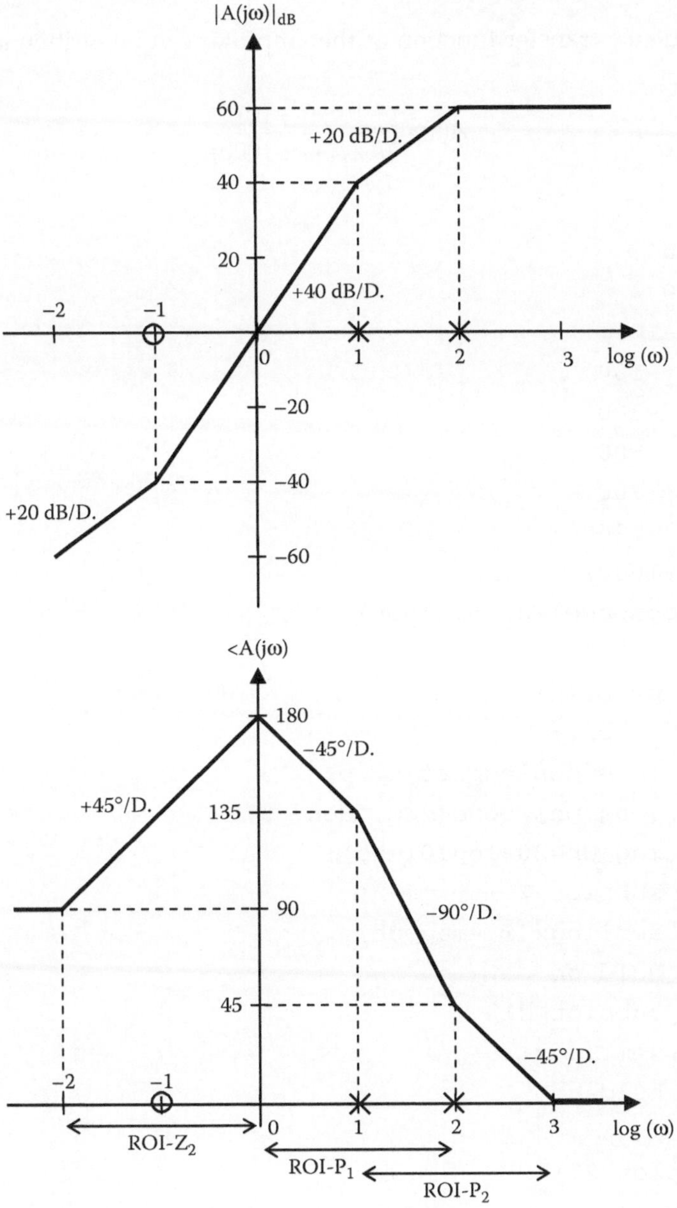

FIGURE EX. 7.12(b)

Solution:
Low-frequency transfer function of the amplifier can be written as:

$$A(s) = \frac{10^2 s(s+10)}{(s+p_1)(s+1000)}.$$

(a)

```
k=100;
z=[0
   -10];
p1=[-100
    -300
    -500
    -700
    -1000];
p2=-1000;
w=logspace(-3, 5, 1000);
for i=1:5
    p=[p1(i)
       p2];
    [num den]=zp2tf(z, p, k);
    [mag ph]=bode(num, den, w);
    mag_db=20*log10(mag);
    subplot 211
    semilogx(w, mag_db)
    hold on
    subplot 212
    semilogx(w, ph)
    hold on
end
subplot 211
xlabel('Frequency (Rad./Sec.)')
ylabel('Magnitude (dB)')
subplot 212
xlabel('Frequency (Rad./Sec.)')
ylabel('Phase (Degree)')
```

The magnitude and phase diagrams obtained from MATLAB simulations are given in Figure Ex. 7.13(a).

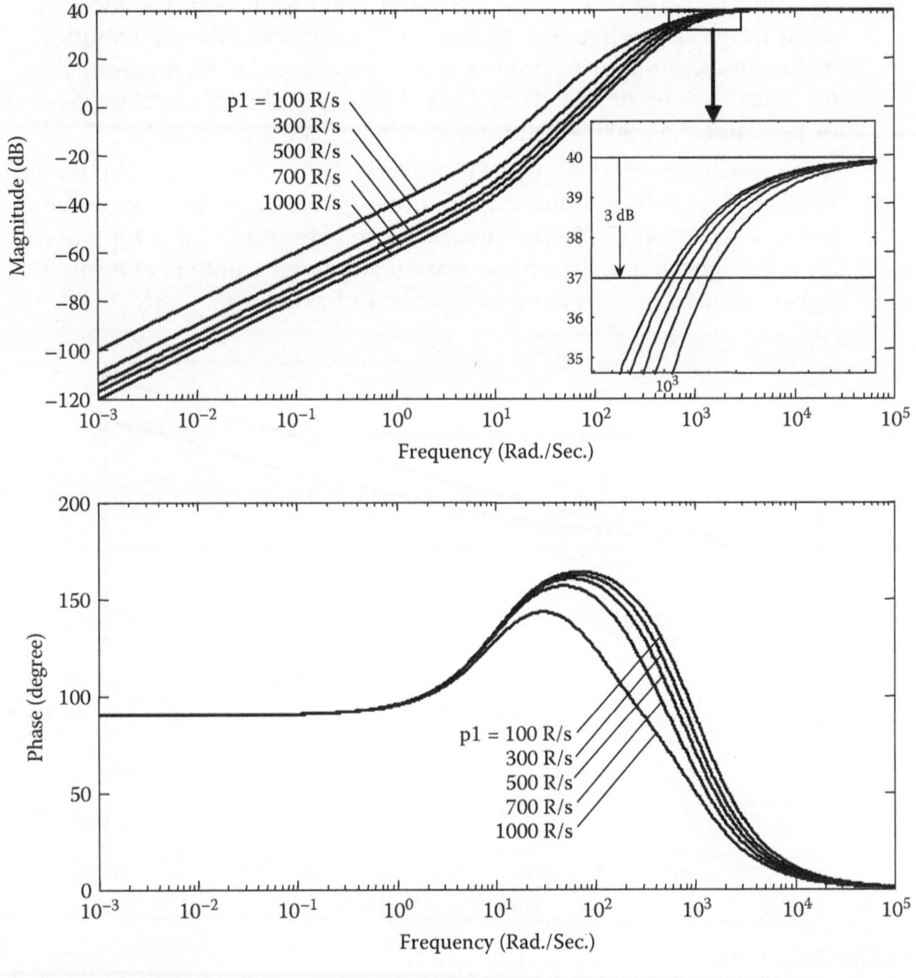

FIGURE EX. 7.13(a)

(b) As indicated in the corner of the magnitude plot, the low cutoff frequencies are determined as the frequencies where the magnitude curves cross the 37-dB line, which is 3 dB lower than the midband gain. The resulted values for the low cutoff frequency are tabulated as follow:

Location of the First Pole (p_1) (Rad./Sec.)	Low Cutoff Frequency (ω_{CL}) (Rad./Sec.)
100	1011
300	1082
500	1197
700	1329
1000	1555

(c) Equation (7.12) gives a fixed value of 1000 Rad./Sec. for the low cutoff frequency, whereas Equation (7.13) considers the contribution of the first pole as it approaches the second pole. As a result, the low cutoff frequency is 1005, 1044, 1118, 1221, and 1414 Rad./Sec. for p_1 = 100, 300, 500, 700, and 1000 Rad./Sec., respectively.

These results are compared in Figure Ex. 7.13(b). The values obtained from MATLAB simulation are the precise values for low cutoff frequency, Equation (7.13) provides a close estimation, and Equation (7.12) is a good approximation only if the largest pole is at a much higher frequency than the other zeros and poles.

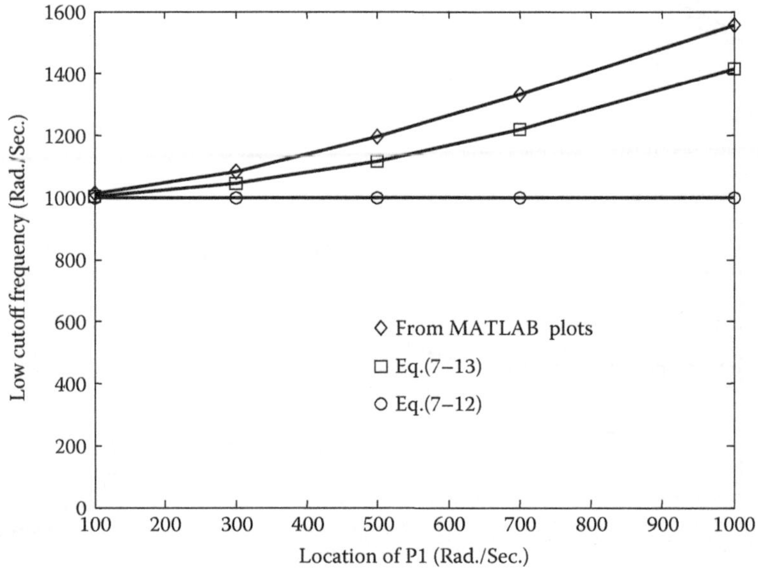

FIGURE EX. 7.13(b)

Example 7.14

A common-emitter amplifier is presented in Figure Ex. 7.14(a). Assuming that $V_{CC} = 5$ V, $R_1 = 4\ k\Omega$, $R_2 = 1\ k\Omega$, $R_S = 60\ \Omega$, $R_E = 200\ \Omega$, $R_C = 3\ k\Omega$, and $C_1 = 1\ \mu F$,

(a) Draw the amplifier's magnitude and phase plots at low and mid-band frequencies (1Hz–100 kHz).

(b) Find the midband gain and low cutoff frequency.

(c) Obtain the amplifier's time response to a 10-Hz square wave (much lower than the low cutoff frequency) with a 1-mV peak-to-peak amplitude and a 0-V DC offset.

(d) Repeat (c) for a 50-kHz square wave (much higher than the low cutoff frequency) with the same amplitude and DC offset.

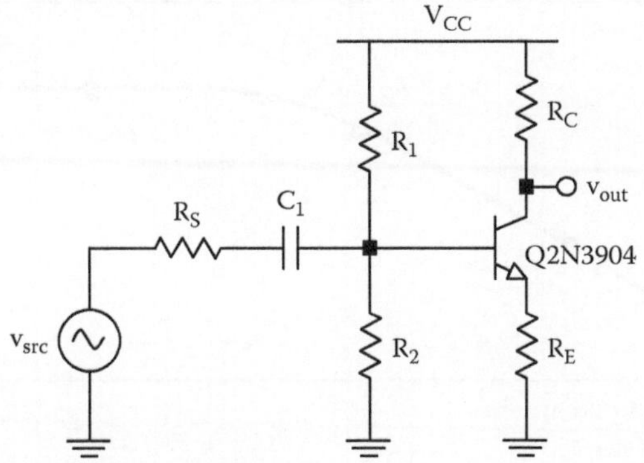

FIGURE EX. 7.14(a)

Solution:

(a)

```
VDC VDD 0 5
vs SRC 0 DC 0 AC 1
R1 VDD B 4k
R2 B 0 1k
RC VDD OUT 3k
RE E 0 0.2k
RS SRC A 0.06k
C1 A B 1u
Q1 OUT B E Q2N3904
.LIB nom.lib
.AC DEC 100 1 1e8
.PROBE
.END
```

(b) As indicated on the Bode plot in Figure Ex. 7.14(b), the midband gain is 5, and the low cutoff frequency is at 477 Hz. It is important to note that the phase plot, which is originally supposed to be caused by one zero and one pole at low frequencies, starts from –90° and approaches –180° as the frequency increases. In fact, a –180° difference exists between this phase plot and the one discussed earlier for such an amplifier, which is caused by the negative gain. Therefore, the midband gain is –5.

(c) To perform the time-domain simulation required for this part of the example, the signal source is replaced with a 10-Hz, 1-mV(p-p) square wave:

```
vs SRC 0 pulse(-0.5m 0.5m 0 1n 1n 50m 100m)
```

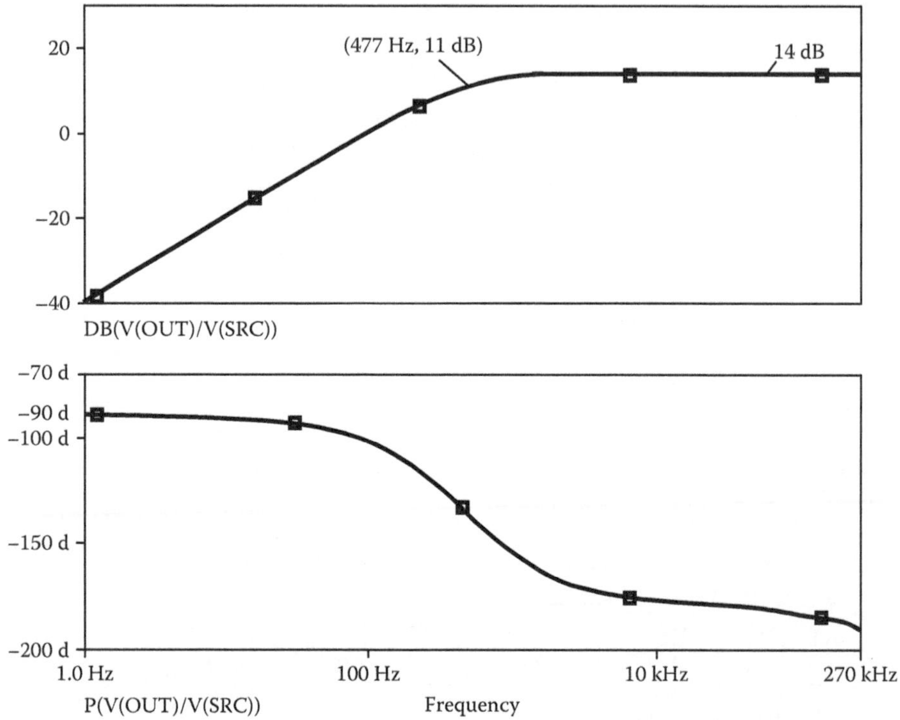

FIGURE EX. 7.14(b)

and a transient analysis is performed with the following specifications:

```
.TRAN 1u 350m 0 1u
```

The waveforms resulting from simulation are illustrated in Figure Ex. 7.14(c).

(d) Now, the signal source is changed to a 50-kHz, 1-mV(p-p) square wave:

```
vs SRC 0 pulse(-0.5m 0.5m 0 1n 1n 10u 20u)
```

and a transient analysis is performed with the following specifications:

```
.TRAN 10n 3m 0 10n
```

The waveforms resulting from simulation are illustrated in Figure Ex. 7.14(d).

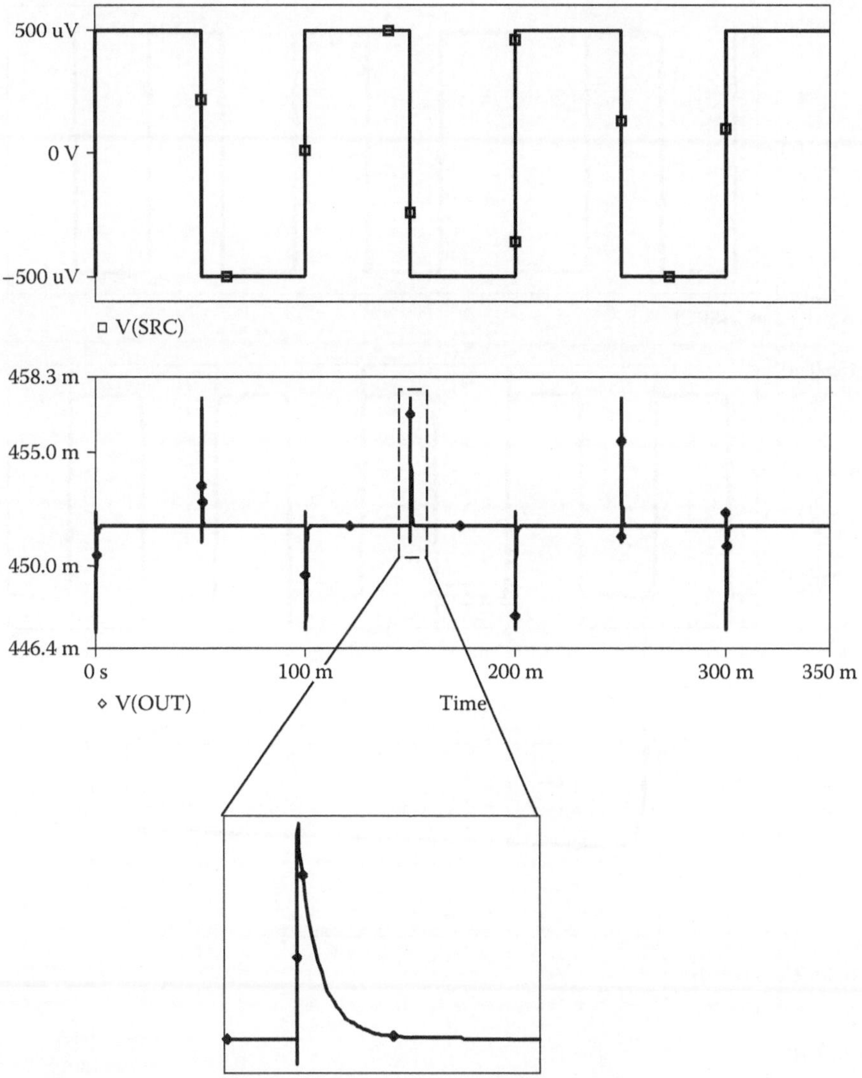

FIGURE EX. 7.14(c)

Example 7.15

Add a 0.1-μF bypass capacitor to the Common-Emitter amplifier of Example 7.14.

(a) Draw the amplifier's Bode plot and compare it to that of the amplifier of Example 7.14.

(b) Determine the new midband gain and low cutoff frequency.

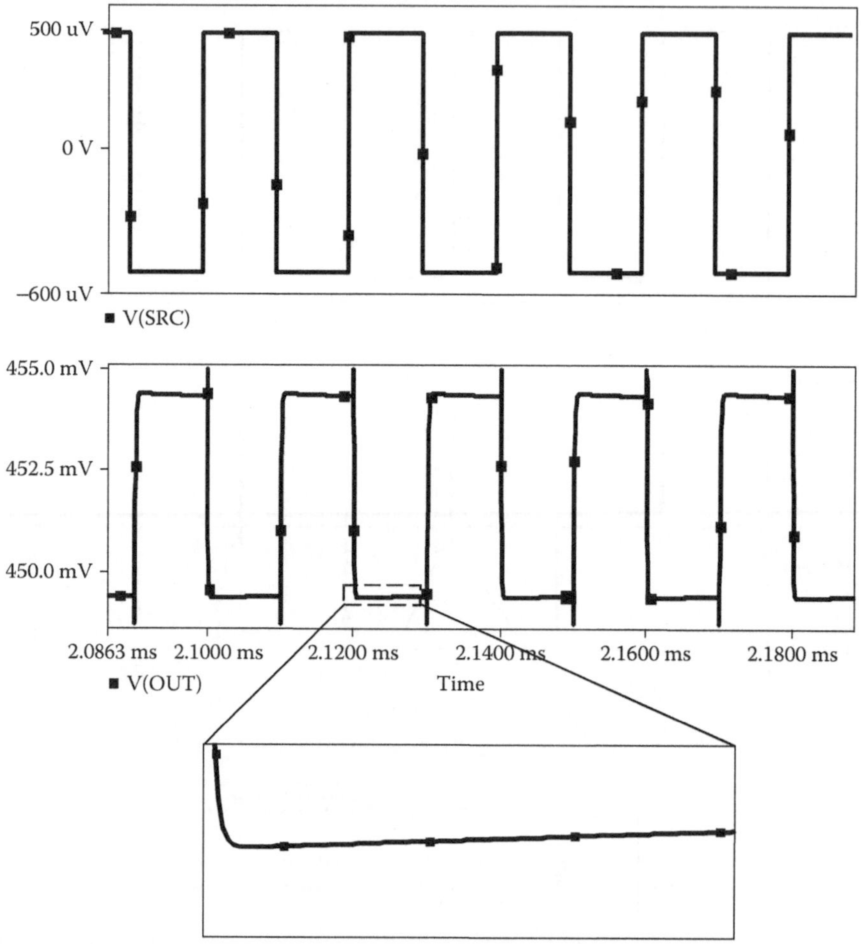

FIGURE EX. 7.14(d)

Solution:

(a)

```
VDC VDD 0 5
vs SRC 0 DC 0 AC 1
R11 VDD B1 4k
R21 B1 0 1k
RC1 VDD OUT1 3k
RE1 E1 0 0.2k
RS1 SRC A1 0.06k
C11 A1 B1 1u
```

```
Q1 OUT1 B1 E1 Q2N3904
R12 VDD B2 4k
R22 B2 0 1k
RC2 VDD OUT2 3k
RE2 E2 0 0.2k
RS2 SRC A2 0.06k
C12 A2 B2 1u
C22 E2 0 0.1u
Q2 OUT2 B2 E2 Q2N3904
.LIB nom.lib
.AC DEC 100 1 1e8
.PROBE
.END
```

(b) Based on the Bode plots in Figure Ex. 7.15, adding the bypass capacitor causes the midband gain to grow from 14 dB to 27.5 dB, and the low cutoff frequency increases from 477 Hz to 34.3 kHz.

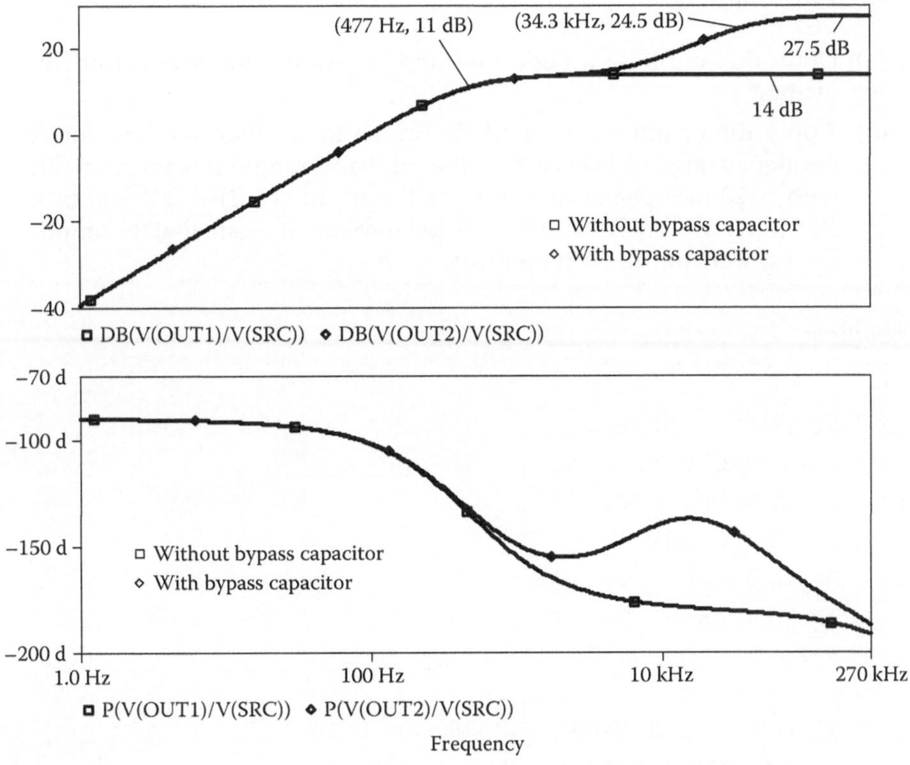

FIGURE EX. 7.15

Example 7.16

For the amplifier in Figure Ex. 7.16(a), assuming that $V_{TH} = 0.8$ V, $k' = 20\ \mu A/V^2$, W = 10 μm, L = 1 μm, $R_1 = 40\ k\Omega$, $R_2 = 10\ k\Omega$, $R_{SRC} = 60\ \Omega$, $R_S = 2\ k\Omega$, $R_D = 600\ k\Omega$, $C_1 = 1\ \mu F$, and $V_{DD} = 5$ V,

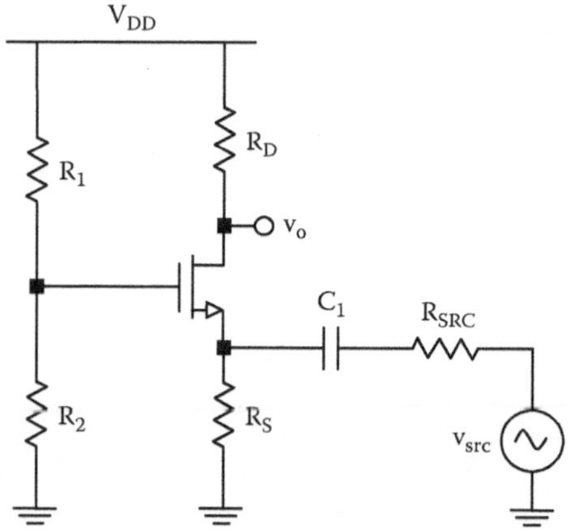

FIGURE EX. 7.16(a)

(a) Draw the amplifier's Bode plot and determine the low cutoff frequency.

(b) Apply the combination of 10-Hz (much lower than the low cutoff frequency) and 10-kHz (within the midband range) sine waves, both with a 20-mV peak amplitude at the input ($v_{src}(t) = 20^{mV} sin(2\pi \times 10^{Hz} \times t) + 20^{mV} \sin(2\pi \times 10^{kHz} \times t)$) and measure the gain that the amplifier exhibits for each component.

Solution:

(a)

```
VDC VDD 0 DC 5
VSRC SRC 0 DC 0 AC 1
RD VDD OUT 600k
R1 VDD G 40k
R2 G 0 10k
RS S 0 2k
RSRC SRC AA 0.06k
C1 AA S 1u
M1 OUT G S S NMOSmodel w=10u l=1u
.model NMOSmodel NMOS(vto=0.8 kp=20u)
.AC DEC 100 1 1e6
```

```
.PROBE
.END
```

As depicted in the magnitude and phase diagrams in Figure Ex. 7.16(b), the low cutoff frequency is at 83 Hz.

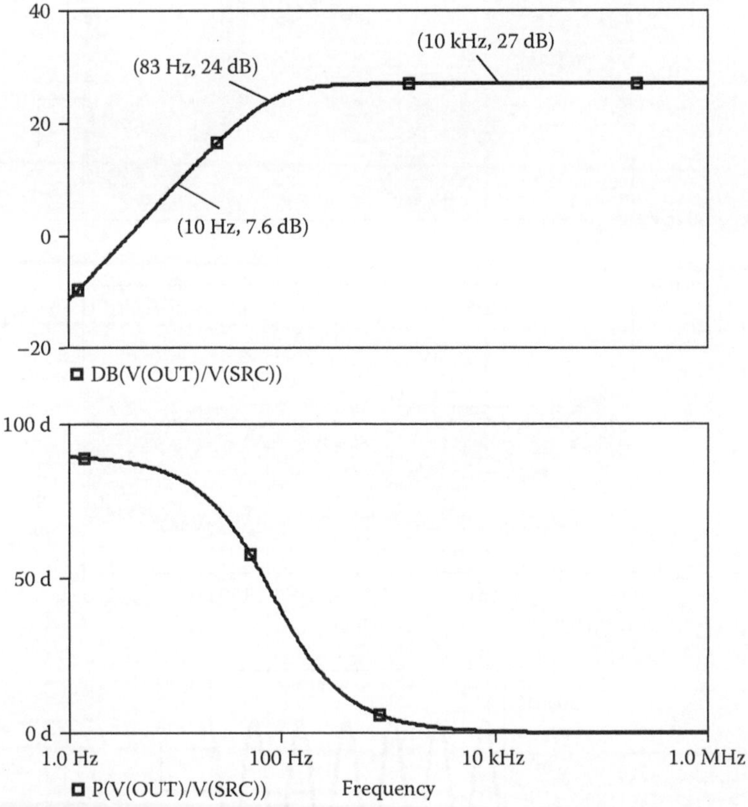

FIGURE EX. 7.16(b)

(b) To run a time-domain simulation for the input signal with the two previously mentioned frequency components, the input signal is replaced with

```
V1 SRC in1 SIN(0 0.02 10k)
```

and

```
V2 in1 0 SIN(0 0.02 10),
```

and a transient analysis is requested by

```
.TRAN  0 200m 0 500n
```

instead of AC sweep. To measure the intended gains, the time-domain response of the amplifier to $v_{src}(t)$ is simulated in Figure Ex. 7.16(c). Based on the output waveforms, the 10-Hz component has a peak-to-peak amplitude of 96 mV and the 10-kHz component appears at the output with 885-mV peak-to-peak amplitude. As a result,

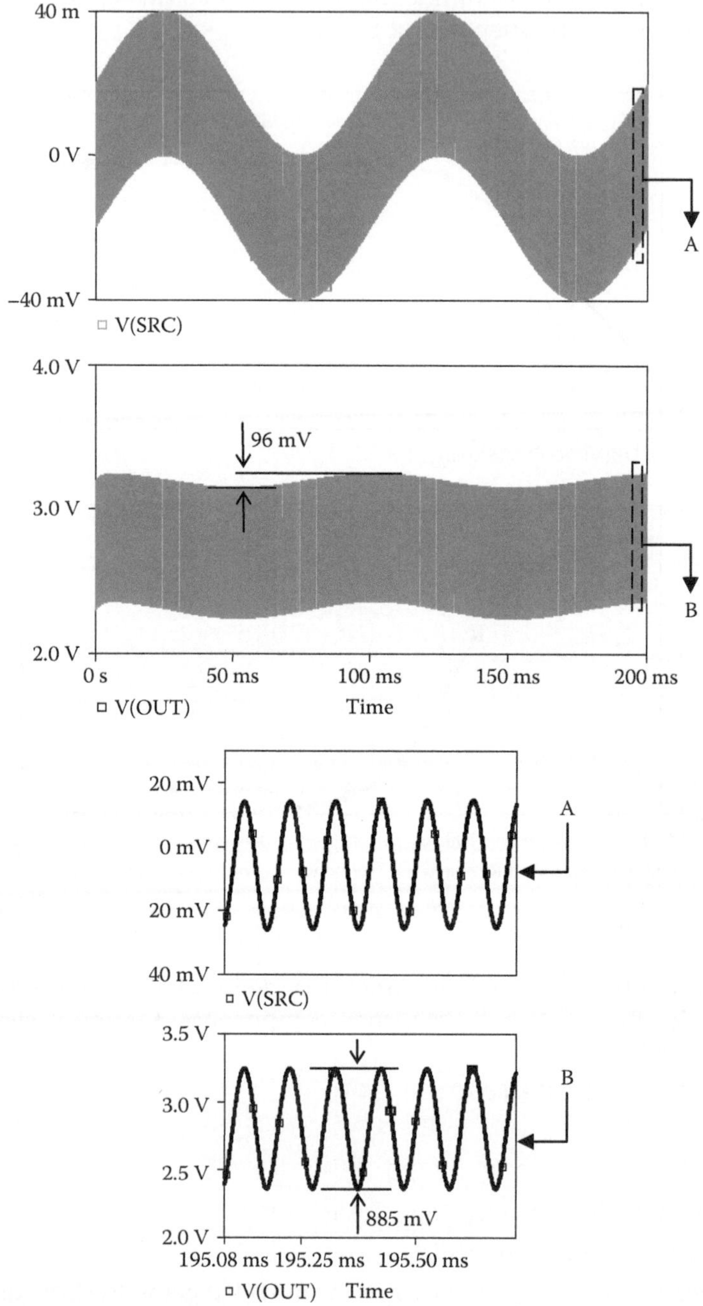

FIGURE EX. 7.16(c)

$$A_{VS}\Big|_{10Hz} = \frac{96mV(p-p)}{40mV(p-p)} = 2.4$$

and

$$A_{VS}\Big|_{10kHz} = \frac{885mV(p-p)}{40mV(p-p)} = 22.2,$$

and in decibels:

$$A_{VS}\Big|_{10Hz} = 20\log|2.4| = 7.6dB$$

and

$$A_{VS}\Big|_{10kHz} = 20\log|22.2| = 27dB.$$

These gain values are in close agreement with the values that are read from the frequency plots in Figure Ex. 7.16(b).

7.9 Problems

7.1 Assuming $|V_{BE}(ON)| = 0.7$ V, $V_A \rightarrow \infty$, and $\beta = 100$ for the transistors, and C_1-$C_3 = 1\ \mu F$ for each of the circuits in Figure Prob. 7.1,

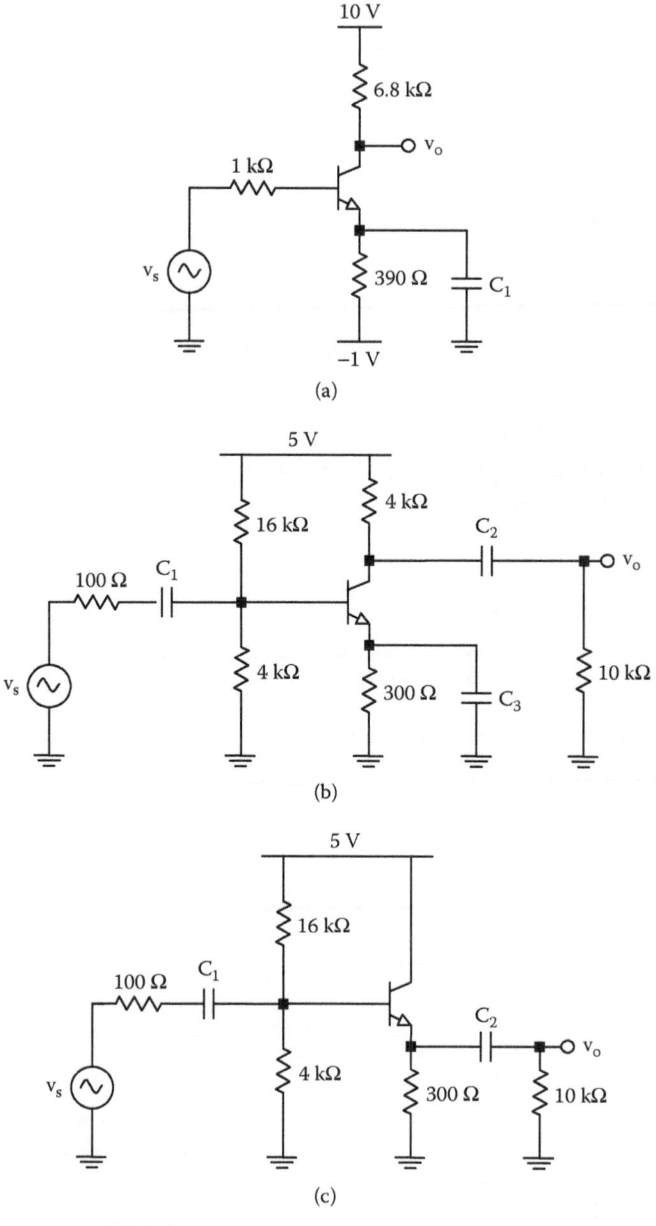

FIGURE PROB. 7.1

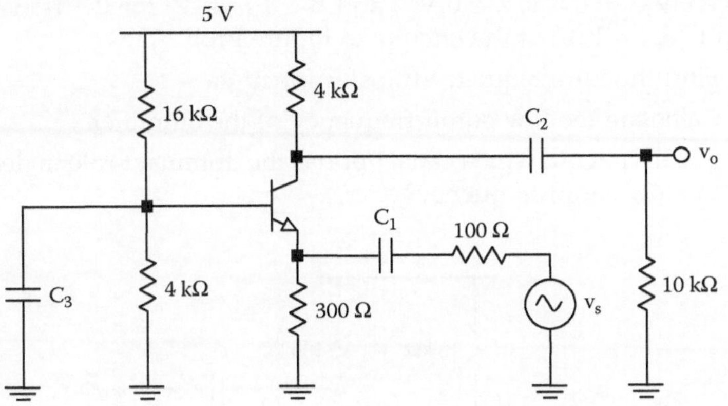

(d)

FIGURE PROB. 7.1 (continued)

(a) Find the low-frequency transfer function.

(b) Calculate the low cutoff frequency of the amplifier.

(c) In each circuit, which capacitor has the dominant role in defining the low cutoff frequency?

7.2 Given $|V_{TH}| = 0.7$ V, $\lambda \cong 0$ V^{-1}, and $\beta = 1\ mA/V^2$ for the transistors, and C_1-C_3 = 1 μF in the circuits in Figure Prob. 7.2,

(a) Find the low-frequency transfer function.

(b) Calculate the low cutoff frequency of the amplifier.

(c) In each circuit, which capacitor has the dominant role in defining the low cutoff frequency?

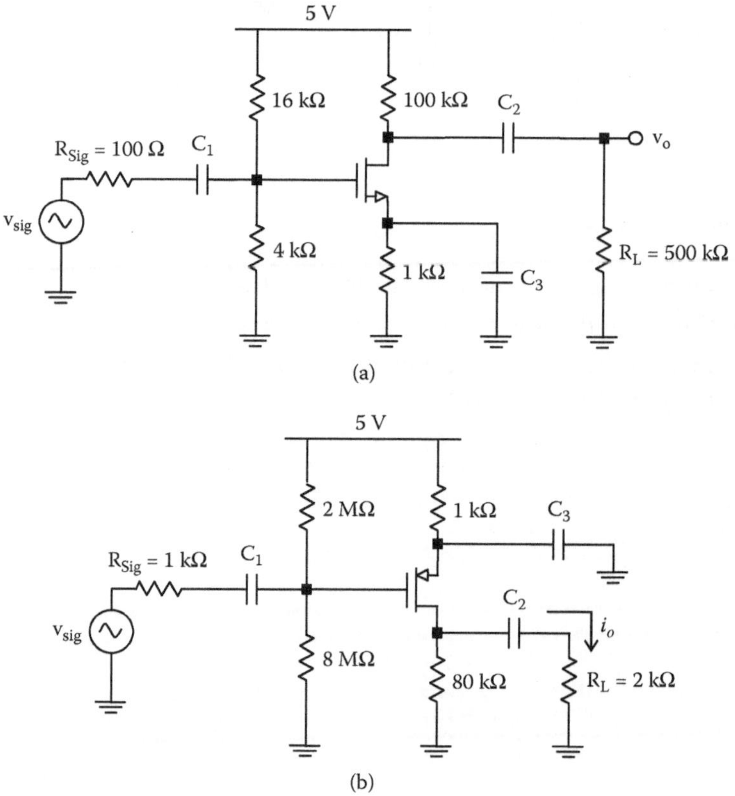

FIGURE PROB. 7.2

7.3 In the amplifier in Figure Prob. 7.3, the load resistance, R_L, may vary from 40 $k\Omega$ to 100 $k\Omega$. Assuming V_{TH} = 0.7 V, and $\beta = 2\ mA/V^2$ for the transistor, determine C_1 and C_2 for a low cutoff frequency of 1 kHz, independent of the load resistance.

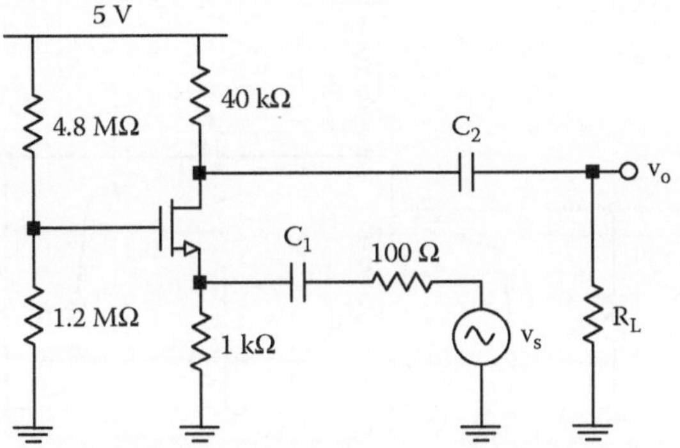

FIGURE PROB. 7.3

7.4 For each of the circuits in Figure Prob. 7.4, it is assumed that the output resistance of the transistor is much larger than the resistors used in the circuit.

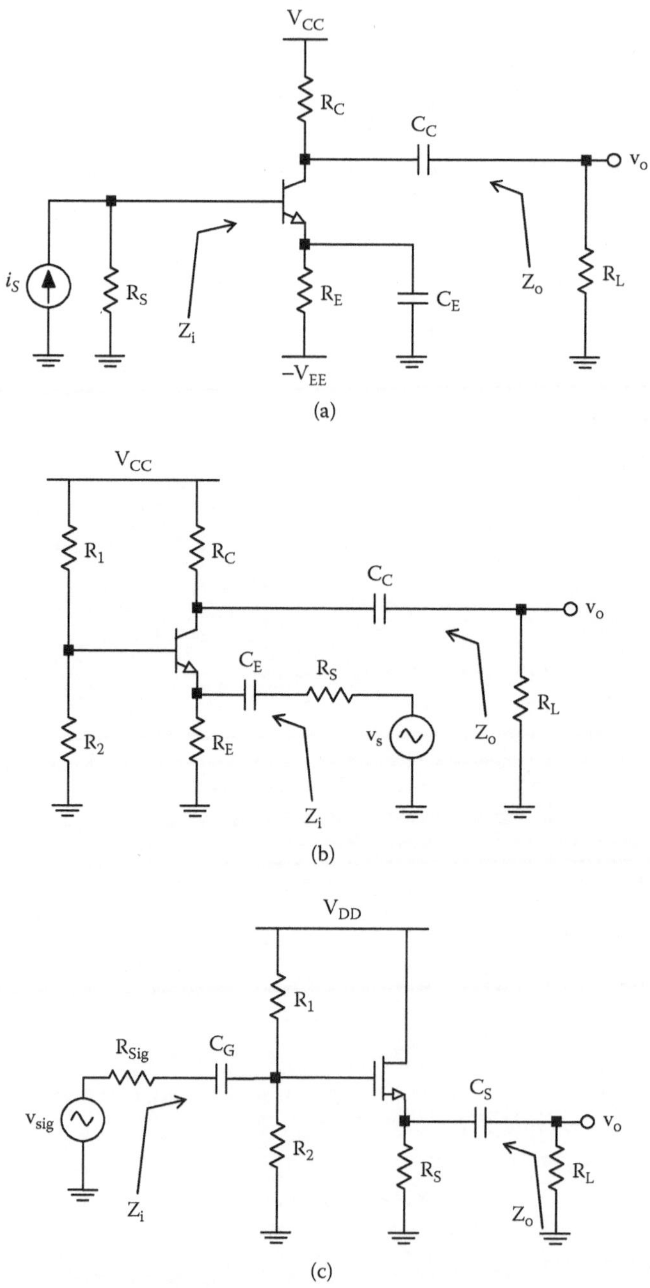

FIGURE PROB. 7.4

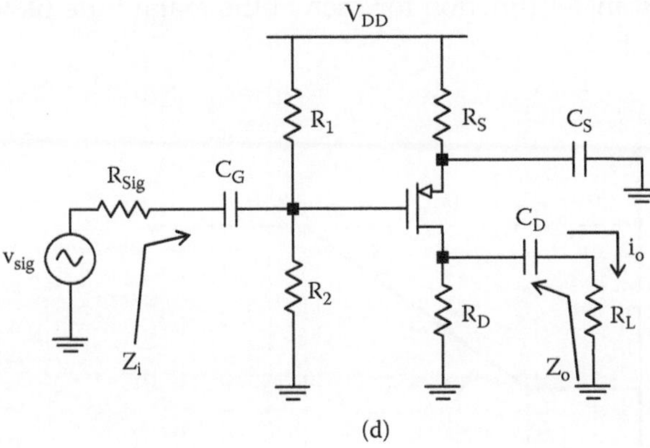

FIGURE PROB. 7.4 (continued)

(a) Replace each capacitor with its impedance in the s-domain and find the input and output impedances of the amplifier, Z_i and Z_o.

(b) Rewrite the obtained impedances in terms of the midband impedance, zeros, and poles with the "low-frequency" format introduced for low-frequency transfer functions in this chapter.

7.5 Draw the magnitude and phase plots for the following low-frequency transfer functions:

(a) $A(s) = \dfrac{10^4(s+10)}{(s+1000)}$

(b) $A(s) = \dfrac{10^3 s}{(s+100)}$

(c) $A(s) = \dfrac{10^4 s(s+1000)}{(s+10)(s+5000)}$

(d) $A(s) = \dfrac{-2000s^2}{(s+10^2)(s+10^3)}$

7.6 Draw the magnitude and phase plots for an amplifier with the following specifications at low and midband frequencies:

(a) $A_0 = 100$, $z = 0$Rad/Sec, and $p = 2$kRad/Sec

(b) $A_0 = 10^5$, $z_1 = 0$Hz, $z_2 = 100$Hz, $p_1 = 1$kHz, and $p_2 = 5$kHz.

7.7 Find the transfer function for each of the magnitude plots drawn in Figure Prob. 7.7.

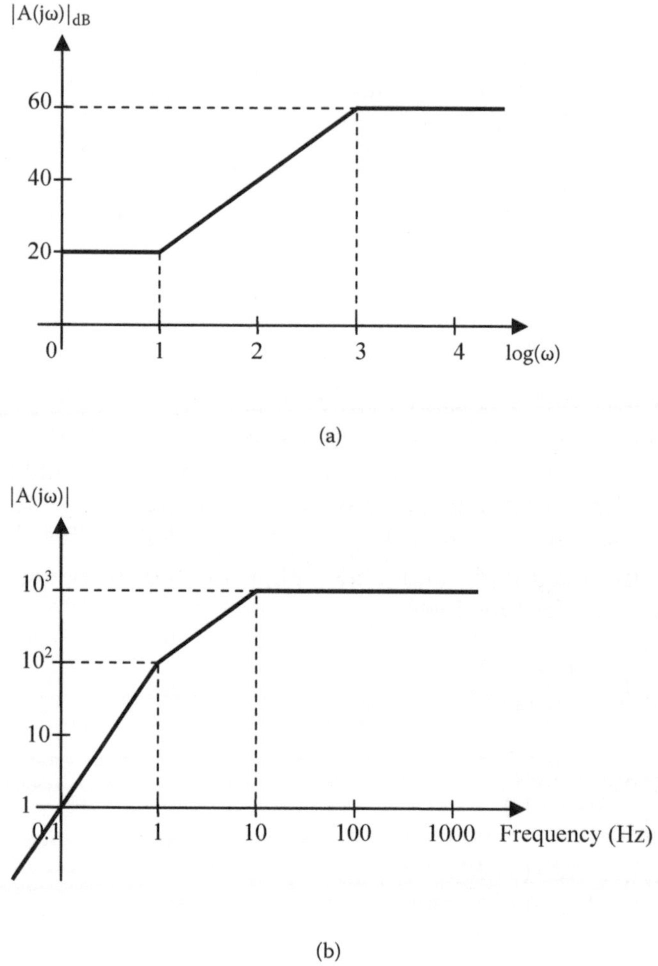

FIGURE PROB. 7.7

7.8 A voltage amplifier has a midband gain of $A_V = v_o/v_i = 1000$, a zero at 0Hz, and a low-frequency pole at 1kHz.

(a) Draw the amplifier's magnitude and phase plots.

(b) Find the amplifier's output when an input voltage of $v_i(t) = 1^{mV}\text{Sin}(2\pi \times 10^{Hz}t) + 0.5^{mV}\text{Sin}(2\pi \times 10^{kHz}t)$ is applied.

Chapter 8

Analysis of Amplifiers at High Frequencies

8.1 Introduction

After learning how to analyze an amplifier at low frequencies in the previous chapter, it is now time to deal with the analysis of small-signal amplifiers at high frequencies. In this chapter, we will study the amplifier's behavior when the frequency approaches and even passes the upper extreme of the operational frequency band. Furthermore, the impacts of the amplifier's high-frequency behavior on its time response will also be studied.

8.2 Basic Concepts at High Frequencies

As explained in the previous chapter, an amplifier exhibits a relatively constant (frequency-independent) gain at midband frequencies. As the frequency increases, the impedances of the transistors' (internal) parasitic capacitances decrease. At some point, these impedances will not be large enough to be ignored anymore. This is where the amplifier's frequency-dependent behavior starts to appear, and is in fact the beginning of the "high-frequency" range. Because of their locations, if the internal capacitors are replaced with finite impedances (instead of open circuits) the gain will drop. It was also mentioned that the frequency at which the gain drops to 70% of the midband gain ($A_0\sqrt{2}/2$ to be exact), is known as the high cutoff frequency, f_{CH}. Similar to the external ones, usually each internal capacitor, contributes to the amplifier's transfer function by a *zero* and a *pole*. The frequency at which an internal capacitor comes out of the open-circuit state is indeed its pole, and its zero is where it enters the short circuit state. Unlike the external capacitors,* an internal capacitor's *region of activity* occurs at high frequencies, starts with

* As mentioned in the previous chapter, in some cases external capacitors with very small capacitances are used to dominantly determine the high-frequency behavior of amplifiers. In these cases, such external capacitors are treated in the same way as the internal capacitors are.

its pole and ends with its zero. Starting from midband frquencies and going toward higher frequencies, the gain remains nearly constant up to the point where *the first internal capacitor comes out of the open-circuit state.*

8.3 Amplifiers at High Frequencies

The small-signal models that have been used so far are good enough to represent a transistor's behavior at low and midband frequencies. For a transistor fabricated in the real world, some nonideal capacitive behaviors occur. These parasitic capacitances are the main contributing factors that limit both the transistor's and the circuit's operation at high frequencies. Therefore, it is of critical importance to develop a model for the transistor to model its parasitics as well as its main function. Figure 8.1 presents the simplified transistor models that are equipped with only two of the most important parasitic capacitances. Because of their small capacitances (hundreds of femto Farads [fF] up to a few pico Farads [pF]), the parasitic capacitors can be easily taken as open circuits at low and midband frequencies. This is why they have not appeared in the small-signal transistor models so far.

As any other linear time-invariant (LTI) system, an amplifier's linear behavior in frequency domain is formulated by its transfer function. The transfer function that models an amplifier's frequency-dependent behavior

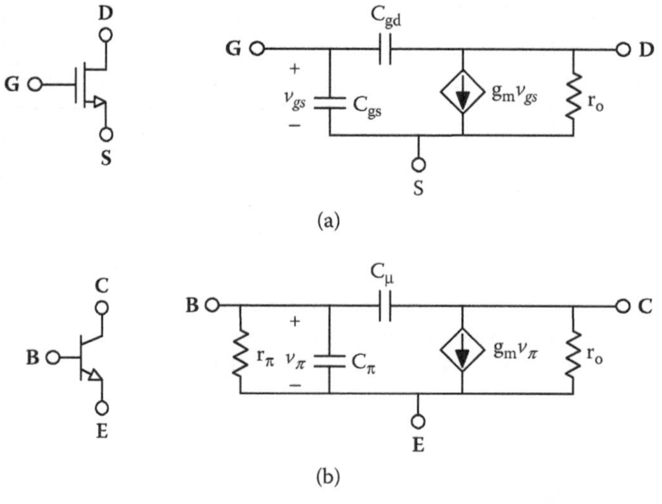

FIGURE 8.1
Simplified high-frequency models for (a) MOS and (b) bipolar transistors.

at midband and high frequencies is called the *high-frequency transfer function*, which is in general written as:

$$A(s) = \frac{A_0 \prod_{i=1}^{N} \left(1 + \frac{s}{z_i}\right)}{\prod_{i=1}^{N} \left(1 + \frac{s}{p_i}\right)} \tag{8.1}$$

where A_0 is the amplifier's midband gain, N is the number of *independent* internal capacitors in the circuit,* and z_i and p_i are the zero and pole for the i-th capacitor. Two capacitors will be independent if they are not either in series or parallel. Therefore, to figure out the number of independent capacitors, series and or parallel combinations of capacitors should first be replaced with their equivalent capacitor.

The zeros can be either at finite frequencies, or at infinity. There will be no term associated with the zeros at infinity written in the transfer function. The number of zeros that are located at infinity will be the difference between the orders of the transfer function's numerator and denominator. As will be seen in the next section, this determines the final slope of the magnitude plot and the final phase of the amplifier when the frequency approaches $+\infty$.

8.4 Plotting an Amplifier's Response at High Frequencies

The general transfer function represented in Equation (8.1) exhibits a magnitude of A_0 and 0° of phase before any one of the zeros and poles start affecting the amplifier's behavior. Then, as the frequency increases, the magnitude and phase curves pass the amplifier's poles and zeros, and their slopes are affected accordingly. Because for each capacitor the pole occurs first, it pulls the magnitude and phase plots down, and then if the zero for that capacitor is at a finite frequency, all it can do is pull up the magnitude and phase plots again. On the magnitude plot, the pole adds a slope of –20 dB/D. first, and then the zero adds 20 dB/D. to the slope. On the phase plot, the pole adds some negative phase with a slope of –45°/D. first, and then the zero adds the same amount of positive phase with a slope of 45°/D. As a result, for the high-frequency transfer function introduced in Equation (8.1), the magnitude will be either decrementing or constant, and the phase, which can have both increments and decrements, will always be negative or, at most, zero.

* The number of capacitive loops in the circuit also contributes to the order of the transfer function. This rarely happens in amplifier circuits; thus, it is not discussed in this book.

Example 8.1 compares the magnitude and phase plots for first-order, high-frequency transfer functions when the zero is both at a finite frequency and at infinity.

Example 8.1

Using MATLAB draw the Bode plots for high-frequency transfer functions:

$$A_1(s) = \frac{10^3\left(1+\dfrac{s}{10^7}\right)}{1+\dfrac{s}{10^4}}$$

and

$$A_2(s) = \frac{10^3}{1+\dfrac{s}{10^4}}.$$

Solution:

The default format for system representation in MATLAB using either "transfer function (tf)" or "zero-pole-gain (zp)" representations is compatible with the general format described earlier for low-frequency transfer functions. Therefore, to draw Bode plot for a high-frequency transfer function, special care should be taken to have the right midband gain presented to MATLAB. One can either write the numerator and the denominator as polynomials and use the "*tf*" representation, or convert the transfer function into the low-frequency general format:

$$A(s) = \frac{A_0\prod_{i=1}^{N}\left(1+\dfrac{s}{z_i}\right)}{\prod_{i=1}^{N}\left(1+\dfrac{s}{p_i}\right)} = \frac{A_0\prod_{i=1}^{N} p_i}{\prod_{i=1}^{N} z_i}\cdot\frac{\prod_{i=1}^{N}(s+z_i)}{\prod_{i=1}^{N}(s+p_i)}$$

and then use the "*zp*" representation. The resulting magnitude and phase plots are presented in Figure Ex. 8.1.

```
num1=[1e-4 1e3];
den1=[1e-4 1];
num2=1e3;
den2=[1e-4 1];
w=logspace(1, 9, 100);
```

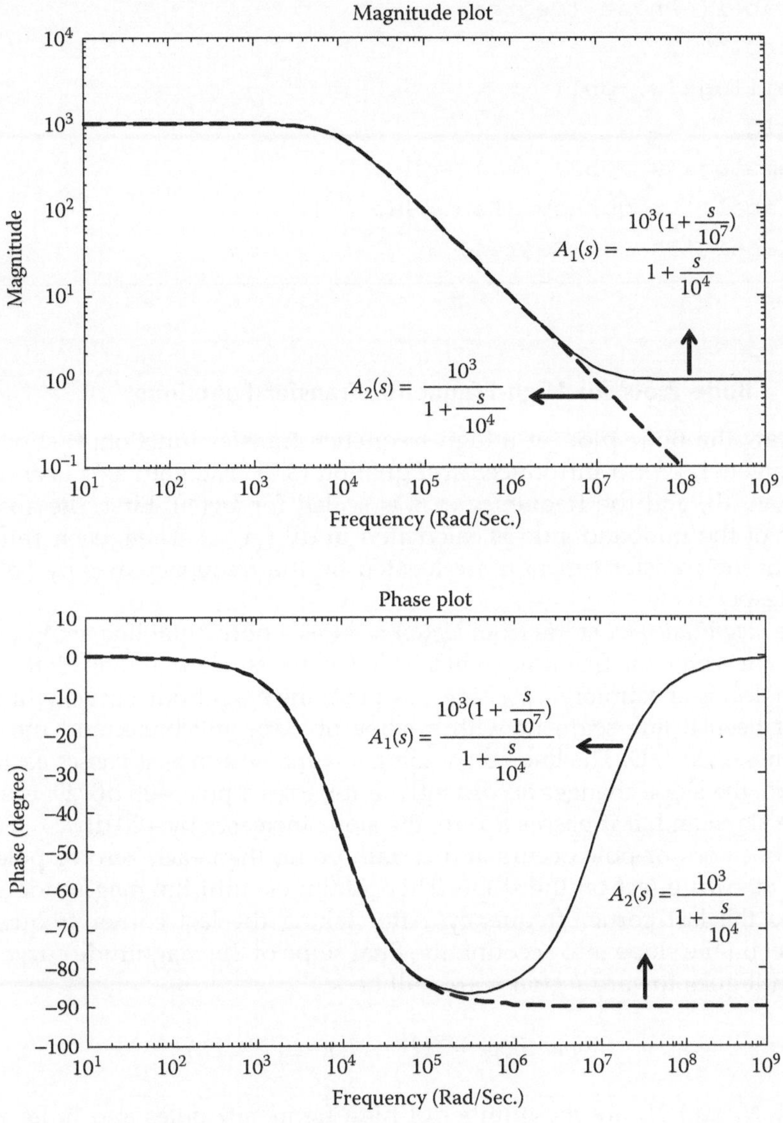

FIGURE EX. 8.1

```
[mag1, ph1]=bode(num1, den1, w);
[mag2, ph2]=bode(num2, den2, w);
subplot 211
loglog(w, mag1)
hold
loglog(w, mag2, '--')
xlabel('Frequency (Rad./Sec.)')
```

```
ylabel('Phase (Degree)')
subplot 212
semilogx(w, ph1)
hold
semilogx(w, ph2,'--')
xlabel('Frequency (Rad./Sec.)')
ylabel('Phase (Degree)')
```

8.4.1 Bode Plots for High-Frequency Transfer Functions

To draw the Bode plot for a high-frequency transfer function, it should be in the general form introduced in Equation (8.1). The gain axis is assumed to be in dB, and the frequency axis is scaled for $log(\omega)$. First, the absolute value of the midband gain is calculated in dB ($|A_0|_{dB}$). Then, each zero and pole of the transfer function are located on the frequency axis by "o" and "x", respectively.

The *magnitude plot* starts from $log(\omega) = -\infty$ as a horizontal line at $|A_0|_{dB}$ up to the point where the first pole (which is in fact the smallest among all the zeros and poles) is encountered. Passing this point the magnitude curve, which has so far been a line segment with a slope of zero, will break and the slope becomes –20dB/D. The line keeps going toward $+\infty$ and as it passes each pole or zero, the slope changes accordingly. If it passes a pole, –20 dB/D. is added to the slope and, if it passes a zero, the slope increases by +20 dB/D. If more than one zero or pole occurs at a certain point, then each zero or pole will have its own effect on the slope. This continues until the magnitude curve passes the last corner frequency. After taking the last corner frequency's impact on the slope into account, the final slope of the magnitude curve with which it goes toward $log(\omega) = +\infty$ will be

$$Final\ Slope = (N_P - N_Z)(-20\ dB/D.) \tag{8.2}$$

where N_P and N_Z are the number of high-frequency poles and finite zeros, respectively.

Example 8.2

Draw the magnitude plot for an amplifier with a midband gain of 10^4, two high-frequency poles at 10 kRad/Sec and 100 kRad/Sec, and a high-frequency zero at 1 MRad/Sec.

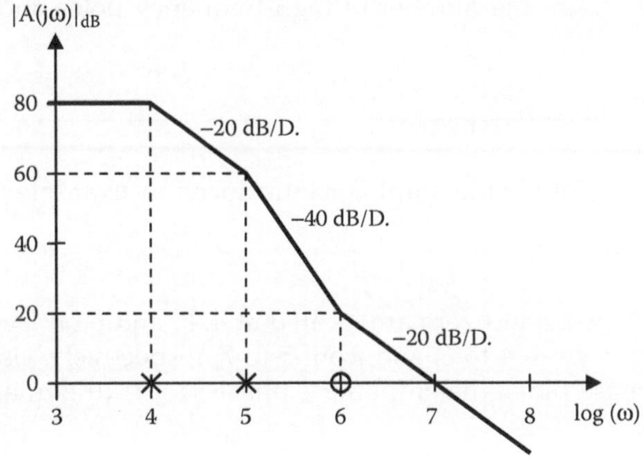

FIGURE EX. 8.2

Solution:
The amplifier's high-frequency transfer function is:

$$A(s) = \frac{10^4\left(1+\dfrac{s}{10^6}\right)}{\left(1+\dfrac{s}{10^4}\right)\left(1+\dfrac{s}{10^5}\right)}.$$

Its midband gain is 80 dB (= $20log_{10}(10^4)$), and the poles and zero are located at 4, 5, and 6 on log(ω) axis, respectively (see Figure Ex. 8.2). Magnitude curve starts from $log(\omega) = -\infty$ at 80dB with slope 0. After passing the first pole at $log(\omega) = 4$, the slope becomes −20 dB/D. At $log(\omega) = 5$, the slope becomes −40dB/D. as the second pole is passed. Going toward higher frequencies, when passing the zero at $log(\omega) = 6$ the slope returns to −20dB/D. With this slope, the magnitude curve goes toward $log(\omega) = +\infty$.

The *phase plot* for a high-frequency transfer function is drawn on the same axes described for low-frequency transfer functions in the previous chapter. First, zeros and poles and the associated ROIs are located on the frequency axis. Then, the phase curve is drawn as a horizontal line segment at 0° with slope 0 starting from $log(\omega) = -\infty$ up to the point where the first ROI starts. The influence of zeros and poles on the slope of the phase plot is similar to what was already explained in the previous chapter. When exiting the last ROI, the phase is supposed to be:

$$\textit{Final Phase} = (N_Z - N_P) \times 90^\circ \tag{8.3}$$

where N_P and N_Z are the number of high-frequency poles and finite zeros, respectively.

Example 8.3

Draw the phase plot for the amplifier introduced in Example 8.2.

Solution:

The amplifier's poles and zero are located at 4, 5, and 6 on $log(\omega)$ axis with ROIs from 3 to 5, from 4 to 6, and from 5 to 7, respectively. Using the rules for drawing phase plots, the amplifier's phase plot is drawn as depicted in Figure Ex. 8.3.

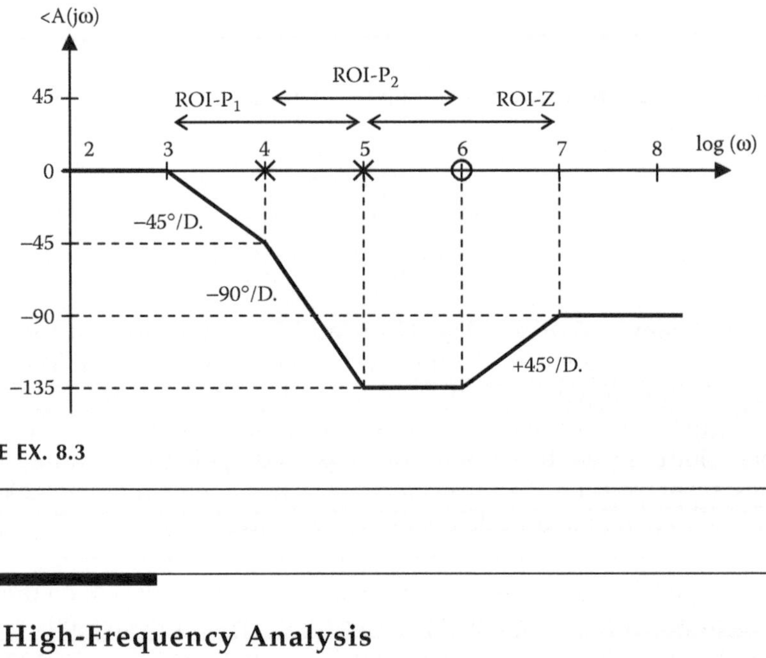

FIGURE EX. 8.3

8.5 High-Frequency Analysis

The high-frequency transfer function for an amplifier can be determined either by *AC analysis methods* or by *high-frequency inspection*. Basically, the classic approach to analyze an amplifier at high frequencies is similar to the AC analysis methods explained earlier. First, the external capacitors should be replaced with short circuits. This is because all the external capacitors are shorted at much lower frequencies as we enter the midband frequency range. Second, each transistor is replaced with its high-frequency small-signal model. Each capacitor is then represented by its impedance in the complex frequency, or s-domain. After an appropriate circuit schematic is obtained by:

- Setting DC independent sources to zero,
- Shorting the external capacitors, and
- Representing the internal capacitors with their impedances in the s-domain,

the amplifier is simply analyzed using the previously described AC analysis methods for its gain or other AC specifications.

Example 8.4

For the amplifier in Figure Ex. 8.4(a), assuming that $R_G = R_1 \| R_2 = 100\ k\Omega$, $R_{SRC} = 1\ k\Omega$, $R_S = 100\ \Omega$, $R_D = 10\ k\Omega$, $r_o = 10\ M\Omega$, $g_m = 10\ mA/V$, $C_1 = C_2 = 1\ \mu F$, and $C_{gs} = C_{gd} = 0.1\ pF$,

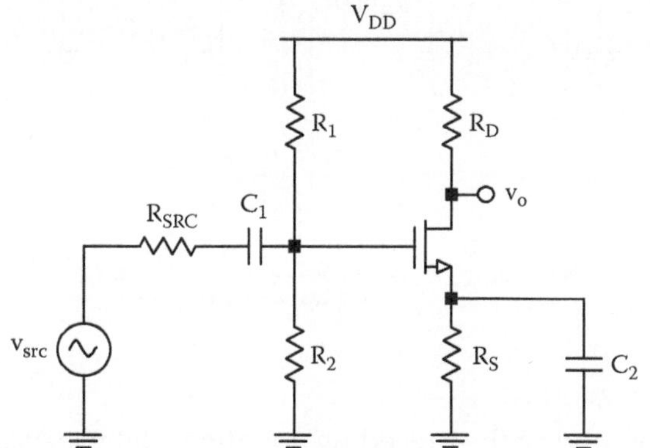

FIGURE EX. 8.4(a)

(a) Find the high-frequency transfer function.
(b) Draw the amplifier's Magnitude plot.

Solution:

(a) First, to obtain the appropriate circuit schematic for high frequencies, the independent DC source V_{DD} is set to zero, the external capacitors (C_1 and C_2) are shorted, and the transistor is replaced with its high-frequency small-signal model (see Figure Ex. 8.4(b)). Then, as seen in Figure Ex. 8.4(c), the circuit is simplified as much as possible to facilitate the analysis. Now, the high-frequency transfer function $A_{VS}(s) = V_O(s)/V_{SRC}(s)$ is found using a nodal analysis as follows:

$$\text{KCL1: } \frac{v_i - v_{src}}{R_{SRC}} + \frac{v_i}{Z_G(s)} + \frac{v_i - v_o}{1/(C_{gd}s)} = 0,$$

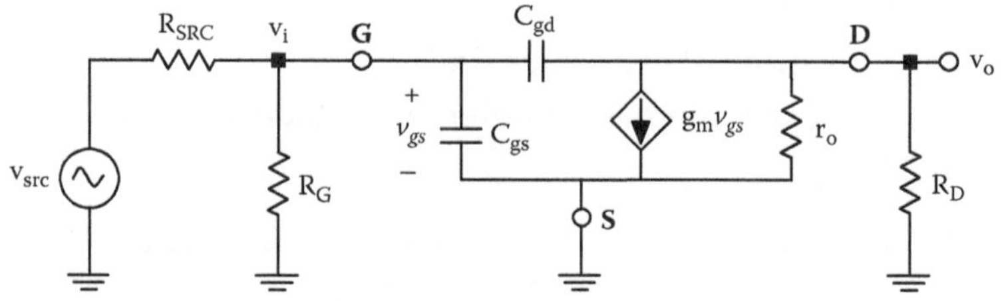

FIGURE EX. 8.4(b)

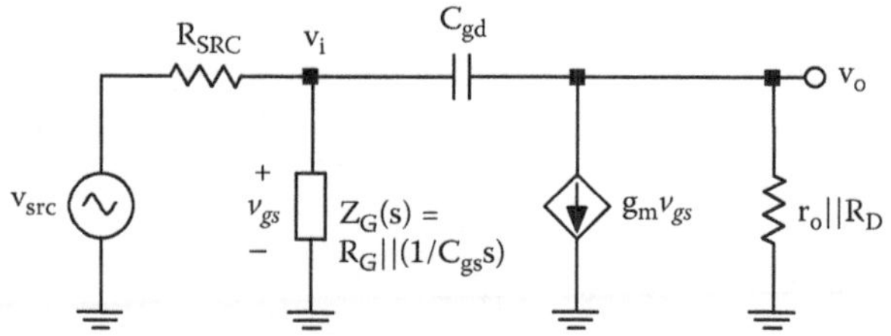

FIGURE EX. 8.4(c)

and

$$\text{KCL2: } g_m v_{gs} + \frac{v_o}{r_o \| R_D} + \frac{v_o - v_i}{1/(C_{gd}s)} = 0,$$

where $v_{gs} = v_i$.

To simplify solving the preceding equations, let us replace the circuit parameters with their numeric values:

$$\text{KCL1: } \frac{v_i - v_{src}}{1} + \frac{v_i}{100/(1+10s)} + \frac{v_i - v_o}{10/s} = 0,$$

and

$$\text{KCL2: } 10v_i + \frac{v_o}{10} + (0.1s)(v_o - v_i) = 0.$$

It should be noted that if the resistances are in $k\Omega$, the capacitances in pF, and the transconductances in mA/V, pole and zero frequencies will be obtained in GRad/Sec. Using KCL2, v_i can be expressed in terms of v_o:

$$v_i = \frac{-(0.1+0.1s)}{10-0.1s}.v_o.$$

Now, after rewriting KCL1 as:

$$\left(1+\frac{1+10s}{100}+\frac{s}{10}\right).v_i - v_{src} - \frac{s}{10}v_o = 0,$$

and replacing v_i we have:

$$-\frac{1+s}{10(10-0.1s)}(1.01+0.2s)v_o - v_{src} - 0.1sv_o = 0,$$

or

$$-\left(\frac{(1+s)(1.01+0.2s)}{100-s}+0.1s\right).v_o = v_{src}.$$

The intended transfer function then will be:

$$\frac{v_o}{v_{src}} = \frac{-(100-s)}{1.01+11.21s+0.1s^2},$$

which can be rewritten based on the general high-frequency format as:

$$\frac{v_o}{v_{src}} = \frac{-99\left(1-\frac{s}{100}\right)}{\left(1+\frac{s}{0.09}\right)\left(1+\frac{s}{112}\right)}.$$

As mentioned earlier, zero and pole frequencies seen in the transfer function are in GRad/Sec. Thus, the amplifier has a midband gain of –99, two poles at 90 MRad/Sec and 112 GRad/Sec, a *negative* zero* at 100 GRad/Sec, and of course another zero at infinity.

(b) The midband gain is approximately –100 or 40 dB. The poles and the finite zero are located at:

$$log|90\ MRad/Sec| = 7.95 \simeq 8,$$

$$log|112\ GRad/Sec| = 11.05 \simeq 11,$$

and

$$log|-100 GRad/Sec| = 11,$$

* Located in the right-hand plane of the S-plane.

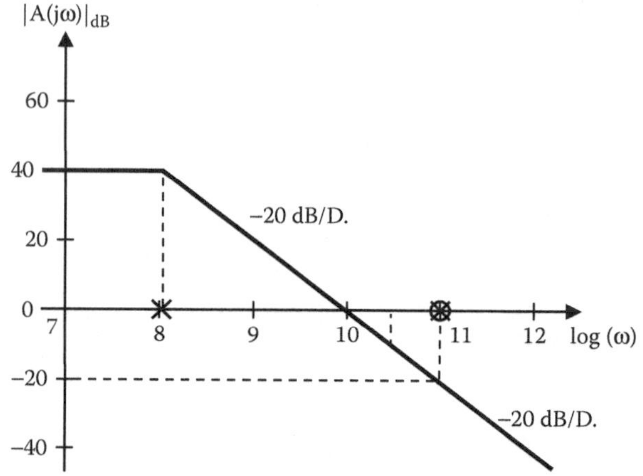

FIGURE EX. 8.4(d)

respectively. A negative zero has the same effect on the magnitude plot as a regular zero, but its influence on the phase plot is similar to a pole. Since the second pole is very close to the zero on the frequency axis, they almost cancel out each other's effects on the magnitude plot. Thus, the magnitude plot will be drawn as depicted in Figure Ex. 8.4(d).

8.5.1 High-Frequency Analysis by Inspection

As seen in Example 8.4, because the frequency-dependent behavior of the internal capacitors is of interest in high-frequency analysis, the transistor cannot be treated as a lumped device anymore. As a result, after replacing the transistor with its small-signal, high-frequency model, the resulting circuit will be an electric circuit, and is traditionally preferred to be analyzed by standard analysis methods (e.g., nodal or mesh analysis methods). It was seen that even if the circuit includes one transistor with only two parasitic capacitors, the analysis would be time consuming. With regular analysis methods, the transfer function's numerator and denominator are usually polynomials, and it would not be easy to rewrite them as the product of their prime factors. This way, it cannot be easily decided which circuit elements dominate the amplifier's frequency-domain aspects. The preceding discussion makes us look for a better approach to analyze the amplifiers at high frequencies. A straightforward method, known as the *high-frequency inspection method,* can be used to simplify calculations.

Keeping in mind that each parasitic capacitor creates a pole in the high-frequency transfer function (and probably a finite zero), it is expected that an amplifier with only a few transistors has many poles at high frequencies. An amplifier with three transistors, for instance, will have a high-frequency transfer function of sixth order or higher! Clearly, only one or at most two of these poles will dominate the amplifier's behavior and the rest can be ignored. As for the zeros, they are either at infinity or located at much higher frequencies than the dominant poles. Therefore, even if the transfer function includes all the zeros and poles associated with all the parasitic capacitors, it could be approximated to a first- or second-order transfer function to be more easily used. The pole (or the two poles) of the approximated transfer function will be the amplifier's *equivalent* high-frequency pole(s) and the general high-frequency transfer function of Equation (8.1) is reduced to:

$$A(s) = \frac{A_0}{\left(1+\frac{s}{\omega_{1H}}\right)\left(1+\frac{s}{\omega_{2H}}\right)} \tag{8.4}$$

where A_0 is the midband gain, and ω_{1H} and ω_{2H} are the first and the second equivalent high-frequency poles. Both ω_{1H} and ω_{2H} are determined based on the contributions of all the internal capacitors to the amplifier's behavior in frequency domain. Usually, the second equivalent high-frequency pole is much larger than the first equivalent pole, and thus, the high cutoff frequency is typically determined by the first equivalent pole. In many cases, the second equivalent pole can be ignored and we will end up with a first-order, high-frequency transfer function.

8.5.1.1 The First Equivalent High-Frequency Pole, ω_{1H}

Before continuing, the *high-frequency circuit schematic* for the amplifier is drawn by:

Setting all the independent sources, both DC and AC, to zero
Shorting all the external capacitors
Replacing the transistor(s) with high-frequency, small-signal model(s)

On the high-frequency circuit schematic, the first equivalent high-frequency pole, ω_{1H}, is obtained as:

$$\omega_{1H} = \frac{1}{a_1} = \frac{1}{\sum_{i=1}^{N} C_i R_{ii}^o} \tag{8.5}$$

where N is the number of independent internal capacitors, C_i is the i-th capacitor, and R_{ii}^o is the equivalent resistance that C_i sees when the other (internal) capacitors are opened. It is clear that the capacitor that makes the largest time constant has the most determining role in the amplifier's first equivalent pole or simply high cutoff frequency.

Example 8.5

Assuming that $R_B = R_1 \| R_2 = 10\ k\Omega$, $R_S = 100\ \Omega$, $R_E = 200\ \Omega$, $R_C = 10\ k\Omega$, $\beta = 100$, $g_m = 40\ mA/V$, $r_o = 1\ M\Omega$, $C_1 = C_2 = 1\ \mu F$, $C_\pi = 1\ pF$, and $C_\mu = 0.1\ pF$ for the common-base amplifier in Figure Ex. 8.5(a), find the first-order, high-frequency transfer function.

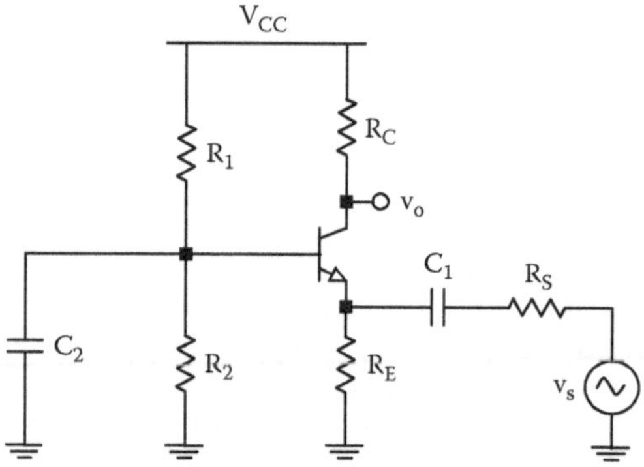

FIGURE EX. 8.5(a)

Solution:

By inspection, the amplifier's midband gain is written as:

$$A_{VS0} = \frac{v_i}{v_s} \cdot \frac{v_o}{v_i} = \frac{R_E \| r_e}{(R_E \| r_e) + R_S} \cdot \frac{R_C}{r_e} = \frac{200^\Omega \| 25^\Omega}{(200^\Omega \| 25^\Omega) + 100} \cdot \frac{10^k}{0.025^k} = 0.18 \times 400 = 72$$

The amplifier's high-frequency circuit schematic is drawn in Figure Ex. 8.5(b) and then simplified as depicted in Figure Ex. 8.5(c). If C_π and C_μ are named C_1 and C_2, respectively, then the first high-frequency pole is obtained as:

$$\omega_{1H} = \frac{1}{a_1} = \frac{1}{\sum_{i=1}^{2} C_i R_{ii}^o} = \frac{1}{C_1 R_{11}^o + C_2 R_{22}^o}.$$

R_{11}^o is the equivalent resistance seen by C_1 when C_2 is opened. Figure Ex. 8.5(d) illustrates how easily this resistance can be found. First, the current source $g_m v_\pi$ and r_o are converted into their Thevenin equivalent. Then, r_o and R_C are merged, and the next step is to convert the resulting combination into their Norton equivalent. Now, notice that the current through the current source is a function of the voltage across it, v_π. Based on Ohm's law, we know

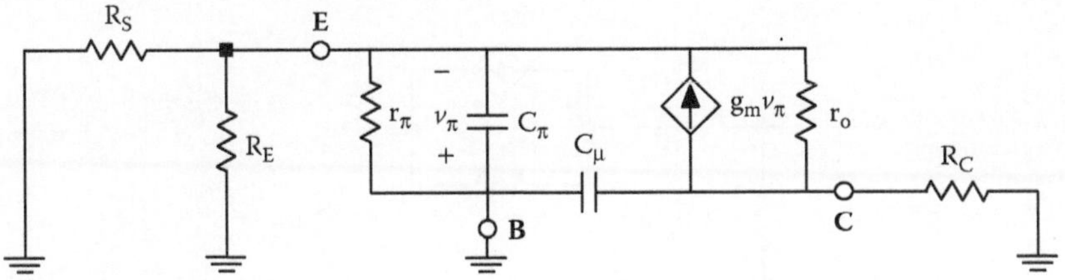

FIGURE EX. 8.5(b)

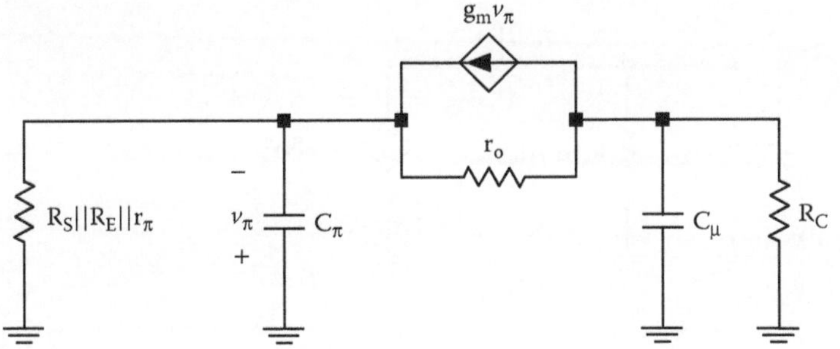

FIGURE EX. 8.5(c)

that when the voltage across an element and the current flowing through it are proportional, this element can be modded by a resistor.

Therefore, the current source can be replaced with a resistor for which the resistance is determined by dividing the voltage across it by the current flowing through it:

$$R = \frac{V}{I} = \frac{v_\pi}{\frac{g_m r_o}{r_o + R_C} v_\pi} = \frac{r_o + R_C}{g_m r_o}.$$

Thus, R_{11}^o is written as:

$$R_{11}^o = \left(R_S \| R_E \| r_\pi\right) \| (r_o + R_C) \| \frac{r_o + R_C}{g_m r_o} \cong \left(R_S \| R_E \| r_\pi\right) \| (r_o + R_C)/g_m r_o$$

$$= \left(100^\Omega \| 200^\Omega \| 2.5^k\right) \| (1^M + 10^k)/(40 \times 1000) \cong 18\Omega.$$

To find R_{22}^o, we should determine the equivalent resistance seen by C_2 when C_1 is opened as depicted in Figure Ex. 8.5(e). R_{22}^o can be considered as the parallel combination of R_C and whatever is seen after that (R_X):

$$R_{22}^o = R_C \| R_X.$$

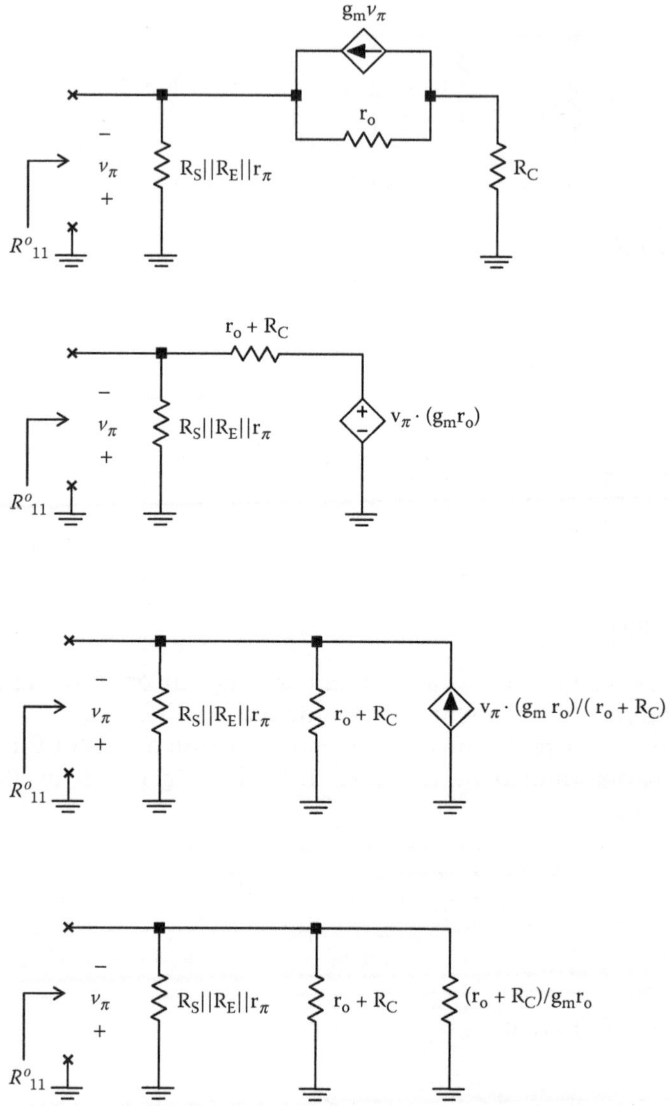

FIGURE EX. 8.5(d)

To find R_X, a test voltage source, V_X, is applied as depicted in Figure Ex. 8.5(f), and using KVL, one can write:

$$\text{KVL: } V_X - r_o(I_X - g_m v_\pi) + v_\pi = 0$$

Knowing from the circuit that:

$$v_\pi = -I_X(R_S \| R_E \| r_\pi)$$

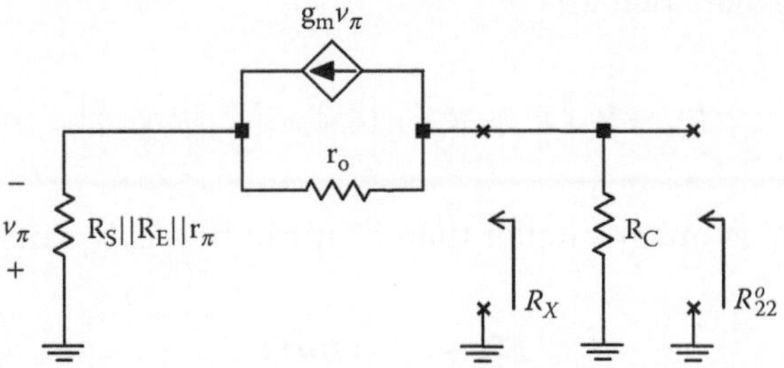

FIGURE EX. 8.5(e)

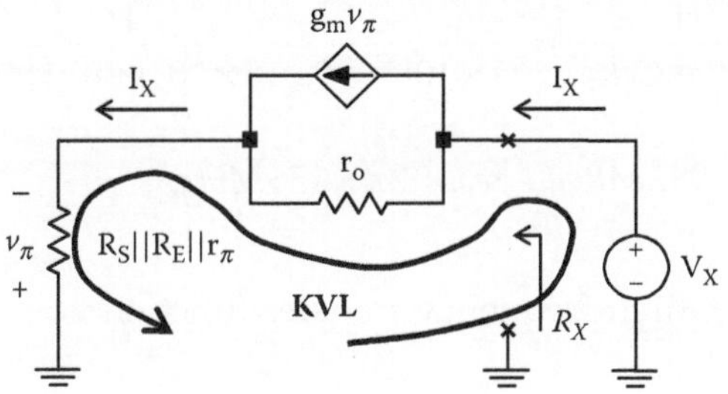

FIGURE EX. 8.5(f)

the KVL equation is rewritten as:

$$V_X = r_o[I_X + g_m I_X (R_S \| R_E \| r_\pi)] + I_X(R_S \| R_E \| r_\pi),$$

and therefore, the resistance R_X is obtained as:

$$R_X = \frac{V_X}{I_X} = r_o + (1 + g_m r_o)\left(R_S \| R_E \| r_\pi\right).$$

As we know, the term $g_m r_o$ for a bipolar transistor is:

$$g_m r_o = \frac{I_C}{V_T} \cdot \frac{V_A}{I_C} = \frac{V_A}{V_T},$$

which is sometimes referred to as the transistor's intrinsic gain. This term is usually much greater than unity and thus:

$$R_X = r_o + g_m r_o\left(R_S \| R_E \| r_\pi\right) = r_o\left[1 + g_m\left(R_S \| R_E \| r_\pi\right)\right].$$

Finally, R_{22}^o is obtained as:

$$R_{22}^o = R_C \| R_X = R_C \| r_o \left[1 + g_m \left(R_S \| R_E \| r_\pi\right)\right],$$

and because R_C is much smaller than r_o, it can be easily said that:

$$R_{22}^o \cong R_C = 10k\Omega .$$

The first equivalent high-frequency pole is then calculated as:

$$\omega_{1H} = \frac{1}{a_1} = \frac{1}{C_1 R_{11}^o + C_2 R_{22}^o} = \frac{1}{(1\times10^{-12}\times0.018\times10^3)+(0.1\times10^{-12}\times10\times10^3)}$$

$$= \frac{1}{1.018^{ns}} = 982MRad/Sec. = 2\pi\times156MHz,$$

and the first-order, high-frequency transfer function is:

$$A_{VS}(s) = \frac{v_o}{v_s} = \frac{A_{VS0}}{1+\dfrac{s}{\omega_{1H}}} = \frac{72}{1+\dfrac{s}{982\times10^6}}.$$

Example 8.6

For the amplifier analyzed in Example 8.4, find the first-order, high-frequency transfer function by inspection.

Solution:

If the input signal source (v_{src}) is set to zero, the amplifier's high-frequency circuit schematic is achieved (see Figure Ex. 8.6(a)).

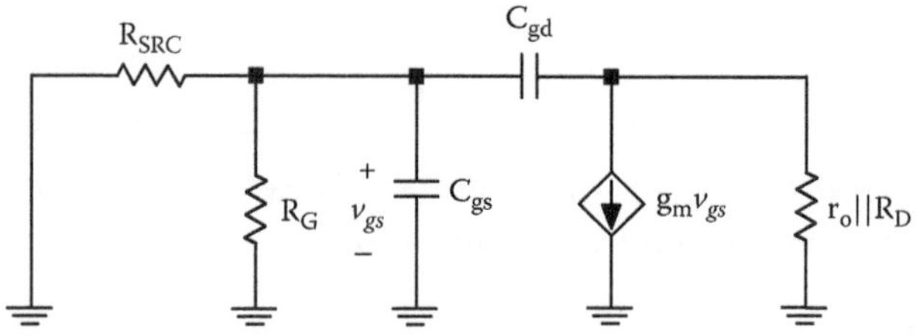

FIGURE EX. 8.6(a)

Let us name C_{gs} and C_{gd} as C_1 and C_2, respectively, and then obtain the first high-frequency pole as:

$$\omega_{1H} = \frac{1}{a_1} = \frac{1}{\sum_{i=1}^{2} C_i R_{ii}^o} = \frac{1}{C_1 R_{11}^o + C_2 R_{22}^o}.$$

R_{11}^o is the resistance seen by C_1 when C_2 is opened. As illustrated in Figure Ex. 8.6(b), this resistance is:

$$R_{11}^o = R_G \| R_{SRC} = 100^k \| 1^k \cong 1\,k\Omega.$$

R_{22}^o is the resistance seen by C_2 when C_1 is opened (see Figure Ex. 8.6(c)). This circuit configuration is frequently encountered when trying to find the resistances seen by C_{gd} in MOS transistors and by C_μ in bipolar transistors.

Figure Ex. 8.6(d) presents the general case, which consists of resistors R_A and R_B connected from nodes A and B to ground, and a current source in parallel with one of the resistors. The current source is a function of the voltage across the other resistor (i.e., GV_A in parallel with R_B or GV_B in parallel with R_A, where G is of a transconductance nature). It can be shown that "*the*

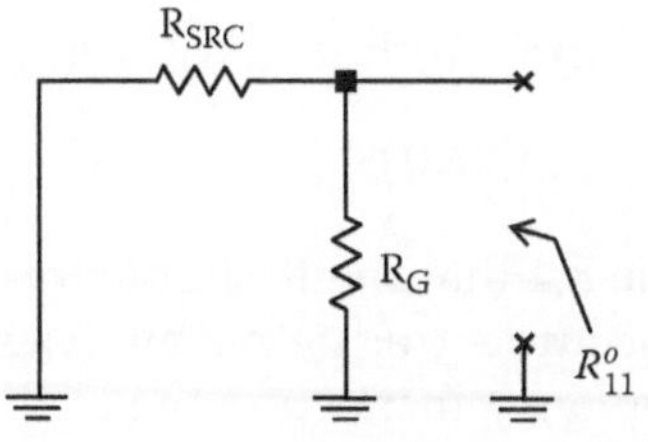

FIGURE EX. 8.6(b)

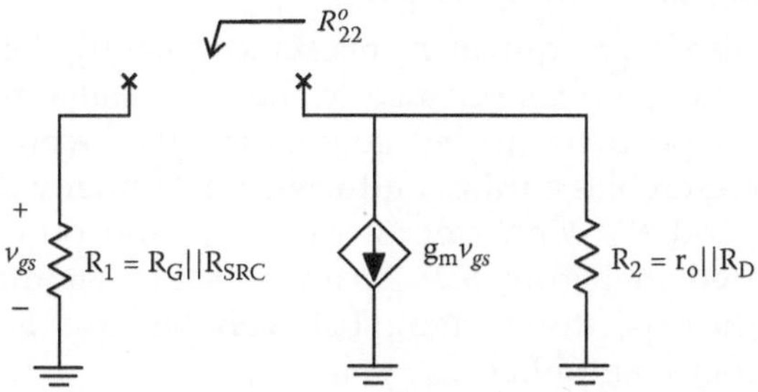

FIGURE EX. 8.6(c)

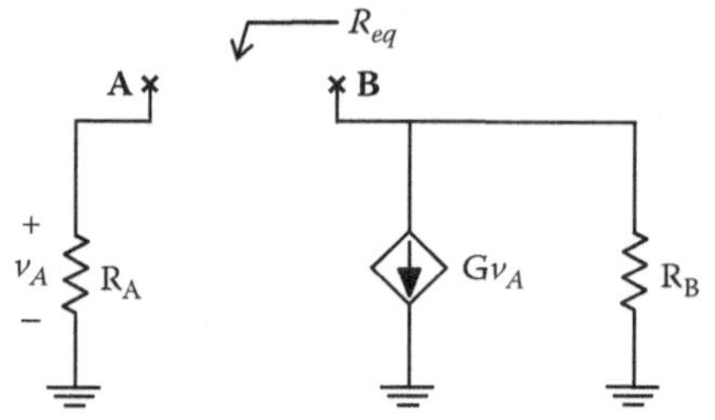

FIGURE EX. 8.6(d)

equivalent resistance (R_{eq}) seen between the nodes A and B is the sum of the two resistances plus the product of the two resistances and the transconductance:

$$R_{eq} = R_A + R_B + G.R_A R_B.\text{''}$$

It is clear that without the dependent current source, the equivalent resistance would be only $R_A + R_B$. Therefore, it can be said that the dependent current source causes the equivalent resistance to be magnified. So, using this formula R_{22}^o in Figure Ex. 8.6(c) is easily obtained as:

$$R_{22}^o = R_1 + R_2 + g_m R_1 R_2 = 1^k + 10^k + (10^m \times 1^k \times 10^k) = 111\ k\Omega.$$

The first equivalent high-frequency pole is then calculated as:

$$\omega_{1H} = \frac{1}{a_1} = \frac{1}{C_1 R_{11}^o + C_2 R_{22}^o} = \frac{1}{(0.1\times10^{-12}\times1\times10^3)+(0.1\times10^{-12}\times111\times10^3)} = \frac{1}{11.2^{ns}}$$

$$= 89.3 MRad/Sec. = 2\pi \times 14.2 MHz.$$

Recalling from Example 8.4 that the first pole for this circuit obtained was 90 MRad/Sec using the accurate traditional analysis method, it is interesting to note that the high-frequency inspection method achieves the same result with less than 1 percent error in this case.

8.5.1.2 Boosting the Time Constant

It was seen that the huge equivalent resistance seen by the capacitance C_{gs} in the previous example was because of the *transconductance amplification*, modeled by the dependent current source. In other words, the circuit in Figure Ex. 8.6(d) resembles a transconductance amplifier with a gain* of –G, and resistors R_A and R_B, connected at the input and output, respectively. Thus, as illustrated in Figure 8.2, it can be said that the time constant associated with the capacitor bridging between the input and output nodes of a transconductance amplifier, C_{Bridge}, is:

* Note the opposite direction of the dependent current source in Figure Ex. 8.6(d) compared with the two-port model for a transconductance amplifier in Chapter 1 (Figure 1.13).

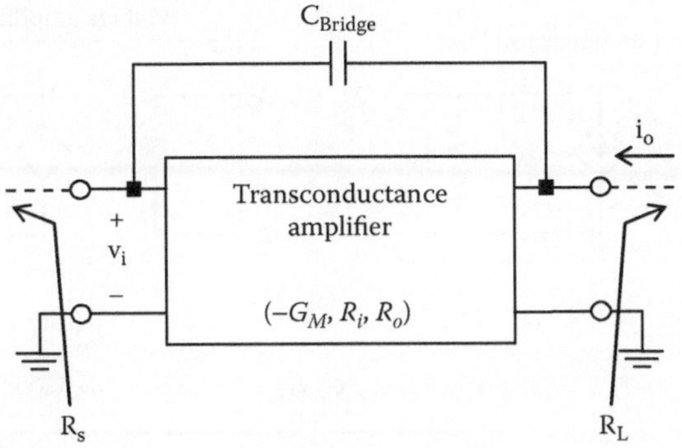

FIGURE 8.2
Boosting the time constant by a transconductance amplifier.

$$TC_{C_{Bridge}} = C_{Bridge}(R_{in,total} + R_{out,total} + G_M R_{in,total} R_{out,total}) \quad (8.6(a))$$

where $R_{in,total}$ and $R_{out,total}$ are the total resistances seen at the input and output nodes comprised of the input and output resistances of the amplifier and the resistances seen into the circuits connected to the input and output terminals of the amplifier, respectively:

$$R_{in,total} = R_i \| R_s, \quad (8.6(b))$$

and

$$R_{out,total} = R_o \| R_L. \quad (8.6(c))$$

If the parallel combination of the dependent current source and the output resistance in the model of the transconductance amplifier in Figure 8.2 is replaced with its Thevenin equivalent, it will be converted to a voltage amplifier, as illustrated in Figure 8.3(a), in which:

$$A_V = G_M R_o.$$

Thus, as shown in Figure 8.3(b) it can be said that a capacitor can also be subject to time constant boosting. For this circuit, Equation (8.6(a)) is rewritten as:

$$TC_{C_{Bridge}} = C_{Bridge}\left[\left(1 + A_V^*\right) R_{in,total} + R_{out,total}\right], \quad (8.7(a))$$

in which $R_{in,total}$ and $R_{out,total}$ are still determined by Equation (8.6(b)) and Equation (8.6(c)). A_V^* is also the amplifier's gain when loaded by R_L:

$$A_V^* = A_V \frac{R_L}{R_L + R_o} \quad (8.7(b))$$

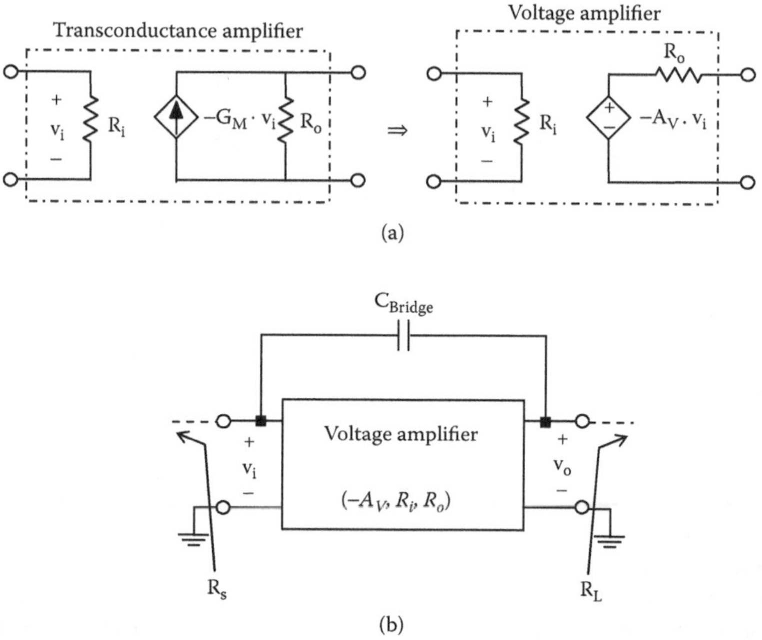

FIGURE 8.3
(a) Converting the transconductance amplifier to a voltage amplifier, (b) boosting the time constant by a voltage amplifier.

Time constant boosting can also be interpreted by *Miller's theorem*. This theorem can be applied to the general case where an impedance "Z" is connected between the input and output of an inverting voltage amplifier with a gain of $-A_V$. This situation is illustrated in Figure 8.4(a). The theorem states that the impedance Z_{Bridge} can be broken into two impedances, Z_{in} and Z_{out}, connected from the input and output nodes of the amplifier to the ground, respectively. The resulting circuit, known as the *Miller equivalent circuit*, is presented in Figure 8.4(b), where:

$$Z_{in} = \frac{Z_{Bridge}}{1 + A_V} \tag{8.8(a)}$$

and

$$Z_{out} = \frac{Z_{Bridge}}{1 + \frac{1}{A_V}}. \tag{8.8(b)}$$

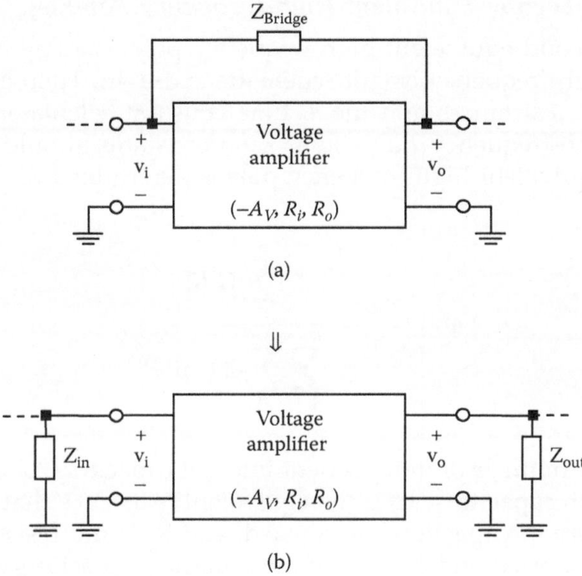

FIGURE 8.4
The Miller equivalent circuit.

For the case of our interest, depicted in Figure 8.3(b), a bridging capacitance, C_{Bridge}, connected across an inverting voltage amplifier with a gain of $-A_V$, can be broken into

$$C_{in} = C_{Bridge}(1 + A_V) \tag{8.9(a)}$$

and

$$C_{out} = C_{Bridge}\left(1 + \frac{1}{A_V}\right), \tag{8.9(b)}$$

connected from the input and output nodes of the amplifier to the ground, respectively. For the amplifiers with a large voltage gain, a large C_{in} appears at the input of the amplifier, resulting in a large time constant. The total time constant associated with C_{Bridge} in Figure 8.3(b) will be:

$$\begin{aligned} TC_{C_{Bridge}} &= C_{in}R_{in,total} + C_{out}R_{out,total} \\ &= C_{Bridge}(1 + A_V)R_{in,total} + C_{Bridge}\left(1 + \frac{1}{A_V}\right)R_{out,total}, \end{aligned} \tag{8.10}$$

which, if $1/A_V$ is neglected due to the large gain, will become identical to Equation (8.7(a)).

8.5.1.3 The Second Equivalent High-Frequency Pole, ω_{2H}

To find the second equivalent high-frequency pole, ω_{2H}, by inspection the amplifier's high-frequency circuit schematic is drawn. Then, in addition to the equivalent resistances and the a_1 time constant calculated for the first equivalent high-frequency pole, other time constants should be obtained. The second equivalent high-frequency pole is determined as:

$$\omega_{2H} = \frac{a_1}{a_2} = \frac{\sum_{i=1}^{N} C_i R_{ii}^{o}}{\sum_{i=1}^{N-1}\sum_{j=i+1}^{N} C_i C_j R_{ii}^{o} R_{jj}^{i}}, \tag{8.11}$$

where N is the number of independent internal capacitors, C_i and C_j are the i-th and the j-th capacitors, R_{ii}^{o} is the equivalent resistance that C_i sees when the other (internal) capacitors are opened, and R_{jj}^{i} is the resistance seen by C_j, when C_i is shorted and the rest of the (internal) capacitors are opened. It is interesting that for any of the terms in the denominator of Equation (8.11), the summation indices i and j can be swapped without any change in the value of the obtained ω_{2H} (i.e., for a given (i, j) pair, one can calculate either $C_i C_j R_{ii}^{o} R_{jj}^{i}$ or $C_i C_j R_{jj}^{o} R_{ii}^{j}$).

As an example, for an amplifier with four independent internal capacitors C_1–C_4, a_2 in Equation (8.11) is calculated as follows:

$$a_2 = \begin{Bmatrix} C_1C_2R_{11}^{o}R_{22}^{1} \\ or \\ C_1C_2R_{22}^{o}R_{11}^{2} \end{Bmatrix} + \begin{Bmatrix} C_1C_3R_{11}^{o}R_{33}^{1} \\ or \\ C_1C_3R_{33}^{o}R_{11}^{3} \end{Bmatrix} + \begin{Bmatrix} C_1C_4R_{11}^{o}R_{44}^{1} \\ or \\ C_1C_4R_{44}^{o}R_{11}^{4} \end{Bmatrix} + \begin{Bmatrix} C_2C_3R_{22}^{o}R_{33}^{2} \\ or \\ C_2C_3R_{33}^{o}R_{22}^{3} \end{Bmatrix}$$
$$+ \begin{Bmatrix} C_2C_4R_{22}^{o}R_{44}^{2} \\ or \\ C_2C_4R_{44}^{o}R_{22}^{4} \end{Bmatrix} + \begin{Bmatrix} C_3C_4R_{33}^{o}R_{44}^{3} \\ or \\ C_3C_4R_{44}^{o}R_{33}^{4} \end{Bmatrix}. \tag{8.12}$$

Because a_1 has been calculated already, each of the terms in Equation (8.12) contains only one parameter to be determined and that is the resistance R_{jj}^{i} or R_{ii}^{j} (whichever is easier to find).

Example 8.7

For the amplifier analyzed in Example 8.6, find the second equivalent high-frequency pole.

Solution:

For reference convenience, the amplifier's high-frequency circuit schematic is given in Figure Ex. 8.7(a). With only two capacitors, the second equivalent high-frequency pole is:

$$\omega_{2H} = \frac{a_1}{a_2},$$

where $a_2 = C_1C_2R_{11}^oR_{22}^1$ or $C_1C_2R_{22}^oR_{11}^2$. It is clear that shorting C_1 eliminates R_G and R_{SRC}. Moreover, this short circuit forces v_{gs} to be zero. This means that the current source g_mv_{gs} will consequently be zero, and thus will be treated as an open circuit. Therefore, we end up with the simplified circuit schematic in Figure Ex. 8.7(b) for R_{22}^1.

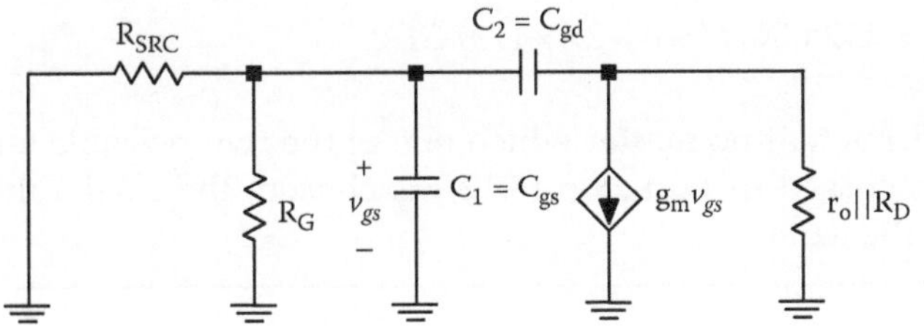

FIGURE EX. 8.7(a)

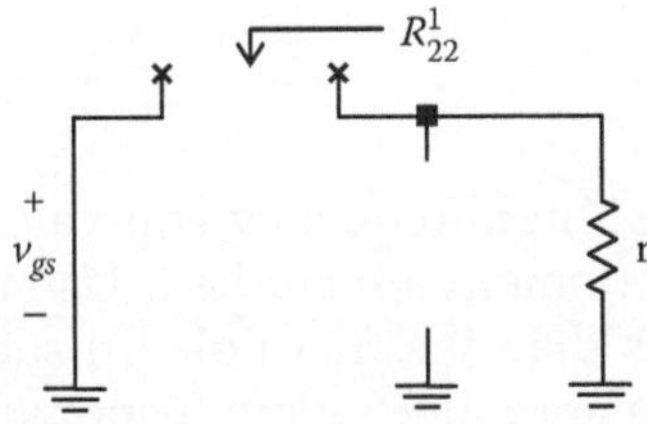

FIGURE EX. 8.7(b)

Therefore, it can easily be said that:

$$R_{22}^1 = r_o \| R_D = 10^{M\Omega} \| 10^{k\Omega} \cong 10\ k\Omega,$$

and the second equivalent high-frequency pole is calculated as:

$$\omega_{2H} = \frac{a_1}{a_2} = \frac{a_1}{C_1C_2R_{11}^oR_{22}^1} = \frac{11.2^{ns}}{0.1^{pF} \times 0.1^{pF} \times 1^{k\Omega} \times 10^{k\Omega}} = \frac{11.2 \times 10^{-9}}{1 \times 10^{-17}}$$

$$= 112\ GRad/Sec = 2\pi \times 17.8\ GHz.$$

Thus, it can be said that for this circuit the high-frequency inspection method returns the same value for the second high-frequency pole as the accurate traditional analysis method does (as seen in Example 8.4).

If instead of R_{22}^{1}, R_{11}^{2} was to be calculated, it can be shown that we would have:

$$R_{11}^{2} = R_{SRC} \| R_G \| \frac{1}{g_m} \| \left(r_o \| R_D \right) \cong 90\Omega,$$

and the second equivalent high-frequency pole would be:

$$\omega_{2H} = \frac{a_1}{a_2} = \frac{a_1}{C_1 C_2 R_{22}^{o} R_{11}^{2}} = \frac{11.2^{ns}}{0.1^{pF} \times 0.1^{pF} \times 111^{k\Omega} \times 90^{\Omega}} = \frac{11.2 \times 10^{-9}}{1 \times 10^{-17}}$$

$$= 112\ GRad/Sec = 2\pi \times 17.8\ GHz.$$

This confirms that no matter which one of the two possible terms for each (i,j) pair (discussed in Equation (8.12)) is chosen, the final value found for ω_{2H} will be the same.

8.6 Time Response

In this section, the effects of high-frequency equivalent poles on the amplifier's response in the time domain are studied. Usually, the second pole is at much higher frequencies than the first pole. In such cases, the first pole dominantly determines the amplifier's time response, and the second pole can be often ignored. To be mathematically exact, for an amplifier with two high-frequency poles, the amplifier can be modeled by a first-order transfer function if the second pole is much greater than the first pole and is located much farther away than the frequency at which the amplifier's gain falls to unity. This frequency, called the *unity-gain frequency,* ω_u or f_u, is one of the important frequency domain specifications of any amplifier. It indicates the maximum frequency that the amplifier can be operated and still performs amplification ($|Gain| > 1$). Figure 8.5 illustrates the magnitude plots for the cases where the amplifier's second high-frequency pole cannot and can be ignored. It should be noted that even in Figure 8.5(a), although both poles determine the unity-gain frequency, the first pole might be the dominant pole when finding the high cutoff frequency of the amplifier.

For the cases where the amplifier can be modeled by a first-order transfer function at high frequencies, in other words, it has a *dominant pole* (ω_{1H}), we have:

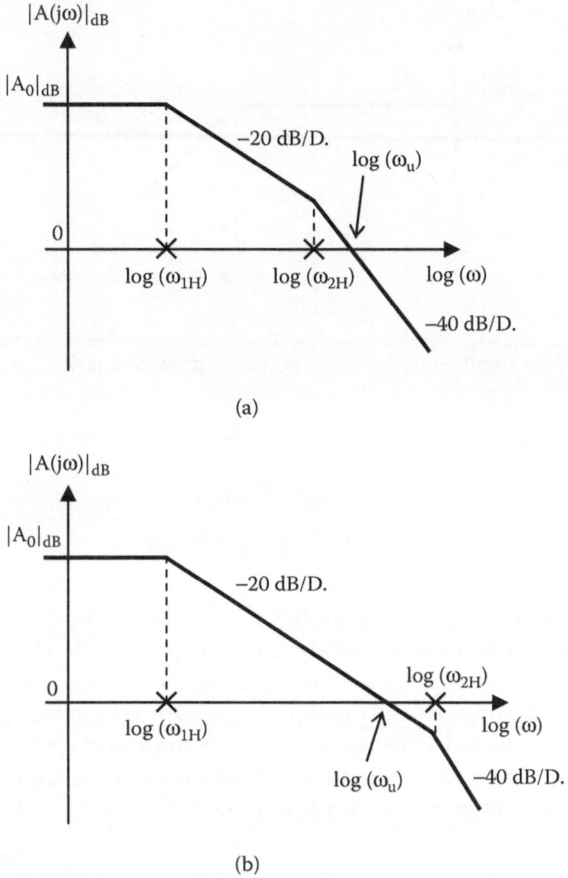

FIGURE 8.5
Magnitude plot for the cases where the second high frequency (a) cannot and (b) can be ignored.

$$A(s) = \frac{A_0}{1 + \dfrac{s}{\omega_{1H}}}, \tag{8.13}$$

and the amplifier's magnitude plot is as depicted in Figure 8.6.

In this section, first, the amplifier's response to a step function is studied and the results are then extended to square wave inputs.

8.6.1 Step Response

The response of an amplifier with the transfer function in Equation (8.13) to the step function

$$x_s(t) = X_I.u(t)$$

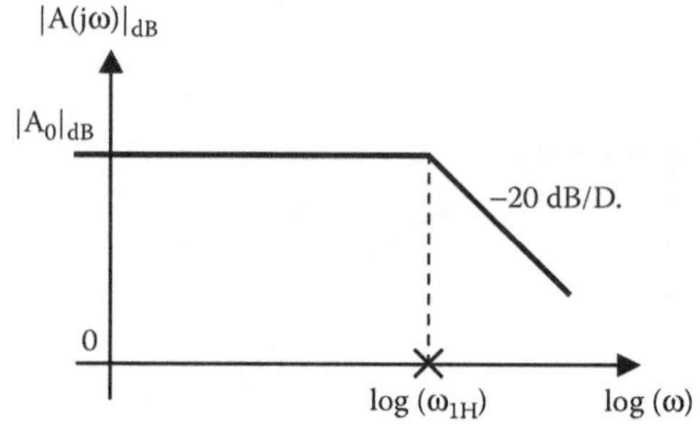

FIGURE 8.6
Magnitude plot for an amplifier with a high-frequency dominant pole.

is formulated as:

$$x_o(t) = A_0 X_I (1 - e^{-\omega_{1H} t}) u(t),$$

and is drawn as depicted in Figure 8.7.

As can be seen, when the input signal is amplified, its sharp edge, which corresponds to high-frequency components, is smoothened. In addition, notice that the flat part of the input signal represented by the lower-frequency components is amplified with the amplifier's midband gain. This is in complete agreement with the amplifier's magnitude plot that resembles the frequency characteristic curve of a low-pass filter.

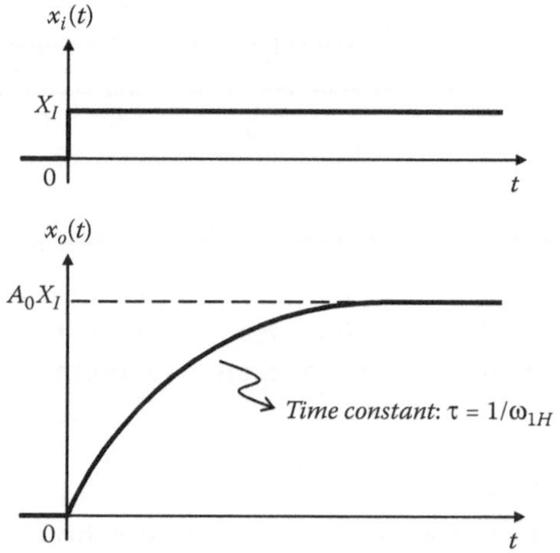

FIGURE 8.7
Step response of an amplifier with a dominant high-frequency pole.

8.6.1.1 Contribution of the Nondominant Pole

To take the impact of the second (i.e., nondominant) pole on the amplifier's time response into account, let us derive the step response of the amplifier in the time domain. If two high-frequency poles are assumed for the amplifier in Equation (8.4), then, based on the basics of systems theory, the amplifier's response to the step function

$$x_s(t) = X_I.u(t)$$

is:

$$x_o(t) = A_0 X_I\left(1 - \frac{\omega_{2H}}{\omega_{2H} - \omega_{1H}} e^{-\omega_{1H}t} + \frac{\omega_{1H}}{\omega_{2H} - \omega_{1H}} e^{-\omega_{2H}t}\right)u(t).$$

Figure 8.8 visualizes the effect of a second pole, which is 10 times larger than the high-frequency dominant pole, on the time response. The step response presented belongs to an amplifier with a midband gain of 1000 and two high-frequency poles at 1 and 10 MRad/Sec. Part (a) compares the step response of the amplifier when the second equivalent high-frequency pole is and is not ignored. The difference between these two responses, which in fact demonstrates the effect of the second pole, is given in part (b). As seen, the error caused by ignoring the second (nondominant) high-frequency pole can easily be neglected.

8.6.2 Response to a Square Wave

Assuming a square wave as a train of alternatively occurring positive and negative step functions, the amplifier's response to a square wave is, in fact, a composition of its responses to positive and negative step functions, as illustrated in Figure 8.9. As seen, each half-cycle of the input square wave in Figure 8.9 is long enough to allow the associated exponential decay to approximately* reach its final value. As explained for amplifiers at low frequencies, two extremes for the signal frequency, $f_S = 1/T_S$, can occur compared with the amplifier's pole, which are worth being studied in detail.

8.6.2.1 $2\pi f_S << \omega_{1H}$

When the signal frequency is much smaller than the amplifier's pole, the exponential decay at the output has enough time to settle down to its steady-state value and stay at that level for a while before the next step occurs at the input. In this case, the circuit's time response will be as depicted in Figure 8.10. As mentioned for low-frequency response, when comparing the

* From mathematical viewpoint, a signal with such an exponential transient regime approaches its final value, but does not *exactly* reach it within a limited period of time. In practice, defining a certain amount of approximation error, it is said that the signal *approximately* reaches its final value if it is close enough to it.

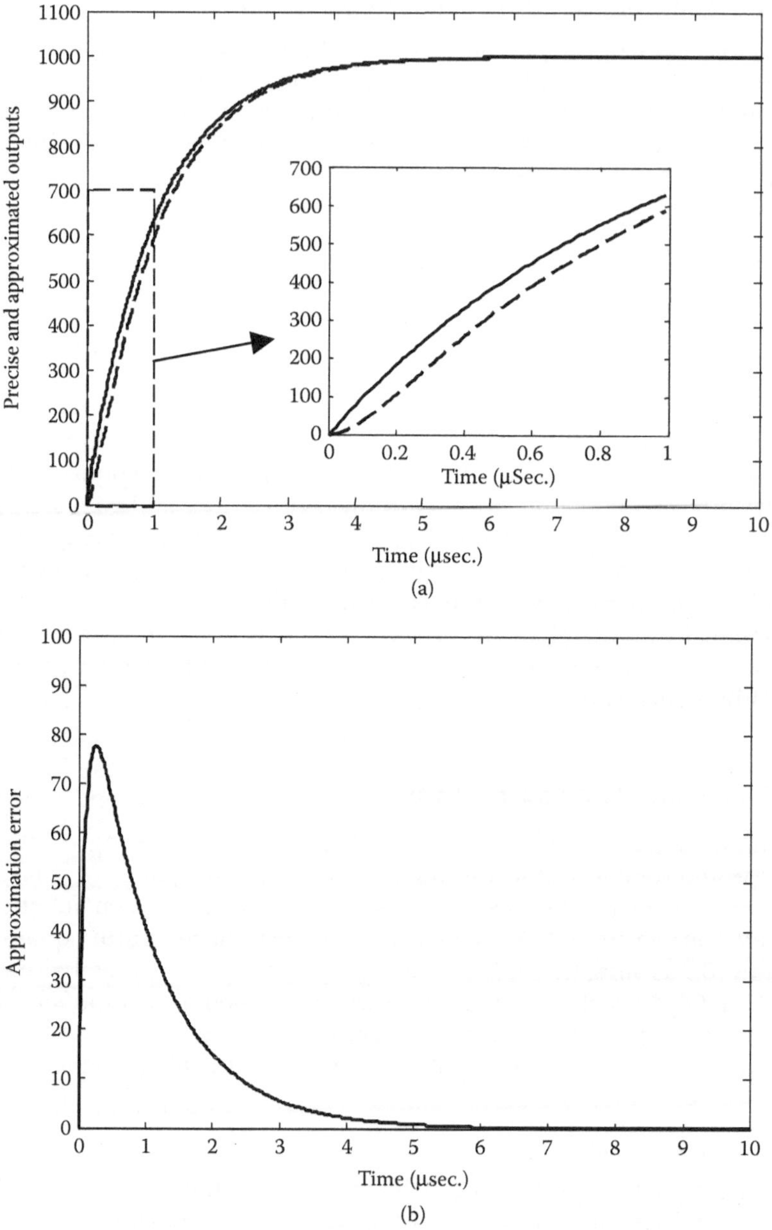

FIGURE 8.8
(a) Step response of the amplifier with (dashed) and without (solid) the second high-frequency pole, (b) pure effect of the second pole.

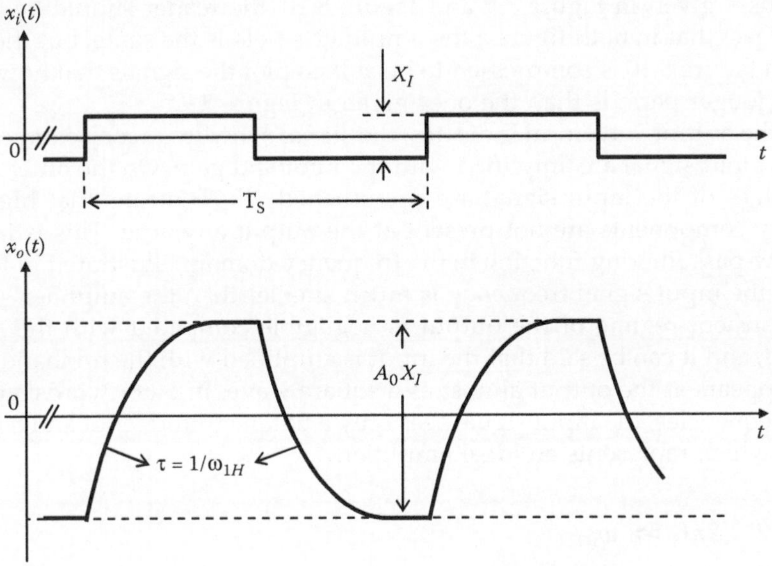

FIGURE 8.9
Response of an amplifier with a dominant high-frequency pole to a square wave at the input.

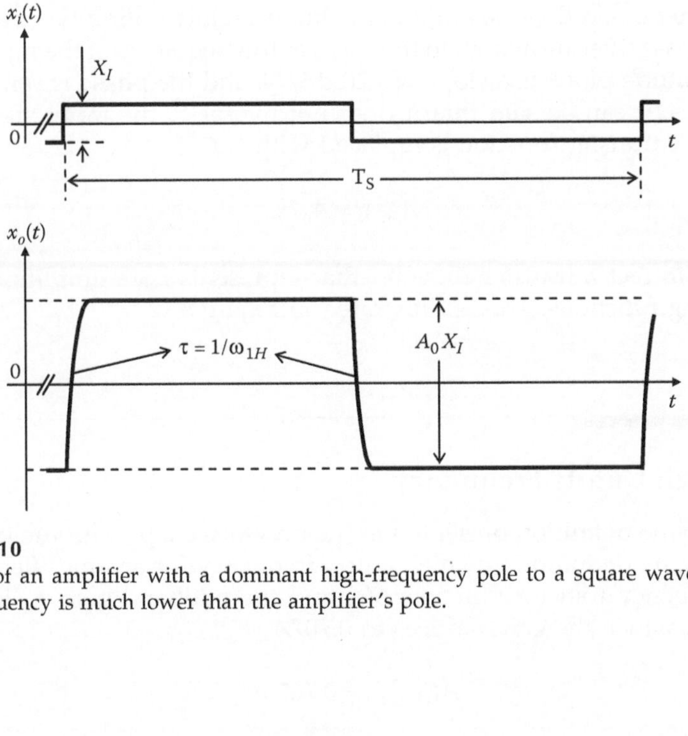

FIGURE 8.10
Response of an amplifier with a dominant high-frequency pole to a square wave when the signal frequency is much lower than the amplifier's pole.

responses given in Figure 8.9 and Figure 8.10, the reader should be aware of the fact that in both figures, the amplifier's pole is the same, but the time axis in Figure 8.10 is compressed to be able to plot the signals that now have much longer periods than the ones given in Figure 8.9.

As the output waveform indicates, the signal's low frequency components of the input signal are amplified with the midband gain. On the other hand, the edges of the input signal are smoothened which means that high-frequency components are not present at the output anymore. This is indeed the low-pass filtering function in the frequency domain, illustrated in Figure 8.6. If the input signal frequency is much smaller than the amplifier's pole, the transient regime of the output is negligible compared with the signal period, and it can be said that the input is amplified with the midband gain, and appears at the output almost as a square wave. In other words, at such low frequencies, the input signal sees only a flat magnitude plot and a phase of 0°, which represents an ideal amplifier.

8.6.2.2 $2\pi f_S >> \omega_{1H}$

The other extreme is where the signal frequency is much higher than the amplifier's pole (at least 10 times, i.e., 1 decade on the logarithmic frequency axis). In this case, the next step occurs at the input before the output finds enough time to exhibit the complete exponential response to the previous step. As a result, the circuit's time response will be as depicted in Figure 8.11. It can be said that the amplifier exhibits an integrating function as well as a low-pass filtering action. In this case, at the frequency of the input signal, the magnitude plot has a slope of –20 dB/D. and the phase is constantly at –90°. Thus, it can be said that it does not matter to the input signal if the amplifier's transfer function is replaced with

$$A^*(s) = A_0/s,$$

which is in fact a transfer function that represents both amplification and integrating functions. This is illustrated in Figure 8.12.

8.7 High Cutoff Frequency

Similar to the definition presented in the previous chapter for the low cutoff frequency, by definition, the high cutoff frequency for an amplifier with the high-frequency transfer function $A(s)$ and the midband gain A_0 is the frequency at which the gain reduces to $0.707A_0$:

$$|A(j\omega_{CH})| = 0.707\,A_0, \tag{8.14}$$

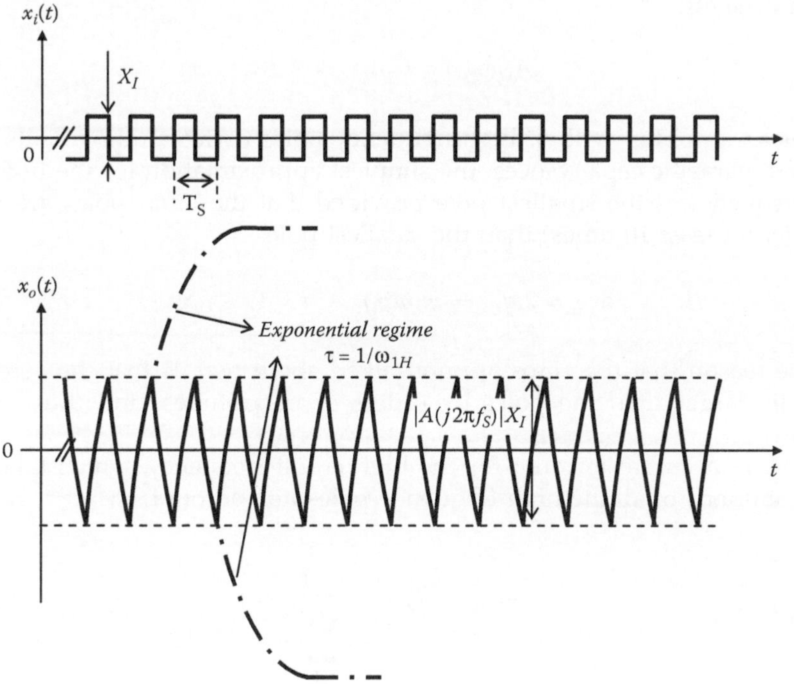

FIGURE 8.11
Response of an amplifier with a dominant high-frequency pole to a square wave when the signal frequency is much higher than the amplifier's pole.

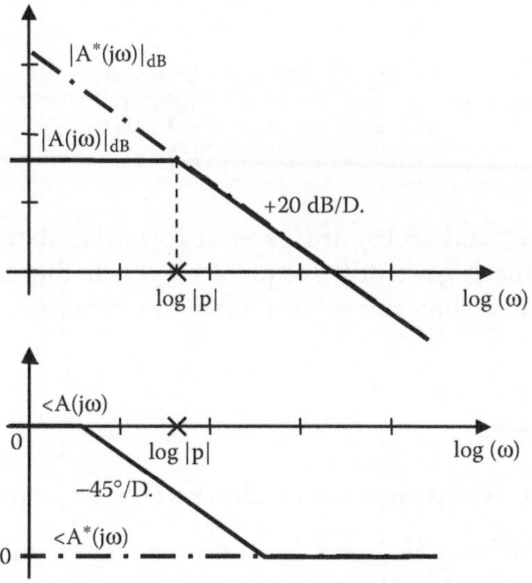

FIGURE 8.12
Approximating the amplifier with an integrating circuit at very high frequencies.

or in decibels:

$$|A(j\omega_{CH})|_{dB} = |A_0|_{dB} - 3\ dB. \tag{8.15}$$

For an amplifier with N high-frequency poles corresponding to its transistors' parasitic capacitances, the simplest approximation for the high cutoff frequency is the smallest pole provided that the other poles are much greater (at least 10 times) than the smallest pole:

$$\omega_{CH} = 2\pi f_{CH} = min(p_i) \qquad i = 1, \ldots, N. \tag{8.16}$$

The reason that the zeros are not talked about here is that the zeros are usually larger than the poles by orders of magnitudes, and thus can be ignored.

A more accurate formula for the high cutoff frequency, which takes the contributions of all the high-frequency poles into account, is:

$$\omega_{CH} = 2\pi f_{CH} = \frac{1}{\sum_{i=1}^{N} \frac{1}{p_i}}, \tag{8.17}$$

which in fact gives the first equivalent high-frequency pole of the amplifier, ω_{1H}, introduced in Equation (8.5). To more accurately determine the amplifier's high cutoff frequency, we have:

$$\omega_{CH} = 2\pi f_{CH} = \frac{1}{\sqrt{\sum_{i=1}^{N} \frac{1}{p_i^2}}}. \tag{8.18}$$

Equations (8.17) and (8.18) are general formulas that always give good estimations for the high cutoff frequency even if the smallest pole is not significantly smaller than the other corner frequencies.

8.8 Complete Response over the Entire Frequency Range

So far, the analysis of amplifiers has been separated to low- and high-frequency ranges. When analyzing an amplifier at low frequencies, it was assumed that the amplifier had no frequency limitation at high frequencies. The resulting transfer function represented the amplifier's behavior at low and midband frequencies. Similarly, when an amplifier was studied at high frequencies, the external capacitors were assumed to be shorted, and, in fact, the associated low-frequency poles and zeros were not taken into account. As the result, the high-frequency transfer function represented the amplifier's behavior at midband and high frequencies. Now, it is time to come up with a complete transfer function that models the amplifier's behavior over the entire frequency range.

8.8.1 Transfer Function and Frequency Response

To find an amplifier's *complete transfer function* using traditional analysis methods, both the internal and the external capacitors are modeled by their impedances, and the circuit is analyzed in the *s*-domain. As an alternative approach, one can first employ the inspection methods described earlier to find the amplifier's low- and high-frequency zeros and poles and also its midband gain. Then, the complete transfer function is written as:

$$A(s) = \frac{x_o}{x_i} = A_0 \cdot \frac{\prod_{i=1}^{N}(s+z_{Li})}{\prod_{i=1}^{N}(s+p_{Li})} \cdot \frac{1}{\left(1+\dfrac{s}{\omega_{1H}}\right)\left(1+\dfrac{s}{\omega_{2H}}\right)} \tag{8.19}$$

where A_0 is the midband gain, z_{Li} and p_{Li} are the low frequency zero and pole for the *i*-th external capacitor, and ω_{1H} and ω_{2H} are the equivalent poles at high frequencies.

The Bode plot for a complete transfer function is drawn with the same rules that have already been explained. The only thing that should be mentioned is that after the transfer function is obtained as shown in Equation (8.19), drawing the magnitude plot first begins from midband frequencies as a line segment at $|Ao|_{dB}$ with slope 0, and then is extended toward both extremes ($\log(\omega) = -\infty$ and $\log(\omega) = +\infty$). The phase plot is also drawn with the same guidelines presented before.

Example 8.8

Draw the Bode plot for an amplifier with a midband gain of 1000, low frequency poles at 10 Rad/Sec and 1 kRad/Sec, low frequency zeros at 0 Rad/Sec and 100 Rad/Sec, and high-frequency poles at 1 MRad/Sec and 10 MRad/Sec.

Solution:

First, finite zeros and poles are located on the frequency axis (see Figure Ex. 8.8). Within the midband frequency range, which is between the last low

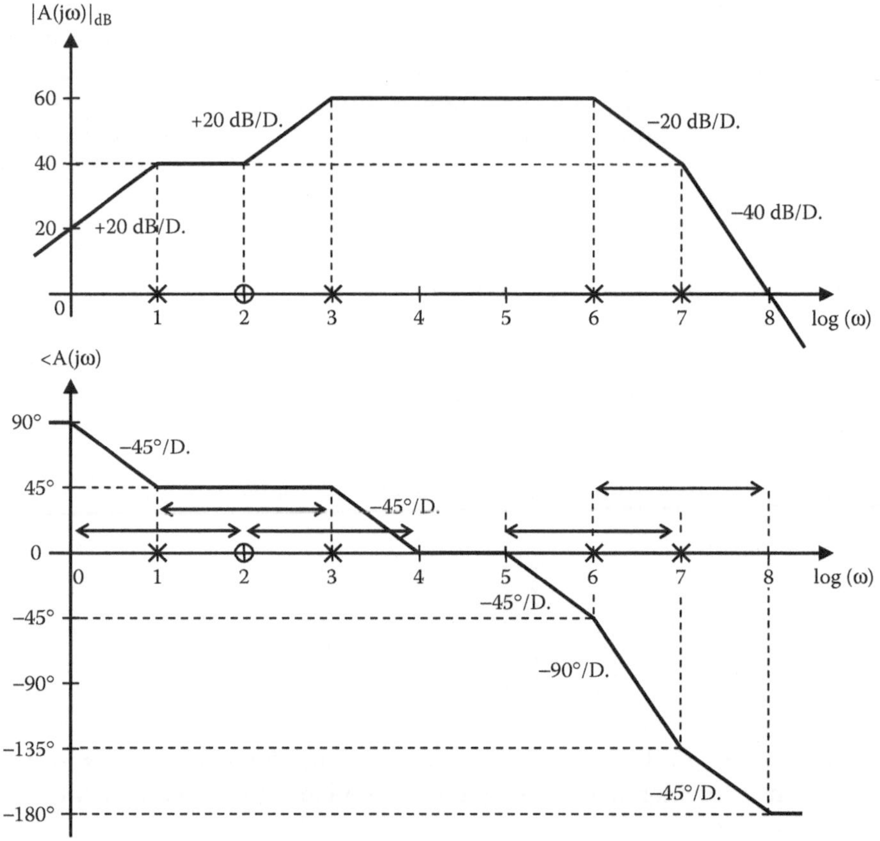

FIGURE EX. 8.8

frequency pole and the first high-frequency pole, the magnitude curve is a flat line segment at 20 *log*|midband gain| = 60 dB. Going toward lower frequencies, the magnitude curve passes a pole, then a zero, and finally another pole. As a result, the curve will approach *log*(ω) with a slope of +20 dB/D. because of having one zero at ω = 0 Rad/Sec (*log*(ω) = −∞). In the other direction, when the curve leaves the midband range toward higher frequencies, it passes two poles and the final slope will be −40 dB/D.

At midband frequencies, the phase plot is at 0° with slope 0. Moving toward low frequencies, the slope of the phase plot changes according to the ROIs of low frequency poles and the finite zero. After exiting the last ROI, the phase plot becomes a horizontal line at 90° as a result of having a zero at ω = 0 Rad/Sec (log(ω) = −∞). With the same approach, the phase plot is drawn when moving from midband frequencies toward high frequencies. The final phase after exiting the last ROI will be −180° as a result of having two zeros at ω = +∞ (log(ω) = +∞).

8.8.2 Time Response

The time response of the amplifiers modeled by their low- and high-frequency transfer functions was studied earlier. Now, if these two transfer functions are merged, the amplifier will be expected to have a time response that reflects both low-frequency- and high-frequency-related phenomena. In fact, the amplifier will try to reject the frequency components that are beyond its two cutoff frequencies (i.e., lower than the low cutoff frequency and higher than the high cutoff frequency). In other words, it can be said that the amplifier generally performs as a band-pass filter. In general, when a step-wise variation occurs at the input, it will be smoothened because of the rejection of high-frequency components, and also if the input stays constant for a relatively long time, it will be treated as a low-frequency component, and will be rejected because of the low-frequency band limitation in the amplifier's frequency response. Step response of an amplifier considering the effects of both low- and high-frequency band limitations is presented in Figure 8.13. In this figure, it is assumed that P_{1L} and ω_{1H} are the low- and high-frequency dominant poles, respectively (i.e., P_{1L} is much larger than the other low-frequency poles and zeros, and ω_{1H} is much smaller than ω_{2H}).

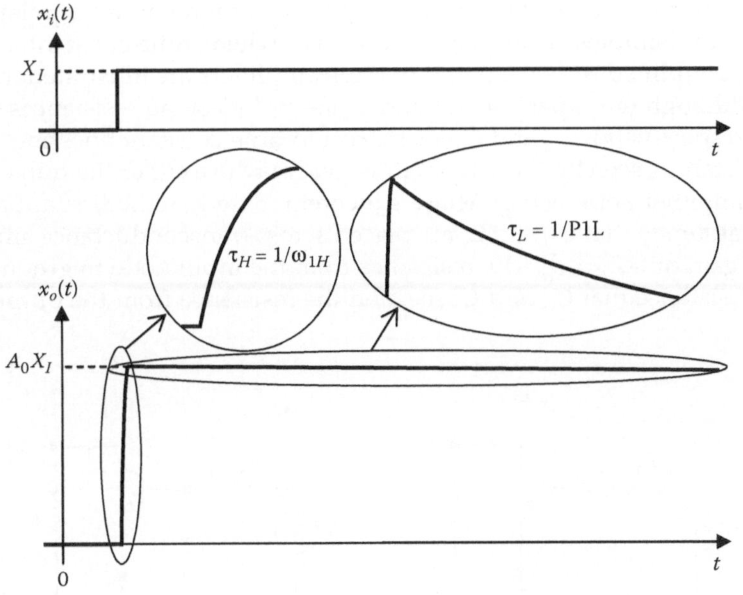

FIGURE 8.13
Step response of an amplifier considering the effects of both low- and high-frequency band limitations.

8.9 Case Studies for High-Frequency Analysis of Multistage Amplifiers

This section is intended to present examples of the application of the concepts studied so far using the qualitative analysis of amplifiers at high frequencies.

When analyzing multistage amplifiers at high frequencies, there are too many capacitors to consider, and it becomes very time-consuming to include the effects of all these capacitors. If one can determine the capacitors that dominate the high-frequency behavior of the circuit, then the other capacitors can be ignored. The capacitors that limit the frequency band of the amplifier at high frequencies are the ones that introduce the largest time constants. These capacitors are also of special interest when designing a wideband amplifier.

In addition to the capacitors that directly see large resistances in the circuit, the capacitors that are subject to time-constant boosting can also be the main contributing capacitors at high frequencies. The following case studies help practice this intuitional approach in finding the element/part of the circuit that dominates the high-frequency response of an amplifier.

8.9.1 Case Study 1: The Common-Source/Common-Emitter Amplifier

The common-source MOS amplifier and the common-emitter bipolar amplifier are the simplest examples of the case where time-constant boosting occurs. Simplified AC schematics of these amplifiers are illustrated in Figure 8.14. Although the capacitors C_{gs} and C_π see not-so-small resistances (R_S and $R_S \| r_\pi$, respectively), C_{gd} and C_μ are subject to time-constant boosting. To find the resistances seen by these capacitors, one may use either the transconductance amplifier approach or Miller's theorem. Based on the first approach, it can be assumed that C_{gd} and C_μ are placed across transconductance amplifiers with a gain of $-G_M = g_m$. The resistance from the input node to ground is the same resistance that C_{gs} and C_π see, and the resistance from the output node

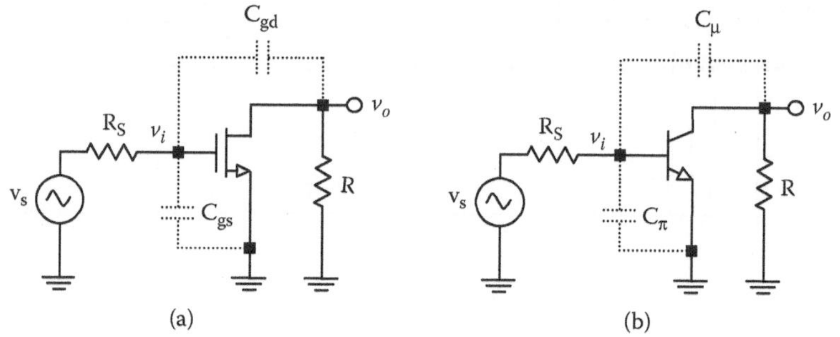

FIGURE 8.14
Time-constant booting in (a) CS amplifier and (b) CE amplifier.

to ground is $R\|r_o$. Therefore, using Equation (8.6) the time constants associated with C_{gd} and C_μ are:

$$TC_{Cgd} = C_{gd}[R_S + (R\|r_o) + g_m R_S(R\|r_o)]$$

and

$$TC_{C\mu} = C_\mu[(R_S\|r_\pi) + (R\|r_o) + g_m(R_S\|r_\pi)(R\|r_o)],$$

which are much larger than those of C_{gs} and C_π. Thus, it can be said that the high cutoff frequency of CS and CE amplifiers is generally determined by C_{gd} and C_μ, respectively.

Using Miller's theorem, it can also be said that the bridging capacitors C_{gd} and C_μ are connected between two nodes of the signal path where the signal is amplified with a negative and usually large voltage gain of $-g_m(R\|r_o)$. As a result, the large Miller equivalent capacitance at the input of the amplifiers:

$$C_{Mi} = C_{gd}[1 + g_m(R\|r_o)]$$

and

$$C_{Mi} = C_\mu[1 + g_m(R\|r_o)]$$

will dominantly contribute to the high cutoff frequency of the MOS and bipolar cascode amplifiers, respectively.

To conclude, it can be said that the CS and CE amplifiers suffer from the fact that a bridging parasitic capacitance is connected between the input and output of the amplifier, subject to the time-constant boosting effect caused by the gain of the amplifier. The larger the gain of the amplifier, the smaller high cutoff frequency it will have.

8.9.2 Case Study 2: The Cascode Amplifier

The cascode amplifier is an example of how having a good understanding about a circuit can help improve its performance.

Figure 8.15 presents simplified AC schematics of MOS and bipolar versions of the cascode amplifier, in which the bridging parasitic capacitors of the first stages are indicated with dotted lines. The main role of the stacked transistor (M_2 or Q_2) in a cascode amplifier can be explained as follows. The stacked transistor acts as a very small load resistance for the first stage, lowering its gain as much as possible. This way, the time-constant boosting caused by the gain of the first stage will be as low as possible. The stacked transistor also acts as a current buffer (in CG or CB configuration), taking the AC current produced by the first transistor (M_1 or Q_1) to a node far away from the bridging parasitic capacitor. There, the AC current is delivered to a large resistance to provide the amplified voltage. This way, the cascode amplifier will have a high cutoff frequency much larger than a single-stage CS or CE amplifier with the same gain.

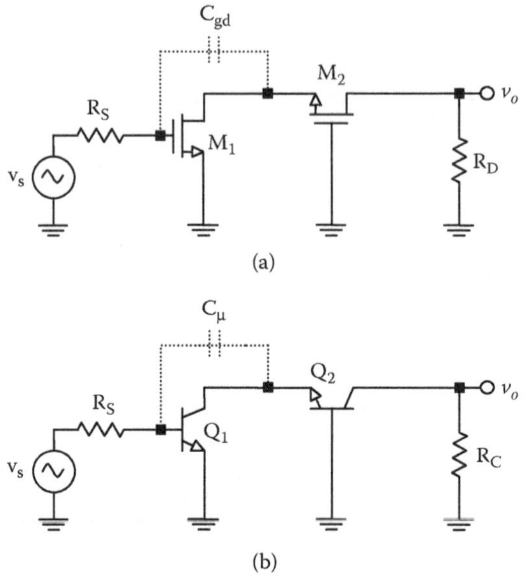

FIGURE 8.15
Cascode amplifier: (a) MOS version, (b) bipolar version.

8.9.3 Case Study 3: A Multistage Amplifier

Circuit schematic of a three-stage CMOS amplifier is illustrated in Figure 8.16. A cascode differential amplifier is used as the input stage, and a source follower is used as the output stage to provide low output resistance. The middle stage is a common-source amplifier, which along with the first stage provides a high overall voltage gain. Three current sources are used in the circuit, sharing the same reference branch composed of M_{12} and R_{REF}. M_{13} is the tail current source for the input stage, and M_{14} and M_{15} serve as the active loads for the middle and output stages.

To figure out where in the circuit the high cutoff frequency is determined, let us inspect the circuit. Transistors M_1 and M_2 are used with CS configuration in the input stage, but because of their cascode connections with M_3 and M_4, they are unlikely to limit the bandwidth. The node where the gate of M_{10} and the drains of M_4 and M_6 are connected, has a very high resistance associated with it, thus making it a possible node that limits the bandwidth. In addition, consider the node where the gate of M_{11} and the drains of M_{10} and M_{14} are connected. Although it indicates a lower equivalent resistance than the previous node, it is still a high-resistance node, thus another one of the possible band-limiting nodes. A much more dominant contribution than either of these two nodes comes from the bridge between them, across which the transistor M_{10} behaves as a CS amplifier. This makes the bridging capacitance C_{gd10} subject to time-constant boosting with a large negative

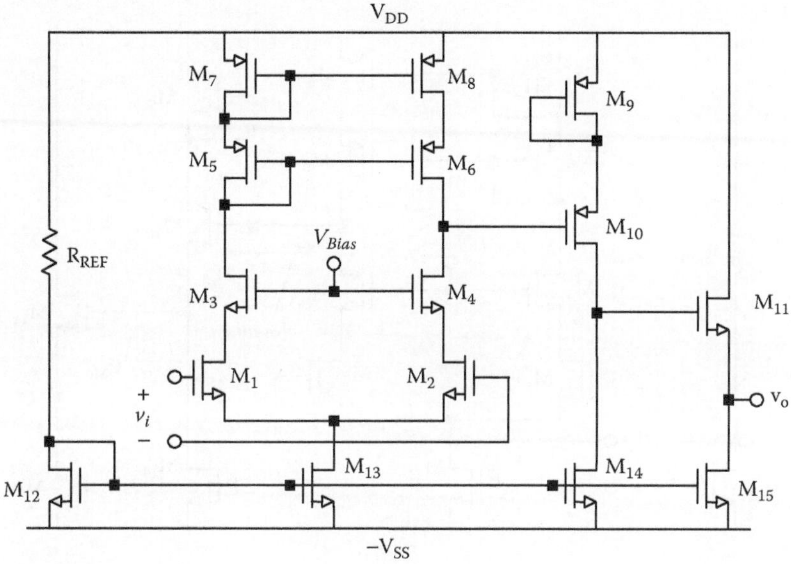

FIGURE 8.16
A multistage amplifier.

voltage gain across it and huge resistances connected from both of its ends to ground.

As depicted in Figure 8.17, applying Miller's theorem, C_{gd10} is broken into two equivalent capacitances:

$$C_{Mi} \cong C_{gd10} \cdot |A_{V10}|$$

and

$$C_{Mo} \cong C_{gd10},$$

where A_{V10} is the gain of the middle stage:

$$A_{V10} = \frac{-r_{o14} \| r_{o10}\left(1+\frac{g_{m10}}{g_{m9}}\right)}{\frac{1}{g_{m10}}+\frac{1}{g_{m9}}},$$

The high cutoff frequency of the circuit can then be approximated to:

$$f_{CH} = \frac{1}{2\pi C_{Mi} R_{eq,g10}},$$

where

$$R_{eq,g10} = r_{o6}(1+g_{m6}r_{o8}) \| r_{o4}(1+g_{m4}r_{o2}).$$

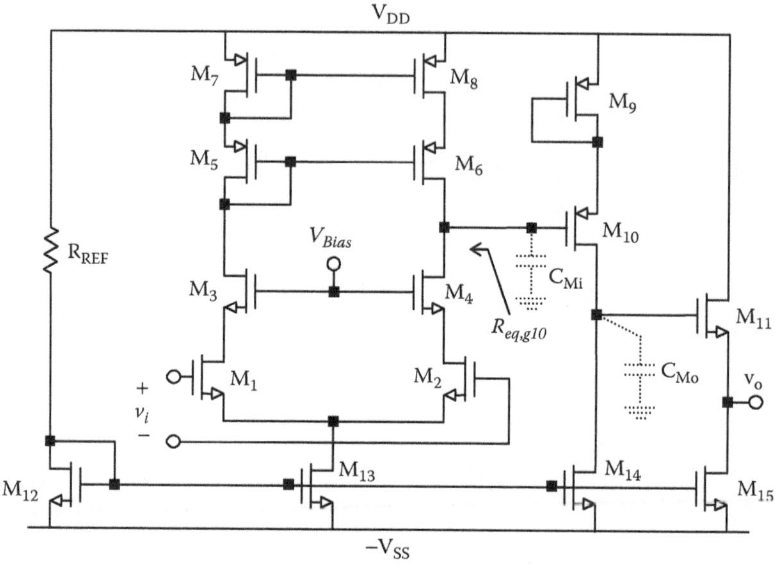

FIGURE 8.17
Applying Miller's theorem on C_{gd10}.

8.10 Simulation Examples

In this section, simulation examples are provided to enhance the reader's insight and understanding about the topics discussed throughout this chapter.

Example 8.9

In the common-source amplifier in Figure Ex. 8.9(a), the transistor has $C_{gd} = C_{gs} = 0.03\ pF$, $g_m = 10\ mA/V$ and $r_o = 1\ M\Omega$. The infinite capacitance for the bypass capacitor means that its zero and pole are both located at 0 Hz. In other words, it is assumed that the bypass capacitor does not contribute to the frequency response at all.

(a) Draw the amplifier's Bode plot from 1 Hz to 1 GHz. Repeat the simulation with $C_{gd} = 0.3$ pF to determine which one of the parasitic capacitors has the dominant contribution in the high cutoff frequency of the amplifier.

(b) Now change the gate-source capacitance to 5pF, and repeat the simulation.

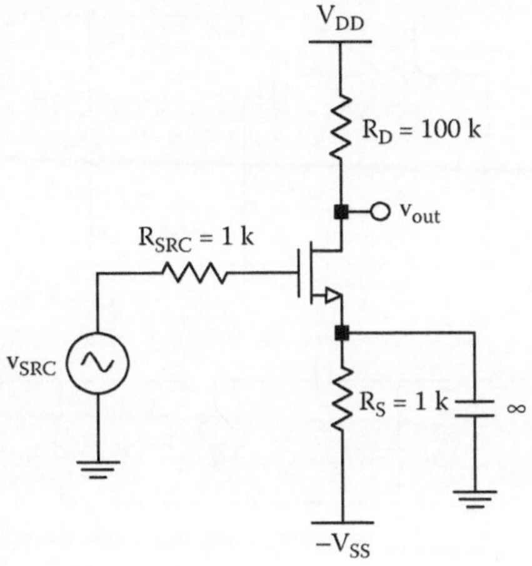

FIGURE EX. 8.9(a)

Solution:

(a) With the information that is given for the transistor, and to be able to easily change the parasitic capacitances, the best way to simulate the circuit is that the AC schematic of the circuit is drawn, and then the transistor is replaced with its small-signal model. This is presented in Figure Ex. 8.9(b).

```
VSRC SRC 0 AC 1
RSRC SRC G 1k
XTr OUT G 0 MOST PARAMS: PCGD={CGD}, PCGS={CGS}
RD OUT 0 100k
.PARAM CGD 0.03p CGS 0.03p
.STEP PARAM CGD LIST 0.03p 0.3p
.AC DEC 100 1 1E19
.PROBE
.SUBCKT MOST D G S PARAMS: PCGD={CGD}, PCGS={CGS}
CGS G 0 {PCGS}
CGD G D {PCGD}
Gm D 0 G 0 10m
ro D 0 1meg
.ENDS
.END
```

Figure Ex. 8.9(c) presents the Bode plots resulted from simulation, in which the high cutoff frequency for $C_{gd} = C_{gs} = 0.03$ pF is 5.30 MHz. As

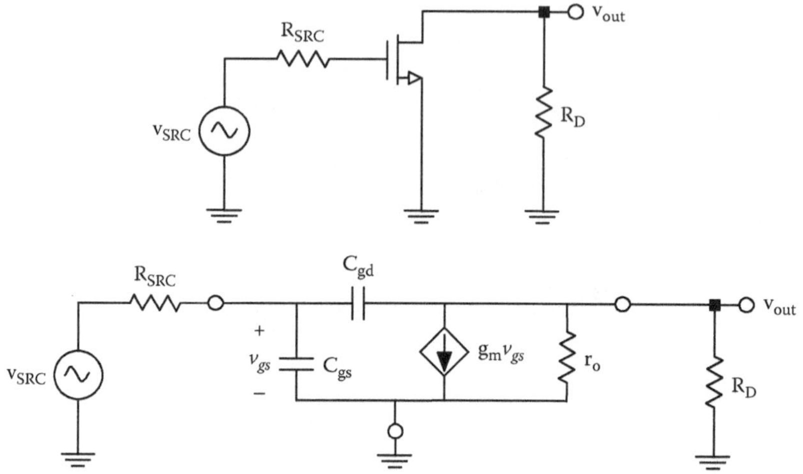

FIGURE EX. 8.9(b)

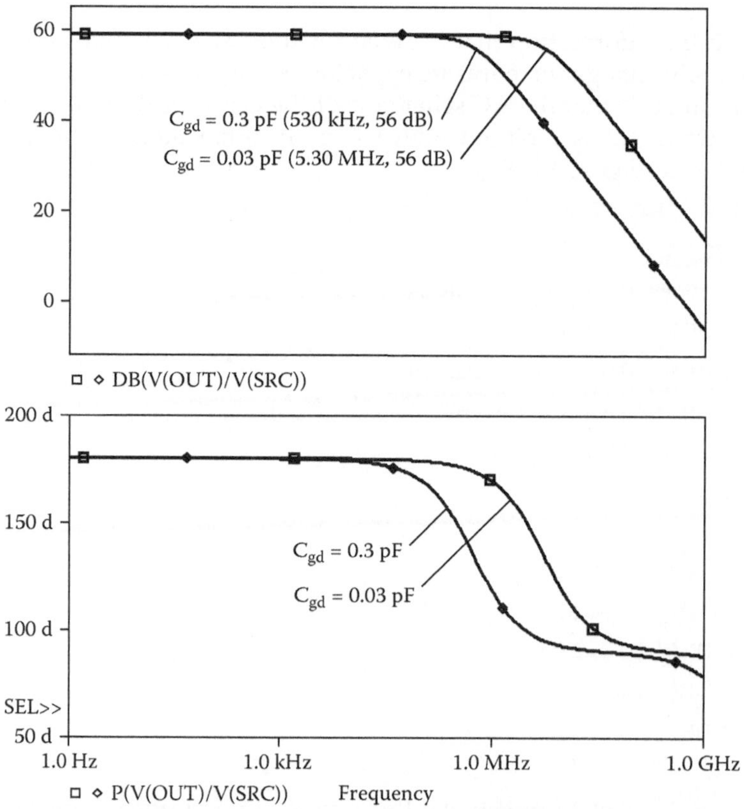

FIGURE EX. 8.9(c)

indicated, increasing C_{gd} by a factor of 10 lowers the high cutoff frequency by the same factor (530 kHz). This means that the high cutoff frequency is dominantly determined by this parasitic capacitance.

(b) The simulations in part (a) demonstrate that the effects of the gate-source parasitic capacitance, C_{gs}, take place at much higher frequencies. To observe this, C_{gs} is made much larger to make its frequency-domain behavior happen at relatively lower frequencies. Figure Ex. 8.9(d) presents the Bode plots for C_{gs} = 5 pF. As indicated in both magnitude and phase plots, a second pole is occurring at much higher frequencies than the first pole, which is definitely caused by the gate-source capacitance.

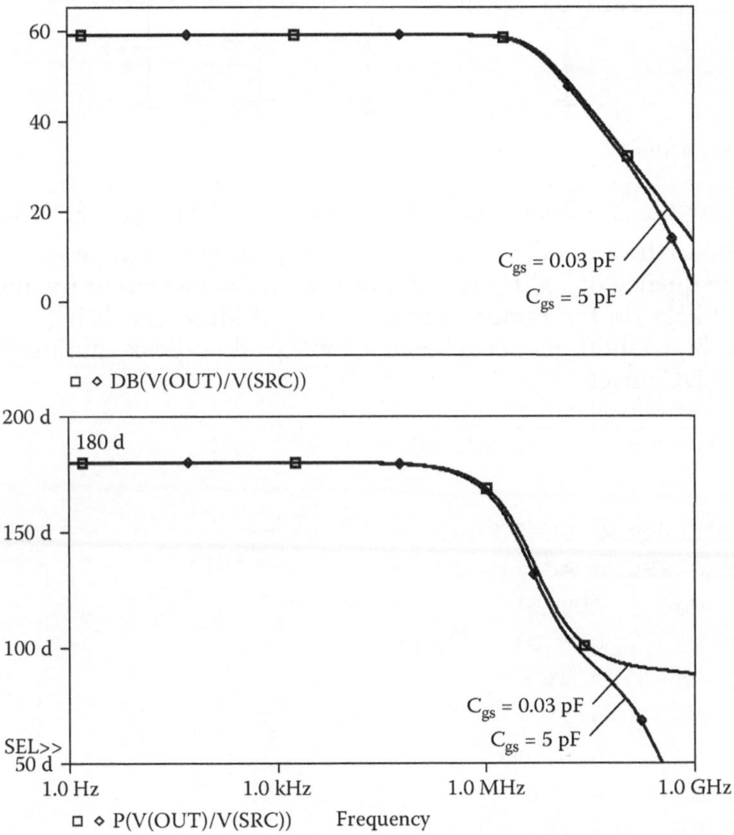

FIGURE EX. 8.9(d)

Example 8.10

A common-emitter amplifier is depicted in Figure Ex. 8.10(a). Assuming that $V_{CC} = 5$ V, $R_1 = 4\ k\Omega$, $R_2 = 1\ k\Omega$, $R_S = 60\ \Omega$, $R_E = 200\ \Omega$, $R_C = 3\ k\Omega$, and $C_1 = C_2 = 1\mu F$,

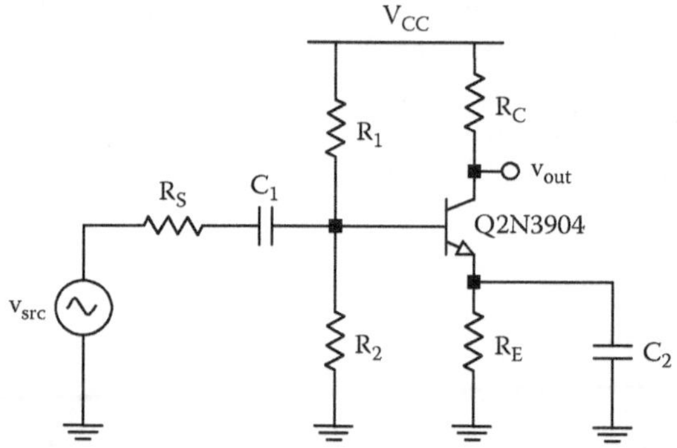

FIGURE EX. 8.10(a)

(a) Draw the amplifier's magnitude plots from 10 Hz to 100 MHz.

(b) Obtain the amplifier's time-domain response to a square wave with a frequency of 100 Hz (much lower than the low cutoff frequency), 100 kHz (in the midband range), and 10 MHz (much higher than the high cutoff frequency) with a 1-mV peak-to-peak amplitude and 0-V DC offset.

Solution:

(a)

```
VDC  Vcc 0 DC 5
VSRC SRC 0 AC 1
RC Vcc   OUT 3k
RE E     0 0.2k
R1 Vcc   B 4k
R2 B     0 1k
RS SRC   A 0.06k
C1 B     A 10u
C2 E     0 10u
Q1 OUT   B E Q2N3904
.lib nom.lib
.AC DEC 100 10 1e8
.PROBE
.END
```

(b) The amplifier's magnitude plot and its response to three square-wave inputs at different frequencies are illustrated in Figure Ex. 8.10(b). As seen, depending on where the input signal is located on the frequency axis, the signal may be amplified, or its derivative or integral is taken.

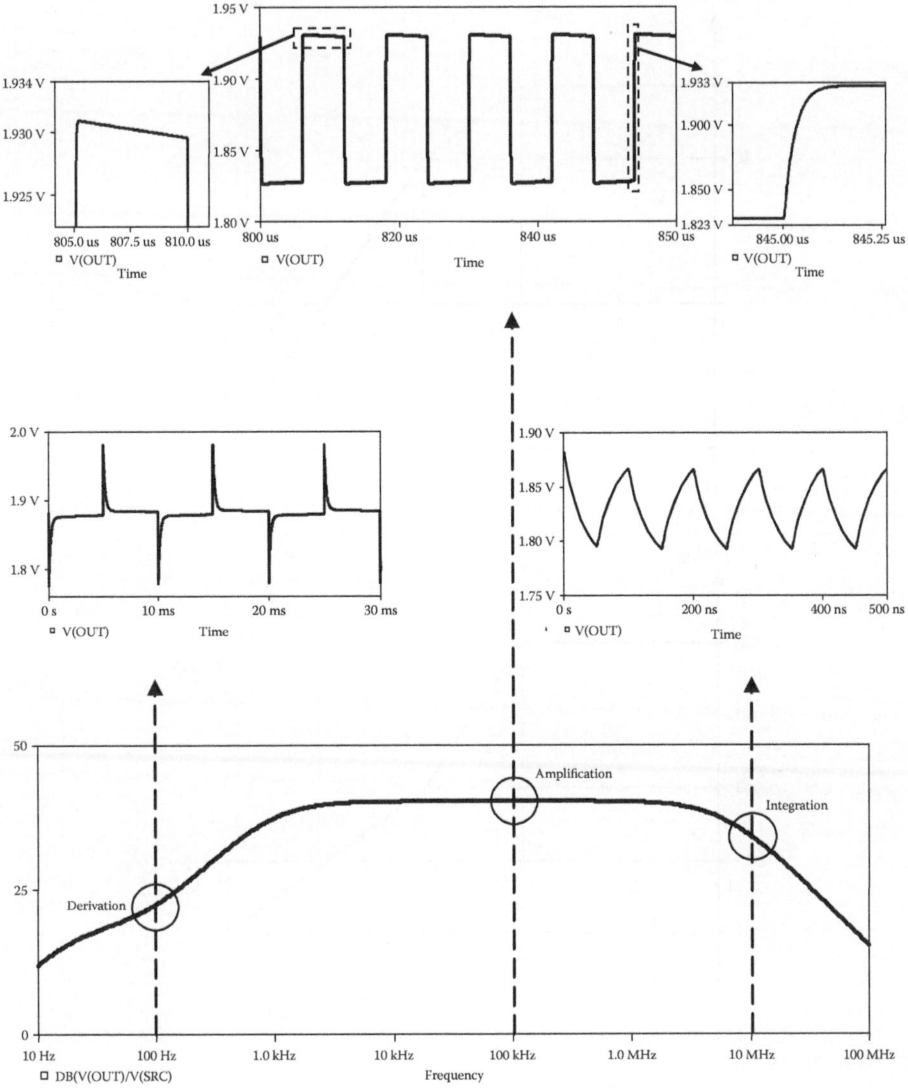

FIGURE EX. 8.10(b)

8.11 Problems

8.1 Find the transfer function for each of the Bode plots drawn in Figure Prob. 8.1.*

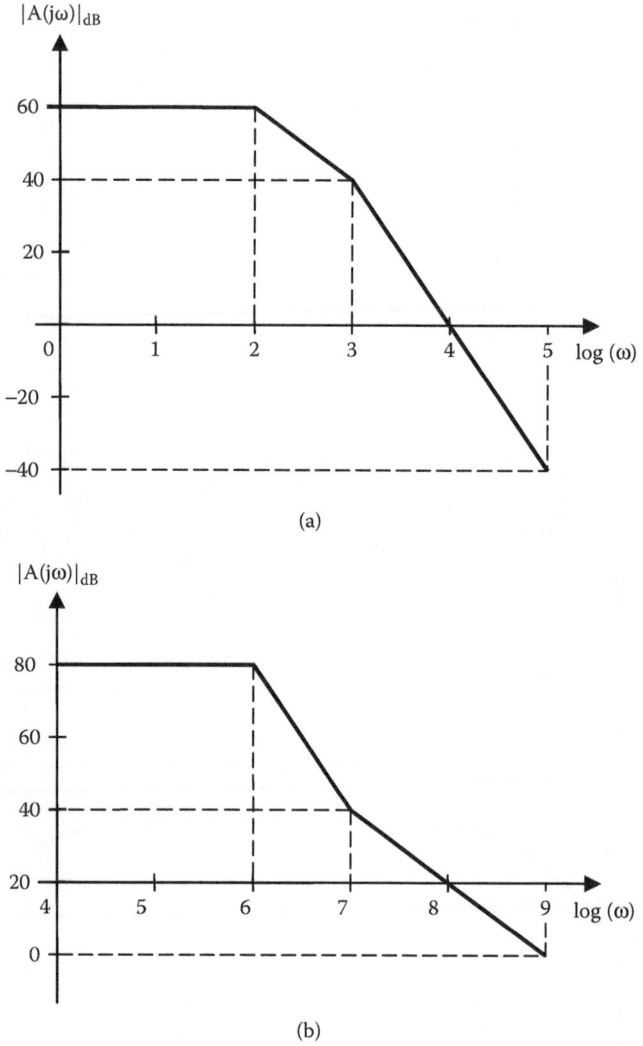

FIGURE PROB. 8.1

* It is assumed that the transfer function has a general form of Equation (8.1) with positive Z_i and P_i values.

8.2 To experimentally obtain the frequency response of an amplifier, a sine wave with a 2-mV peak-to-peak amplitude is applied as the input. The input frequency is manually changed from 100 kHz to 10 MHz. The output peak-to-peak amplitude and the phase difference between the input and the output are then measured for each input frequency, as given in the following table.

Frequency (MHz)	$v_{out,p\text{-}p}$ (mV)	$\Delta\phi$ (degree)
0.1	624	–2.9
0.3	618	–8.5
0.5	606	–14
0.7	590	–19
1.0	559	–27
1.4	512	–35
1.7	476	–40
2.0	442	–45
2.3	410	–49
2.7	372	–54
3.0	347	–56
3.5	310	–60
4.0	279	–63
5.0	232	–68
6.0	198	–72
8.0	152	–76
10.0	123	–79

Using these measured values,

(a) Draw the magnitude and phase plots.

(b) Find the high cutoff frequency.

(c) Derive a first-order transfer function for the amplifier.

8.3 Given $|V_{BE}(ON)| = 0.7$ V, $|V_{CE}(Sat.)| = 0.2$ V, $\beta = 100$, $V_A = 100$ V, $C_\pi = 0.2\ pF$, and $C_\mu = 0.05\ pF$ for the transistor in each of the circuits in Figure Prob. 8.3,

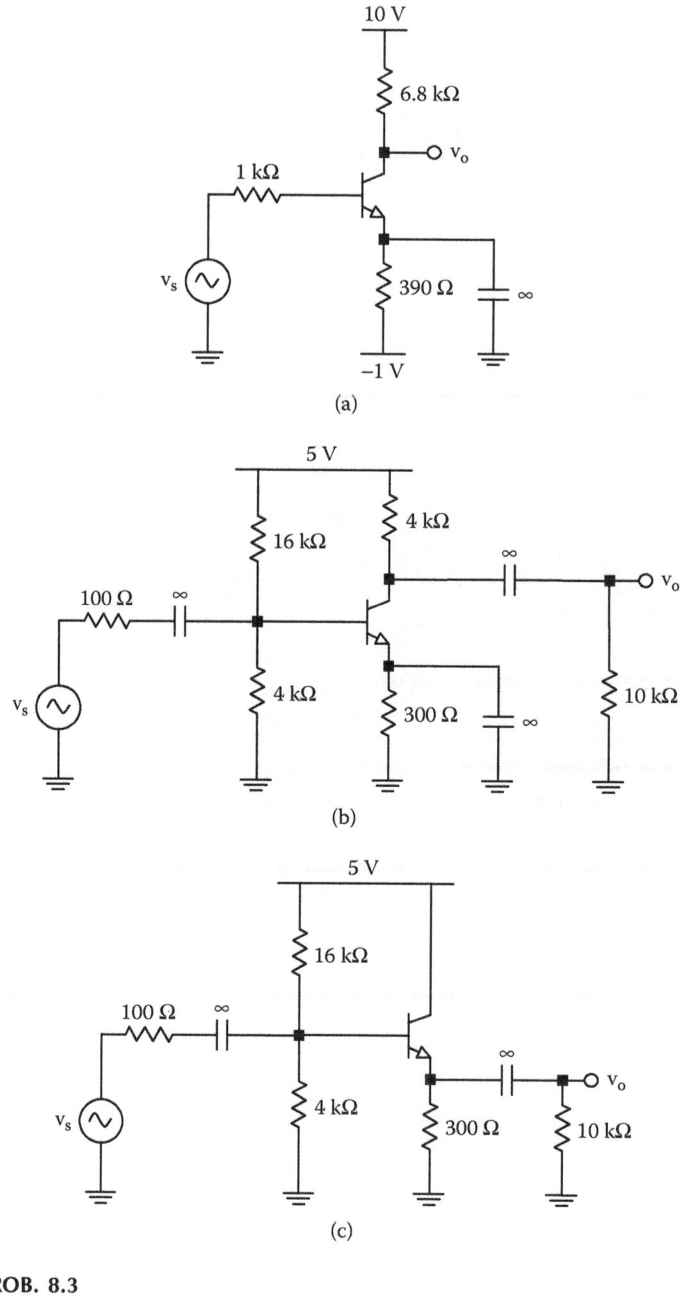

FIGURE PROB. 8.3

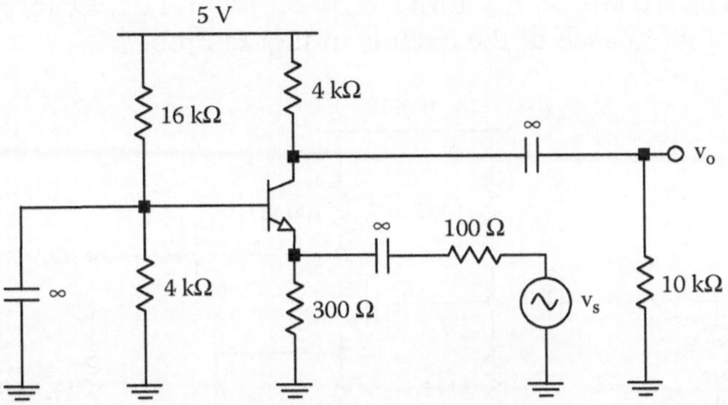

(d)

FIGURE PROB. 8.3 (continued)

(a) Find the second-order high-frequency transfer function.
(b) Calculate the high cutoff frequency of the amplifier.
(c) In each circuit, which capacitor has the dominant role in defining the high cutoff frequency?

8.4 Noting that the circuits in Figure Prob. 8.3 are the same circuits previously analyzed at low frequencies in Problem 7.1 in the previous chapter, form the overall transfer function for each of the amplifiers using low- and high-frequency transfer functions obtained in Problems 7.1 and 8.3, respectively.

8.5 Given $g_m = 5\ mA/V$, $r_o = 1\ M\Omega$, $C_{gs} = 0.1\ pF$, and $C_{gd} = 0.05\ pF$ for the transistors in each of the circuits in Figure Prob. 8.5,

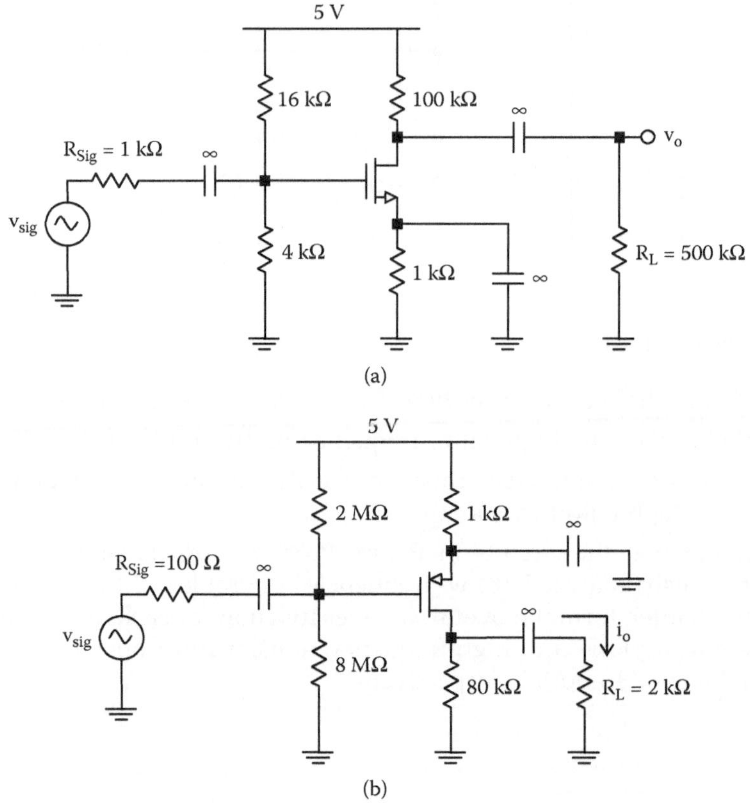

FIGURE PROB. 8.5

(a) Find the second-order, high-frequency transfer function.

(b) Calculate the high cutoff frequency of the amplifier.

(c) In each circuit, which capacitor has the dominant role in defining the high cutoff frequency?

8.6 For the current amplifier in Figure Prob. 8.6 let $\beta_1 = 15\ mA/V^2$, $\beta_2 = 10\ mA/V^2$, $C_{gs1} = 0.05\ pF$, $C_{gd1} = 0.02\ pF$, $C_{gs2} = 0.5\ pF$, $C_{gd2} = 0.2\ pF$, $\lambda_1 = \lambda_2 = 0.01\ V^{-1}$.

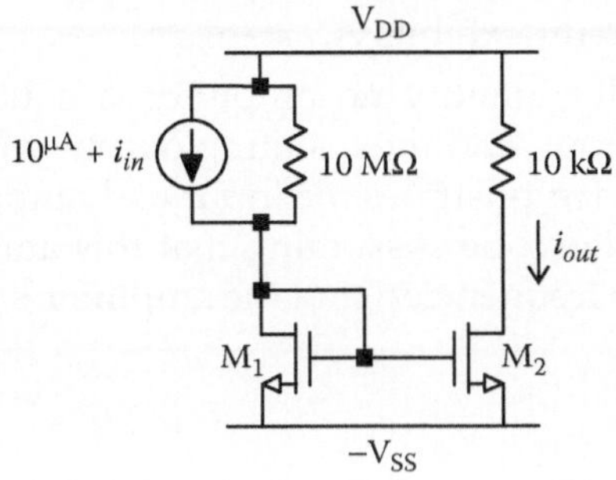

FIGURE PROB. 8.6

(a) Find the first and second equivalent high-frequency poles.

(b) Sketch the Bode plot.

8.7 The amplifier in Figure Prob. 8.7 is the MOS version of the circuit analyzed in Problem 4.8 from Chapter 4. In this circuit, it is expected to have similar gains with opposite amplitudes from the input to the outputs, provided that the resistors in the source and the drain are the same:

$$R_D = R_S \Rightarrow A_{v1} = -A_{v2}$$

where $A_{v1} = v_{o1}/v_i$ and $A_{v2} = v_{o2}/v_i$. Given $g_m = 10\ mA/V$, $r_o = 1\ M\Omega$, $C_{gs} = 0.2\ pF$, and $C_{gd} = 0.1\ pF$ for the transistors, and $R_{sig} = 1\ k\Omega$, $R_G = 1\ M\Omega$, and $R_D = R_S = 10\ k\Omega$,

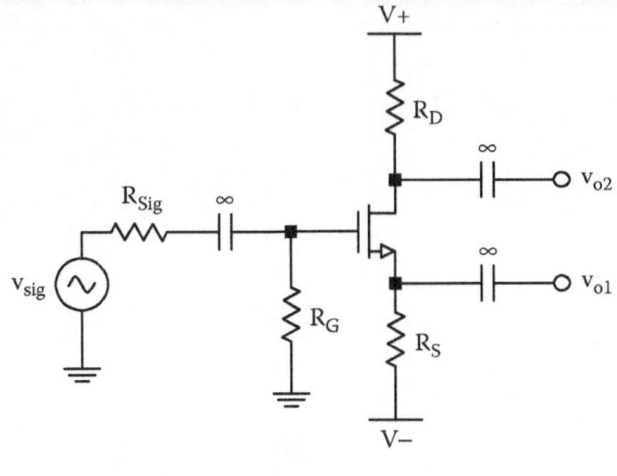

FIGURE PROB. 8.7

(a) Find the high-frequency transfer functions for both amplification paths.

(b) Draw the magnitude plots for the transfer functions obtained in part (a) and determine the maximum frequency of operation for which the circuit exhibits $A_{v1} = -A_{v2}$.

8.8 To experimentally identify an amplifier, a square wave signal is applied as the input, and the resulting output waveform is studied. Figure Prob. 8.8 demonstrates the input and output signals recorded by a digital oscilloscope. Assuming that this amplifier has a dominant pole at high frequencies, find the amplifier's gain and dominant pole.

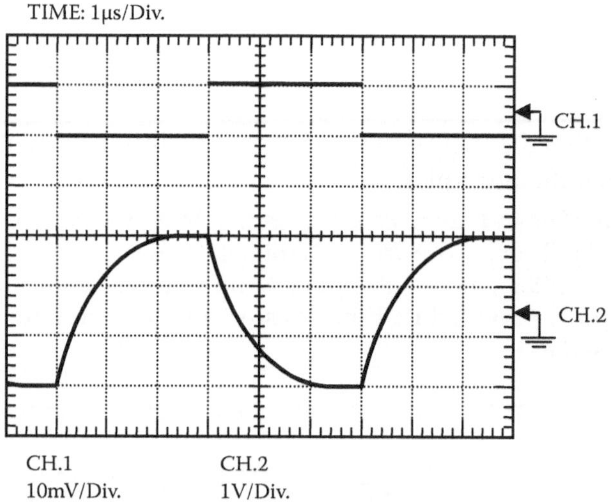

FIGURE PROB. 8.8

References

1. *Merriam-Webster Online Dictionary,* s.v. "signal," http://mw1.merriam-webster.com/dictionary/signal/ (accessed May 17, 2007).
2. Kreyszig, E., *Advanced Engineering Mathematics*, 9th ed., John Wiley & Sons, Hoboken, NJ, 2005.
3. Huelsman, L.P., *Basic Circuit Theory,* 3rd ed., Prentice Hall, Upper Saddle River, NJ, 1991.
4. Streatman, B., and Banerjee, S., *Solid-State Electronic Devices*, 6th ed., Prentice Hall, Upper Saddle River, NJ, 2005.
5. Neudeck, G.W., *Modular Series on Solid-State Devices, Vol. III: The Bipolar Junction Transistor,* 2nd ed., Addison-Wesley, Boston, 1989.
6. Pierret, R.F., *Modular Series on Solid-State Devices, Vol. IV: Field-Effect Devices,* 2nd ed., Addison-Wesley, Boston, 1990.
7. Gray, P.R., Hurst, P.J., Lewis, S.H., and Meyer, R.G., *Analysis and Design of Analog Integrated Circuits*, 4th ed., John Wiley & Sons, Hoboken, NJ, 2001.
8. Allen, P.E., and Holberg, D.A., *MOS Analog Circuit Design*, 2nd ed., Oxford University Press, Oxford, UK, 2002.
9. Horowitz, P., and Hill, W., *Art of Electronics*, 2nd ed., Cambridge University Press, Cambridge, UK, 1989.
10. Oppenheim, A.V., Wilsky, A.S., and Nawab, S.H., *Signals and Systems*, 2nd ed., Prentice Hall, Upper Saddle River, NJ, 1996.
11. Dorf, R.C., and Bishop, R.H., *Modern Control Systems*, 10th ed., Prentice Hall, Upper Saddle River, NJ, 2004.

Index

C

L

M

N

O

P

S

T

V